Social Capital and Information Technology

Social Capital and Information Technology

edited by Marleen Huysman and Volker Wulf

The MIT Press
Cambridge, Massachusetts
London, England

© 2004 Massachusetts Institute of Technology

All rights reserved. No part of this book may be reproduced in any form by any electronic or mechanical means (including photocopying, recording, or information storage and retrieval) without permission in writing from the publisher.

This book was set in Sabon by SNP Best-set Typesetter Ltd., Hong Kong
Printed and bound in the United States of America.

Library of Congress Cataloging-in-Publication Data

Social capital and information technology / edited by Marleen Huysman and Volker Wulf.
  p. cm.
A selection of revised papers from a workshop organized by the editors and held in Amsterdam on 21–22 May 2002, with the addition of some invited papers by social researchers.
Includes bibliographical references and index.
ISBN 0-262-08331-0 (hc : alk. paper)
  1. Social capital (Sociology)—Congresses. 2. Information technology—Social aspects—Congresses. 3. Information networks—Social aspects—Congresses. 4. Knowledge management—Congresses. 5. Organizational learning—Congresses. I. Huysman, Marleen. II. Wulf, Volker.

HM708.S635 2004
303.48′33—dc22   2003068592

10 9 8 7 6 5 4 3 2 1

Contents

Preface   vii

1 Social Capital and Information Technology: Current Debates and Research   1
   Marleen Huysman and Volker Wulf

I Social Capital in Civic Engagement   17

2 Trust, Acceptance, and Alignment: The Role of IT in Redirecting a Community   21
   Anna-Liisa Syrjänen and Kari Kuutti

3 The Effects of Dispersed Virtual Communities on Face-to-Face Social Capital   53
   Anita Blanchard

4 Find What Binds: Building Social Capital in an Iranian NGO Community System   75
   Markus Rohde

5 How Does the Internet Affect Social Capital?   113
   Anabel Quan-Haase and Barry Wellman

II Social Capital in Knowledge Sharing   133

6 The Ties That Share: Relational Characteristics That Facilitate Information Seeking   137
   Rob Cross and Stephen P. Borgatti

7 Exploring the Eagerness to Share Knowledge: The Role of Social Capital and ICT in Knowledge Sharing   163
   Bart van den Hooff, Jan de Ridder, and Eline Aukema

8   Design Requirements for Knowledge-Sharing Tools: A Need for Social Capital Analysis   187
    Marleen Huysman

9   Explaining the Underutilization of Business-to-Business E-Commerce in Geographically Defined Business Clusters: The Role of Social Capital   209
    Charles Steinfield

10 The Impact of Social Capital on Project-Based Learning   231
    Mike Bresnen, Linda Edelman, Sue Newell, Harry Scarbrough, and Jacky Swan

III Applications of IT   269

11 Sharing Expertise: The Next Step for Knowledge Management   273
    Mark S. Ackerman and Christine Halverson

12 Pearls of Wisdom: Social Capital Building in Informal Learning Environments   301
    Robbin Chapman

13 Expertise Finding: Approaches to Foster Social Capital   333
    Andreas Becks, Tim Reichling, and Volker Wulf

14 Fostering Social Creativity by Increasing Social Capital   355
    Gerhard Fischer, Eric Scharff, and Yunwen Ye

Contributors   401
Index   403

Preface

This volume represents a collection of articles on the interplay between two rather independent fields of research: information technology and social capital. With this fine collection, the book contributes to a need to incorporate a relational perspective within the field of applied computer science research.

The idea for this book goes back to summer 2001, when we met in the Boston area as visiting researchers: Marleen Huysman at Harvard Business School and Volker Wulf at MIT. The intellectual environment along with the fortunate lack of daily work routines inspired us to think about the value of social networks and relationships in the research on information technology and, in particular, knowledge management. At the time, Wulf was finishing a coedited book on expertise management. That work argues that the currently accepted approach to the concept of knowledge management is largely driven by technological and managerial concerns, and as a result, it overlooks the importance of informal social relationships. On the other side of the Charles River, Huysman was completing a book on knowledge sharing in practice. In it, she discusses the traps of current approaches to knowledge management. Although we come from two different backgrounds—Huysman is a sociologist by training, and Wulf is a computer scientist—we felt the need to incorporate within the present discourse on knowledge management a focus on the more relational aspects that motivate (and hinder) knowledge sharing.

During our stay in the Boston area, we had the pleasure of meeting with various scholars, thinkers, and writers in related fields of research. Among them, Larry Prusak, the director of the IBM Institute for

Knowledge Management, was pivotal. The inspiring discussions convinced us more and more to include the concept of social capital into the research agenda for technology-supported knowledge sharing.

At the end of our respective visits to the United States, we finally met each other. We realized that although we had taken different paths, we had reached a similar state of mind: more effort is needed to explore the potentials of the concept of social capital for future research on topics such as technology-enhanced knowledge sharing. Whereas the concept of social capital was already introduced and appropriated in various disciplines, it was time for researchers to also think about its potential for entering a new area of technology-supported knowledge sharing. In order to give our ideas more body as well as to garner support for them, we decided to organize a lunch meeting at the European Conference on Computer Supported Cooperative Work (ECSCW) conference in September in Bonn. The purpose of the meeting was to see whether there was an opportunity to build a "community" with researchers sharing similar interests. The group of about twenty like-minded scholars who came encouraged us to start a networking process among researchers in the field of social capital and information technology.

On 21–22 May 2002, we invited an international group of thirty-five researchers to a workshop in Amsterdam to present papers, debate various controversies within this field, and become part of this emerging research community. Given the novelty of the field, the topics of discussion were extremely diverse. Also, the absolute lack of closure in the field created a tolerant and positive atmosphere. At the end of the workshop, we offered to coedit a book on the topic. In order to guarantee a high-quality volume, all workshop papers were peer reviewed, and about half of the papers were selected for publication in the book.

A first volume on social capital and information technology needs to cover as broad a range of topics as possible. The Amsterdam workshop was overly biased toward knowledge management, while the concept of social capital originated from broader societal issues such as development concerns and civic engagement. Consequently, in addition to the selected workshop papers, authors who had conducted societal-related research on social capital and information technology were also asked to submit. These papers went through a thorough reviewing process as well.

The present book is the result of this journey. Although we especially owe a lot to each other—and thanks to e-mail, we were able to keep each other enthused during the whole journey—we could not have done it without the help of so many others. We thank all those with whom we had the pleasure to meet during our stays in the Boston area and who inspired us to persevere in introducing the concept of social capital into the fields of information systems, knowledge management, and computer-supported cooperative work. In particular we wish to thank: Larry Prusak, Dorothy Leonard, Mark Ackerman, Eric von Hippel, Mitchel Resnick, Wanda Orlikowski, Keneth P. Morse, Eric Lesser, David Millen, Brian Lacey, and Ralf Klamma. We also offer our gratitude to all the participants of the workshop Social Capital and Information Technology for the fruitful discussion of the issues addressed in this book. A special thanks goes to Etienne Wenger for his helpful comments and the insights that he shared with us during the two workshop days in Amsterdam.

We want to acknowledge the Department for Information Systems, Vrije Universiteit Amsterdam, the Fraunhofer-FIT, Sankt Augustin, and the Netherlands Organization for Scientific Research for sponsoring the Amsterdam workshop. The state government of North Rhine Westphalia generously supported Wulf's research stay at MIT. We also wish to extend our appreciation to Robert Prior of The MIT Press who encouraged us to go ahead with this book project. As well, we owe a lot to the reviewers of the individual chapters: Elisabeth Davenport, Giorgio DeMichelis, Kari Kuutti, Eric Lesser, Bernhard Nett, Larry Prusak, and Chip Steinfield. We thank Marion Kielmann, University of Siegen, for her help in preparing the final manuscript. Finally, of course, thanks to all the contributors to this volume.

1
Social Capital and Information Technology: Current Debates and Research

Marleen Huysman and Volker Wulf

Social capital has recently gained importance in a variety of research fields. The concept was originally introduced by sociologists and political scientists. Lately, scholars in organizational and management sciences have shown an increased interest in the concept. *Social capital* refers to network ties of goodwill, mutual support, shared language, shared norms, social trust, and a sense of mutual obligation that people can derive value from. It is understood as the glue that holds together social aggregates such as networks of personal relationships, communities, regions, or even whole nations.

Social capital is about the value derived from being a member of a society or community. By being a member, people have access to resources that are not available to nonmembers. Research on social capital varies as to the type of resources that are gained from being member of a social network or community. For example, resources range from those that aid potential career moves or entrepreneurial start-up processes to services in so-called developing countries. Social capital has also been adopted recently by researchers interested in the topic of organizational learning and knowledge management.

Although the concept of social capital has a much longer existence (cf. Hanifan 1916), it has become a prominent topic of discussion over the last two decades. Its emergence within the fields of political science and sociology started as a critic of the narrow analytic perspective on economic activities that is immanent in the neoclassical school of macroeconomic thinking (e.g., Bourdieu 1986; Granovetter 1985; Uzzi 1997). A basic assumption of mainstream economic analysis sees the economy as an increasingly separate, differentiated sphere in modern society, with economic transactions no longer defined by social or kinship obligations but by rational calculations of individual gains. It is argued that industrial societies are distinct from preindustrial ones because the social dimensions of economic activities are subordinated under atomic market transactions. With regard to research methodology, such an understanding marginalizes the analysis of sociological conceptions of economic activities.

Social capital is a notion that challenges this reductionist view. Drawing on the capital metaphor, it allows us to study the social aspects of economic activities. Yet the recent interest in the concept of social capital emerged from rather divergent philosophical traditions. To put it starkly, the actual discussion on social capital can be routed back to either the marxist or communitarian tradition. The marxist conception of social capital is provided by French sociologist Pierre Bourdieu (1986). The communitarian tradition stems from U.S. social scientists such as Amitai Etzioni (1993, 1995) and Robert Putnam (1993, 2000).

The Marxist versus Communitarian Tradition

Bourdieu (1980, 1986) perceives social capital as a specific form of capital, and he contends that it has to be studied in relation to economic and cultural forms of capital. Bourdieu (1985) defines social capital as "the aggregate of the actual or potential resources which are linked to possession of a durable network of more or less institutionalized relationships of mutual acquaintance and recognition" (248). Like all forms of capital, social capital is accumulated labor. It has its own capitalists, who accumulate social capital in the form of relationships, networks, and contacts: "The network of relationships is the product of investment strategies, individual or collective, consciously or unconsciously aimed at establishing or reproducing social relationship, which are directly usable in the short or long term" (249). Bourdieu is specifically interested here in the way the different forms of capital shape the social world, especially those aspects of a class struggle and class nature. Whereas the upper classes take their high level of social capital for granted, lower classes usually are aware of their scarce resources in terms of social capital—for example, the lack of collective bargaining power or access to career jobs.

Against this conflict perspective stands the tradition of U.S. communitarianism. In this tradition, social capital is community centered. Communities, in turn, are seen as voluntaristic social units that promote the harmonic development of organizations and society as a whole. The community concept was studied not just from an "objective" sociological perspective but also to provide the society—in practice, mainly the United States—with a normative, organizational vehicle for revitalizing democracy. Advocates of this community view, known as communitarians, protest against the decline of social trust and the loss of civic engagement, and seek to shore up the moral, social, and political foundations of society (Etzioni 1995). This emphasis on unity and collectivism is in line with the communitarian perspective that surrounds the discussion on "communities of practice" (Lave and Wenger 1991; Wenger 1998). From a communitarian standpoint, it is the community instead of the individual or organization that structures action and provides the key frame of reference. This perspective argues that we know what we know through our relationships with others in the community. The communitarian view also stresses the need to take social responsibility to support the community rather than striving to satisfy individual needs only. This communitarian stance has been adopted by many social scientists interested in social capital, and it largely overshadows the topics—power, class struggle, and class conflict—that characterize the marxist tradition. In this discussion, the term *social capital* underwent quite some shift in meaning.

The Emergence of Meaning

Let us have a closer look at the emergence of meaning. While not using the term social capital explicitly, Granovetter (1985) works out the concept of the embeddedness of social action. He maintains that "the anonymous market of neoclassical models is virtually nonexistent in economic life and that transactions of all kinds are rife with the social connections described" (495). Granovetter criticizes the limited analytic perspective of institutional economists, especially Williamson's (1975) work. He also shows how personal relations and networks of such relations generate trust and discourage malfeasance, undermine formal organizational structures, and shape interorganizational transactions. As such, the embeddedness of social action offers a valid alternative explanation for institutionalization in economic life.

Meanwhile, there are many case studies that have proven the importance of social networks in explaining economic behavior. Loury (1977) attributes racial income differences to different levels of connection to the labor market and access to relevant information. Portes and Sensenbrenner (1993) investigate the effect community participation has on the economic condition of Puerto Ricans in New York and Latin American minorities in Miami. Uzzi (1997) shows how social networks shape interorganizational cooperation in the New York textile industries.

On a theoretical level, Coleman, Burt, and Portes (1998) provide significant contributions to the discussion on social capital. Coleman defines social capital rather vaguely as a "variety of entities with two elements in common: They all consist of some aspect of social structure, and they facilitate certain action of actors—whether persons or cooperated actors—within this structure" (1988, S98). Burt understands social capital as "friends, colleagues, and more general contacts through whom you receive opportunities to use your financial and human capital" (1992, 9). While differing in the scope of their definitions, both of these authors highlight the close interaction between social and human capital. The argument which was already developed by Bourdieu (1986) becomes increasingly more important in knowledge-intense economies. In this sense, Cohen and Prusak (2001) suggest that social capital serves as the main angle to improve knowledge management in organizations. These considerations are the starting point for many of the contributions to this book.

While the analysis of social capital has been grounded so far on the relationship among individual actors or among an individual actor and a social aggregate, Putnam (1993, 2000) equates social capital with the level of civic engagement in general. He applies the concept of social capital to cities, regions, and whole nations, and on this basis, understands social capital as the set of properties (e.g., norms, levels of trust, or social networks) associated with a social entity that enables joint activities and cooperation for mutual benefit. Putnam's perspective necessitates the question, which interaction exists between the level of civic engagement and the use as well as appropriation of information technology? Various researchers such as Wellman (2000) and Kraut et al. (1998) have studied this relationship in more depth. In chapter 5 of this book, Annabel Quan-Haase and Barry Wellman show that the Internet influences civic engagement by adding to already existing levels of social capital.

Beyond its usefulness for civic engagement, however, social capital is also an important ingredient for knowledge development in and between organizations (e.g., Cohen and Prusak 2001; Lesser 1999). It is argued that social capital investments inherently serve to motivate organizational members to share knowledge. This motivation is derived from being a member of a community in which shared norms, trust, cognitions, and experiences stimulate goodwill and reciprocity. Especially in knowledge-intensive organizations, this implies helping each other through the act of sharing knowledge. Such a viewpoint on economic institutions might be utopian, and social capitalists might indeed enforce their power over organizational members who are socially deprived. Nevertheless, the topic of social capital inspires many to remain optimistic about managing the key resource of most organizations: knowledge.

The introduction of the concept of social capital, in combination with those of community and communities of practice, have led to a fundamental shift in thinking about managing knowledge (cf. Ackerman, Pipek, and Wulf 2003). While the notion of human capital formed the core premise of the first wave of knowledge management theories, social capital can be seen as crucial for the second wave. *Human capital* relates to individual knowledge, individual capabilities to act on this knowledge, and individual learning. The first wave of knowledge management mainly centered around the issues of how to support the exchange of human capital so as to avoid unnecessary knowledge redundancy as well as to fill the knowledge gaps that exist as a result of mobility, globalization, and distributed work. Most prominent in this first wave of knowledge management were information technological solutions, such as knowledge repositories or content management systems. It has now become accepted that these technical solutions combined with the emphasis on sparking individual knowledge are problematic. What the first wave of knowledge management neglected was the importance of people's motivation to share their knowledge and learn from other people's knowledge. The second wave of knowledge management, although still in its infancy, asserts that communities will foster social capital, thereby increasing people's motivation to share knowledge (Huysman and De Wit 2002).

Investing in social capital means that long-term benefits such as social networks based on mutuality, trust, respect, and appreciation will last much longer than engineered networks such as organizational teams. It is difficult to point to the conditions that stimulate social capital. Nahapiet and Ghoshal (1998) argue that social capital has three dimensions: a structural dimension that includes, for instance, network structures; a cognitive dimension that incorporates shared stories, language, and culture; and a relational dimension that has to do with mutual trust and reciprocity. These dimensions are highly interrelated and difficult to segregate in practice. What makes the search for its characteristics even more difficult is the fact that social capital, like other forms of capital, accumulates when used productively (Fountain 1997).

Darker Sides

The dominance of the communitarian view, applied to both civic engagement as well as organizational action, is also present in this book. Most contributions take it for granted that high levels of social capital have positive effects on the sharing of knowledge and expertise, on community building and the development of creativity. We would fully agree with representatives of the conflict perspective such as Bourdieu (1986) that this volume is consequently biased and too optimistic. Although Mike Bresnen, Linda Edelman, Sue Newell, and Harry Scarbrough explicitly deal with both sides of the coin (see chapter ten), it is necessary to at least mention some of the negative aspects of social capital in this introduction.

Empirical findings from different case studies have revealed dysfunctional behavior within tight-knit social networks. Uzzi (1997) calls such ambivalent effects the "paradox of embeddedness." As reported in the literature, a high level of social capital can create the following problems:

- restrictions imposed on actors who do not belong to the network (Portes 1998; Cohen and Prusak 2001);
- a lack of perception concerning environmental changes outside the network (Cohen and Prusak 2001);
- negative social dynamics within the network and a downward leveling of norms (Portes 1998);
- a dependency on central actors and their loyalty toward the network (Uzzi 1997);
- restrictions on autonomy and individuality resulting from demands for conformity (Portes 1998);
- irrational economic behavior due to a feeling of solidarity toward partners in the network (Portes and Sensenbrenner 1993); and
- irrational economic behavior due to personal aversion (Uzzi 1997).

Information Technology

Most of the research on social capital is conducted by either social, political, economic, or organizational scientists. As a result, the range of topics is broad; it varies with the effects of social capital on individual career moves to regional development. It is thus interesting to note that although the introduction and acceptance of the concept of social capital in various academic disciplines has been significant, with some notable exceptions (e.g., Preece 2002; Resnick 2001; Lesser and Cothrel 2001; Huysman, Wenger, and Wulf 2003) the topic has not gained comparable attention from scholars concerned with information technology (IT) in organizations or society at large. This is, on the one hand, not surprising as we are dealing with rather different disciplines. Although computer scientists and information system scholars are increasingly open to incorporating social science research into their discipline, and vice versa on the part of social scientists, cross-fertilization between the various research fields is still not standard practice. On the other hand, IT scholars' limited interest in the topic of social capital is perplexing in light of today's "networked society." Indeed, the growing academic focus on networks within and between organizations makes research into the relationship between IT and social capital even more important. Since social capital is about connections among people, one has to ask, Is social capital influenced when these connections are supported by IT, and if so, how? In terms of IT development, how does one set up design processes appropriately? How does one design specific functionality to support social capital? Research is also needed in the other direction—namely, to what extent, if any, is social capital necessary in order to develop, customize, and appropriate IT? Moreover, can IT help us better understand the level of social capital within a network or community? The contributions to this book take the following perspectives: how does IT effect social capital, and how can specific technical functionality to foster social capital be designed?

In general, the relationship between IT and social capital seems to be an ambivalent one. High levels of social capital or strong, preexisting social networks, for example, are seen to be a success factor in establishing electronic-based networks (Fukuyama 1995). At the same time, the existence of IT creates networking infrastructure that encourages the formation of social capital (Calabrese and Borchert 1996). Yet high levels of social capital make communication by means of IT unnecessary (Kumar, van Dissel, and Bielli 1998). As researchers have also documented, existing IT possibilities that were intended to support communication infrastructures, do not support or create a sense of community. This ambivalence is present in this book as well, although there is a strong bias toward IT positively influencing social capital. The chapters by Anita Blanchard (chapter 3) and Anabel Quan-Haase and Barry Wellman (chapter 5) both discuss the dual effect of social capital on Internet usage: the Internet both positively and negatively influences social capital.

Overview of the Book

No single discipline is able to claim exclusive ownership of the social capital concept. This also mirrors the multidisciplinarity of the contributions to the book. *Social Capital and Information Technology* covers research from the computer sciences, sociology, communication studies, business economics, and management studies. The aim of this volume is to provide a meeting place for various communities of academics.

All thirteen contributions to this book share an interest in communities as the social entities in which social capital resides, even though the communities covered here vary to a great extent. For example, Blanchard explores on-line as well as off-line communities of athletes. Liisa Syrjänen and Kari Kuutti present research on Finnish dog breeding communities. Quan-Haase and Wellman discuss civic communities in a more general sense, and refer also to personal communities such as local and far-flung friends and kin. Markus Rohde looks at communities within and among various nongovernmental organizations (NGOs) in Iran. Gerhard Fischer, Eric Scharff, and Yunwen Ye examine Internet-based communities, among others, that are active in the open source movement. Robbin Chapman focuses on communities of children in afterschool programs. Charles Steinfield scrutinizes communities at an interorganizational level and discusses business-to-business hubs. Finally, several contributions share an interest in knowledge-processing communities within organizations, such as the chapters by Bart van den Hooff, Jan de Ridder, and Eline Aukema; Rob Cross and Stephen Borgatti; Huysman; and Mark Ackerman and Christine Halverson.

Analyzing the relationship between IT and social capital calls for a sociotechnical research approach, and all of the contributions to this volume have this background in the sense that they all approach technology as being part of its social environment. Several contributions are explicit about this relationship—such as Rohde; Ackerman and Halverson; Syrjänen and Kuutti; Huysman; and Fischer, Scharff, and Ye—while others explore IT in its social context but do not directly refer to the sociotechnical tradition. Only two contributions (Bresnen, Edelman, Newell, and Scarbrough; and Cross and Borgatti) do not discuss IT, focusing only on knowledge sharing in organizations and its pluses as well as minuses for social capital. We have included these two chapters because they add significantly to the conceptualization of social capital and also offer fertile ground for technological design. Cross and Borgatti investigate more fundamental issues of information use in social networks, providing empirical ground for IT design—for instance, for the matchmaking algorithms proposed in Andreas Becks, Tim Reichling, and Wulf's contribution. Bresnen, Edelman, Newell, and Scarbrough discuss the possible downsides of social capital, and as such, offer insights that counterbalance the enthusiasm of most of the other authors here.

Dealing with the issues raised above, the first collection of chapters analyze the effect of social capital and the role of IT on civic engagement. Part two represents another field of research: organizational learning and knowledge sharing. The five contributions each look at how IT might support or hinder knowledge sharing within communities, and as such, how that might influence the degree of social capital.

The last part of the book takes a computer science perspective. It presents computer applications that have the potential to augment social capital among their communities of users. The four chapters study eleven different applications. Most of the applications are designed to promote social capital as a way to overcome spatial or temporal boundaries by making users aware of each other or of artifacts others have created. In contrast, the Envisionment and Discovery Collaboratory presented by Fischer, Scharff, and Ye supports face-to-face conversations within given communities of interest. Among the systems that bridge spatial and temporal boundaries, topic- and member-centered communication spaces are classic examples. While member-centered communication spaces—such as the Babble or Loops systems presented by Ackerman and Halverson—foster social ties in an already well-defined community, topic-centered communication spaces—such as Zephyr in Ackerman and Halverson's contribution or the Experts Exchange by Fischer, Scharff, and Ye—allow people who are not necessarily well known to each other to exchange ideas or find solutions to problems. A key factor in motivating participation in topic-centered communication spaces seems to be the possibility of enhancing one's personal reputation. System design has to take this fact into account.

Beyond pure communication, applications may foster social capital by offering virtual spaces in which to create, develop, and store topic-centered materials. These repositories of materials are typically augmented by communication and annotation functionality. Editing tools support the development of materials, and they may have additional functionality to distill content out of communication spaces. The Answer

Garden, presented by Ackerman and Halverson, is one of the most influential approaches. While the general functionality of these systems may be similar, their concrete implementation is specific with regard to the topic at hand and the application domain. The Answer Garden was mainly built to encourage learning within organizations. Chapman's Pearls of Wisdom offers a rather sophisticated set of functionalities to foster knowledge sharing and social capital building among teenagers in after-school community centers. The systems discussed so far offer places in virtual space where human actors can go to strengthen existing social ties or build new ones.

In another class of applications, the system takes a more active role in suggesting actors with whom one can establish or refresh social ties. Such applications require personal data from the different human actors along with domain-specific algorithms to match actors appropriately. Moreover, they need to offer access to appropriate communication channels. The Expertise Recommender presented by Ackerman and Halverson matches developers in a software company based on their past work activities on certain software modules. Becks, Reichling, and Wulf detail an algorithmic framework that matches actors based on different sets of personal data and corresponding algorithms. They apply the framework to the problem of matching actors with similar interests in e-learning platforms. Finally, Fischer, Scharff, and Ye examine a recommender system that does not propose human actors but software components created by others. They argue that the CodeBroker will help make shared repositories of software components more useful to software developers, and thus, augment the social capital within software development communities.

Research Challenges

During the past two decades, social, political, organizational, and economic sciences have witnessed a rapid introduction of a fairly new concept. The notion of social capital was greeted with much enthusiasm, especially in academia. Until now, however, the concept has not been analyzed in relation to IT. With the collection of papers in this book, we hope to demonstrate that research on this topic is valuable and deserves more attention. We are, of course, aware that this volume is only a first step toward bridging different academic fields. This collection is also by no means a complete sample of possible research topics.

As mentioned earlier, while the concept of social capital was initially approached from both a conflict as well as communitarian perspective, gradually—and perhaps largely due to enthusiasm—the communitarian approach has come to dominate the debate. The collection of papers here should be seen in this light. Future research needs to focus also on the darker sides of social capital. IT is able to connect people, but at the same time it contributes to depriving those who are not connected. Although this growing digital divide is not addressed in this book, it surely needs more research attention.

Another example of a promising research field that is only explored here by Steinfield is social capital in relation to e-business. Given a new dimension by the use of electronic networks, interorganizational cooperation is nowadays often discussed in terms of business-to-business marketplaces, supply chain management, virtual organizations, or strategic alliances. Many failed attempts to implement these approaches can be attributed to the lack of attention to the issue of social capital. We also need to investigate how new types of communities—for example, on-line communities—will change the relationships between producer and consumer. New business models may be created that allow for a closer interaction among consumers and among consumers and producers.

Also, with the exception of the contribution by Syrjänen and Kuutti, this book is biased toward short-term evaluations of the effect of IT on social capital. Until long-term empirical evaluations look at the social capital impacts of IT, discussions about the effects of IT remain speculative. As various researchers (e.g., Barley 1986; Pipek and Wulf 1999) have demonstrated, the appropriation of IT and its consequences need in-depth, long-term analyses. Hence, this new field of research requires more longitudinal studies.

Furthermore, this book does not discuss how IT can be used to analyze social capital. Research done by Noshir Contractor, Dan Zink, and Mike Chan (1998), for instance, provides valuable insight into tools to investigate diverse aspects of social networks in order to study the social capital of, say, organizations. Finally, new technologies such as large-screen displays or wearable devices will offer opportunities to design innovative applications that attempt to impact the level of social capital within social entities.

Given all this, we believe this book offers an interesting collection of research on social capital and IT.

Note

1. In criticizing the cognitivistic mainstream, sociocultural theories of learning emphasize the collective nature of knowledge acquisition. Learning takes place in informally defined social aggregates: so-called communities of practice (cf. Lave and Wenger 1991, Wenger 1998).

References

Ackerman, M., V. Pipek, and V. Wulf, eds. 2003. *Sharing expertise: Beyond knowledge management*. Cambridge: MIT Press.

Barley, S. R. 1986. Technology as an occasion for structuring: Observations on CT scanners and the social order of radiology departments. *Administrative Science Quarterly* 31:78–108.

Bourdieu, P. 1980. Le capital social: Notes provisoires. *Actes de la Recherche en Sciences Sociales* 31:2–3.

Bourdieu, P. 1986. The form of capital. In *Handbook of theory and research for the sociology of education*, edited by J. G. Richardson. New York: Greenwood Press.

Burt, R. S. 1992. *Structural holes: The social structure of competition*. Cambridge: Harvard University Press.

Calabrese, A., and M. Borchert. 1996. Prospects for electronic democracy in the United States: Rethinking communications and social policy. *Media, Culture, and Society* 18:249–268.

Cohen, D., and L. Prusak. 2001. *In good company: How social capital makes organizations work*. Boston: Harvard Business School Press.

Coleman, J. S. 1988. Social capital in the creation of human capital. *American Journal of Sociology* 94:95–121

Contractor, N., D. Zink, and M. Chan. 1998. IKNOW: A tool to assist and study the creation, maintenance, and dissolution of knowledge networks. In *Community computing and support systems, Lecture Notes in Computer Science 1519*, edited by Toru Ishida. Berlin: Springer-Verlag.

Etzioni, A. 1993. *The spirit of community: The reinvention of American society*. New York: Touchstone.

Etzioni, A. 1995. *New communitarian thinking: Persons, virtues, institutions, and communities.* Charlottesville: University Press of Virginia.

Fountain, J. E. 1997. Social capital: A key enabler of innovation in science and technology. In *Investing in innovation: Towards a consensus strategy for federal technology policy*, edited by L. M. Branscomb and J. Keller. Cambridge: MIT Press.

Fukuyama, F. 1995. *Trust: The social virtues and the creation of prosperity.* New York: Free Press.

Granovetter, M. 1985. Economic action and social structure: The problem of embeddedness. *American Journal of Sociology* 91, no. 3:481–510.

Hanifan, L. J. 1916. The rural school community center. *Annals of the American Academy of Political and Social Science* 67:130–138.

Huysman, M. H., and D. De Wit. 2002. *Knowledge sharing in practice.* Dordrecht: Kluwer Academics.

Huysman, M. H., E. Wenger, and V. Wulf. 2003. *Communities and Technologies.* Dordrecht: Kluwer Academics.

Kraut, R. P., V. Lundmark, S. Kiesler, T. Mykhopadhyay, and W. Scherlis. 1998. "Internet paradox: A social technology that reduces social involvement and psychological well-being?" *American Psychologist* 53, no. 9:1017–1031.

Kumar, K., H. G. van Dissel, and P. Bielli. 1998. The merchant of Prato—revisited: Toward a third rationality of information systems. *Merchant Information Systems Quarterly* 22, no. 2:199–226.

Lave, J., and E. Wenger. 1991. *Situated learning: Legitimate peripheral participation.* New York: Cambridge University Press.

Lesser, E. 1999. *Knowledge and social capital: Foundations and applications.* Boston: Butterworth-Heinemann.

Lesser, E., and J. Cothrel. 2001. Fast friends: Virtuality and social capital. *Knowledge Directions* (spring/summer): 66–79.

Loury, G. C. 1977. A dynamic theory of racial income differences. In *Women, minorities, and employment discrimination*, edited by P. A. Wallace and A. M. Le Mond. Lexington, Mass: Lexington Books.

Nahapiet, J., and S. Ghoshal. 1998. "Social capital, intellectual capital, and organizational advantage." *Academy of Management Review* 23, no. 2:242–266.

Pipek, V., and V. Wulf. 1999. *A groupware's life.* In *Proceedings of the sixth European conference on computer-supported cooperative work.* Dordrecht: Kluwer.

Portes, A. 1998. Social capital: Its origin and application in modern sociology. *Annual Review of Sociology* 24:1–24.

Portes, A., and J. Sensenbrenner. 1993. Embeddedness and immigration: Notes on the social determinants of economic action. *American Journal of Sociology* 98:1320–1350.

Preece, J., ed. 2002. Supporting community and building social capital. Special edition of *Communications of the ACM* 45, no. 4.

Putnam, R. 1993. The prosperous community: Social capital and public life. *American Prospect* 13:35–42.

Putnam, R. 2000. *Bowling alone: The collapse and revival of American community*. New York: Simon and Schuster.

Resnick, P. 2001. Beyond bowling together: Sociotechnical capital. In *Human computer interaction in the new millennium*, edited by J. M. Carroll. Boston: Addison Wesley.

Uzzi, B. 1997. Social structure and competition in interfirm networks: The paradox of embeddedness. *Administrative Science Quarterly* 42:35–67.

Wellman, B. 2000. Computer networks as social networks. *Science* 293, no. 14:2031–2034.

Wenger, E. 1998. *Communities of practice: Learning, meaning, and identity*. New York: Cambridge University Press.

Williamson, O. 1975. *Markets and hierarchies*. New York: Free Press.

I
Social Capital in Civic Engagement

Part I of this book covers research papers that look at social capital and IT at the societal level. All four chapters in this section analyze the role of the Internet in general or specific computer applications in terms of community building. The authors investigate the impact that these networked applications have on an already existing sense of community in different social aggregates.

Chapter 2 presents an interesting case study conducted by Anna-Liisa Syrjänen and Kari Kuutti on a community of hunting dog enthusiasts, who breed dogs for big game hunting using an information system over many years. The study follows how this community defined a new direction for itself by developing its information system, and how this orientation spread among its members and became the accepted principle. The trust, acceptance, and alignment needed for cooperation were generated as well as maintained via daily interactions among the community's members. This process was facilitated by the information system. Gradually, the information system became such an integral part of the membership's interactions that it eventually constituted the infrastructure vital to the day-to-day functioning, self-determination, and development of the community. The most significant role of the information system was that it could display relevant, socially explicit knowledge, thereby allowing for the emergence of trust and social capital. The knowledge presented had its tangible roots in reality since it was firmly linked to concrete objects, events, artifacts, and people through social interaction processes. This empirical case highlights that at least in this way, IT can help to generate and maintain the trust and acceptance that are highly valuable elements in long-term cooperation.

In chapter 3, Anita Blanchard empirically examines the effects of a dislocated virtual community on the emergence of social capital in face-to-face (FtF) circumstances. The virtual community, Multiple Sport Newsgroup (MSN), is a newsgroup for athletes. It is a dispersed virtual community because it is not associated with any corresponding physical place. Members participate in this virtual community from all over the world. To understand how participating in MSN affects members' FtF social capital, the networks of social interaction among members both in MSN and FtF, the norms of behavior in MSN and FtF, and the trust members felt for each other in MSN and FtF were examined using different methods. Data were collected using participant observations along with semistructured interviews with leaders, participants, and lurkers in the virtual community. Results show that active participation in MSN can positively affect FtF social capital if virtual and FtF social networks of interactions overlap. Norms of behavior and trust among members move between the virtual and FtF communities, positively influencing both. Members who were active in their FtF communities reported being more strongly attached to the FtF community than to the virtual one. Finally, it was determined that lurkers make up the vast majority of MSN members; lurkers also may not participate in their FtF communities either. So the case study provides insights into the impacts of social capital gained in a virtual community on social interaction in the physical world.

The fourth chapter, by Markus Rohde, looks at how IT might support the development of social capital among NGOs in Iran. Iranian civil society is developing quickly these days. Hundreds of NGOs are engaged in a process of socio-organizational networking, which is mainly coordinated by two resource centers in Tehran. Chapter 4 presents experiences with a project that supported this networking process by sociotechnical means. The concepts of *social capital* and *communities of practice* are applied to provide a theoretical grounding for sociotechnical interventions. The introduction of an Iranian NGO community system—a process model of integrated organization and technology development—is described. Rohde analyses training measures as well as efforts aimed at socio-organizational and technological developments. He presents mechanisms of both participatory system design and coordinating the networking of an e-community. This chapter shows that the processes of IT introduction can be an important trigger for the development of communities within so-called developing countries, and as such, for the development of social capital.

In chapter 5, Anabel Quan-Haase and Barry Wellman provide a more general and conceptual perspective on the relationship between the Internet and the development of social capital at a societal level. They argue that the Internet has been proposed as a key catalyst of social change: utopians predict that the Internet will rebuild tight-knit communities, while dystopians fear that the Internet will lead to isolation and draw people away from engaging in their communities. These views, based on technological determinism, assume that the Internet plays a dominant role in changing society. In this chapter, data from several studies on the uses and effects of the Internet on social capital are reviewed. The authors situate their analysis in recent discussions about the decline in North Americans' social capital, including debates concerning social contact among friends and kin, community participation, and engagement in politics. The evidence gathered by Quan-Haase and Wellman suggests that the Internet occupies a crucial, but not dominant, place in everyday life, connecting friends and kin both near and far. Contrary to technological determinist views, the Internet is adding on to—rather than transforming or diminishing—social capital. Those who use the Internet the most continue to communicate by phone and through face-to-face encounters. Although the Internet helps connect far-flung communities, it also connects local ones.

2
Trust, Acceptance, and Alignment: The Role of IT in Redirecting a Community

Anna-Liisa Syrjänen and Kari Kuutti

The term *social capital* has gained popularity during the last half dozen years. The origins of this popularity can perhaps be traced to the writings of such U.S. sociologists as Robert Putnam and Francis Fukuyama. Initially, the term was used mostly in sociological discussions, but it has recently spread to other fields as well. Thus, social capital has also been connected with the design and use of IT. One of the visible signs that the term is now taken seriously within IT research communities is Putnam's appearance as the keynote speaker at the 2000 Computer Supported Cooperative Work conference, where he gave a talk on *Bowling Alone*, centering on the demise of civic networks (along with social capital) in U.S. society.[1]

One factor that has served to increase interest in concepts like social capital has been the recognition of the fact that the unregulated market mechanism (or what Fukuyama in 1999 called "Free-Market Liberal Democracy"), which has been intensively tested and tried around the world since the 1980s, does not automatically provide the best or even cheapest solution for customers. In particular, the privatization of public services has often led to disaster—for example, when a company with the lowest bid fails to deliver the promised service, the clearing up of the mess afterward increases the costs enormously. If society is reduced to market transactions, it will not work well; it may even start to derail. As a result, interest in the social "glue" that holds societies together has increased considerably. Another related area has been the search for alternative forms of organizing (commercial) relationships with a focus on different forms of networking—after all, some form of social glue is vital for networks as well. As a term, social capital seems to offer promising possibilities for the analysis of such complicated situations.

The concept of social capital is quite hazy and fluid in the sense that different authors give different meanings to it. Putnam defines it as "features of social organization, such as networks, norms, and trust that facilitate coordination and cooperation for mutual benefit" (1993, 35). This is quite a broad definition that lumps together a variety of apparently different issues, and it is not so clear at all how it can be operatively used as an analytic tool.[2] Both Putnam and Fukuyama define social capital as individual assets, although both also discuss its usefulness in decreasing transactions in a market society by decreasing the need for explicit contracting. Because individuals are rarely involved in transactions that require voluminous contracts, it can be assumed that somehow social capital may also be applicable to relationships among companies.

It is interesting to note that somewhat similar concepts of capital have in fact been discussed earlier by Pierre Bourdieu (1986), who introduced the concepts of *cultural capital* and *intellectual capital*. Bourdieu sees cultural forms of capital as a tool in defining social and cultural fields. Cultural capital is possessed personally, and it is to a certain extent interchangeable with "real" capital. Examples of cultural capital include education, social skills, degrees earned, and social networks created during a lifetime. All this capital can be mobilized to certain ends, if necessary. Bourdieu's definition is much stricter than other definitions of social capital, and he has shown that it can be applied analytically—at least within the social fields he himself has defined. As far as we know, his work has not been used to analyze the problems Putnam, Fukuyama, and others are interested in.

Leaving Bourdieu aside, there seem to be two main groups of researchers with an interest in social capital. First, there are the traditional sociologists like Putnam and Fukuyama, whose primary concern lies in analyzing decline of civic society (that of the United States in particular). This group also includes their followers, who are keen on stopping the perceived decline and improving the situation. The other group consists of organizational/economics researchers like Eric Lesser (2000) and Janine Nahapiet and Sumantra Ghoshal (1998), who extrapolate the validity and usefulness of social capital from Putnam's and Fukuyama's individual perspective to an organizational perspective, and whose focus is on measuring, quantifying, and even commodifying social capital at the service of business, for instance. Their concern is not so much how social capital can be used in analyzing something; rather, they center on making it a "tangible" asset for businesses and then increasing it.

Both of these approaches appear with some hitches. The decline of civic society is a many-faceted issue that necessitates analyses at a level deeper than social capital per se. In fact, Bourdieu's ideas about the dynamics of social fields and the role of cultural capital in these dynamics might be applicable for that purpose. As for the operationalization of social capital for organisational purposes, our view is that the jump from individuals to whole organisations is still ungrounded. It may be more sensible at this moment to concentrate instead on narrower perspectives—for instance, that of actual working communities.

In terms of IT, the main driving force behind the interest in social capital seems to arise from the potential of using networks and the Internet in particular in electronic commerce. One of the bottlenecks in the creation of viable e-business lies in generating trust in a shop or business partner that can be accessed only through their Web pages. It is also easy to lose trust in e-commerce: a well-known example is provided by the case of Amazon.com when it transpired that the company had accepted money from book publishers in return for suitable "recommendations" to its customers. Cases such as this one illustrate the need to understand and manage trust in individual transactions over electronic markets. Concomitantly, there is a strong interest in finding out how such transactions could be supported by IT. A more sophisticated variant of this interest stems from the observation that the Internet has enabled the proliferation of various spontaneous network communities that seem, at least in some ways, to resemble the so-called natural communities that researchers like Putnam and Fukuyama regard as ideal models for the development of social capital. It is intriguing to explore whether network communities are governed by similar social relationships as real communities, and whether this phenomenon can be utilized in e-commerce.

In spite of these limitations, we believe that the concept of social capital per se may serve as a working label to point to issues in generating and maintaining the trust, acceptance, and alignment necessary for successful cooperation, while also highlighting that these cannot be separated from their context, domain, and culture. This social capital manifests itself in the form of alignments, conflicts, learning, and the shaping of knowledge, for instance. In this study, we focus on the role of IT in processes and interactions among participants where the social capital manifests itself. We shall not study the topic from the perspective of individuals but adopt the perspective of long-term cooperation, which has primarily other objectives than just financial benefits. Further, we embrace the view presented by Reijo Miettinen (2001) that social capital should not be studied in the abstract, with no regard for the motive and content of the network collaboration. Miettinen calls trust that is created and reproduced in shared projects and activities *activity-based trust*, and contends that it emerges from a concrete historical activity. Our purpose in this chapter is to illustrate one case of such activity-based trust and the role of IT in its emergence.

Research Settings

The work presented in this chapter concentrates on examining how the information system (IS) of a community functions as a material artifact in which the social capital of the community can be rooted, as suggested by Yrjö Engeström and Heli Ahonen (2001). The concept of social capital will be applied to the community under study by describing the use of the community's IS as a tool for both generating social capital and addressing the results of this process. In assuming that evolving trust in the formation of social capital usually will not happen suddenly—to the contrary, it takes time and unfolds "from the inside"—several types of sources are thus needed to highlight the dynamic and ambiguous nature of the topic. Actual informants were selected based on their active use of the IS and/or a few years of experience with the domain through involved participation including some publishing outputs. In addition, they were—and in some cases still are—holding positions that demand continual contact with other members in the community. Regarding the topic, they therefore could have a good overall view of the core practice. It is important to take in account, however, a whole organizational structure containing viewpoints of activities from the grassroots to the management level. In this particular field where it is typical that all participants are involved in several activities, the selection base of informants could be restricted appropriately so that all necessary activities are considered. All this was complemented by the longtime participation and position of one of the authors in the field. The selected informants, then, can be considered as forming a kind of microcosm community in relation to the study of the formation of social capital.

Working with members of the community both retrospectively and in real time has enabled the use of several types of data sets. Data have been gathered through thematic interviews with ten persons whose accuracy the informants checked and commented on afterward. In the context of a field study, note-taking has typically also been used when working with informants. To study how the situation has changed over the years in relation to a wider context, several types of recorded material published between 1958 and 2002 were used as well. These data include fact-based annuals, club magazines, anniversary and historical books, related videos, and so forth, along with the community's database IS and the Web site, which contains factual information on the case community's practices from the 1930s to the present. Related comments on the Web discussion site have also been used in making analyses. Thus, through these outputs, several members of the community contributed to the results indirectly, but the research focused on activities of the actual informants. Regarding data collection, the choice between primary and secondary sources was mainly estimated based on the data artifacts' use in the context of the case community's practice and/or their original purpose.

In accordance with Lesser, Fontaine, and Slusher (2000), *community* is defined as a collection of individuals who interact regularly within a common set of issues, interests, or needs, and this involvement is often voluntary. *Community knowledge* is seen here as local, particular, situational knowledge about a particular activity around a particular field, used as a resource in 'knowing' in order to make the activity happen (Virkkunen and Kuutti 2000; Kuutti 2001). This community knowledge, moreover, contains all kinds of knowledge that can be revealed in one way or another, including human beings as bodily actors. Hence, community knowledge can be shared with others and used in collective sense-making processes. Our work stresses objective, or socially explicit, knowledge (cf. Spender 1996), which as will be shown, can be considered an important incentive for the emergence of trust in the formation of social capital. The IS is viewed here through a wide lens, and it includes information sources, tools, processing, outputs, and organizational arrangements (Hirschheim, Klein, and Lyytinen 1995, 11). This perspective is attached to the concept of *infrastructure* as defined by Susan Leigh Star (1999). Infrastructures, Star contends, have such properties as embeddedness, transparency, reach, or scope; they are learned as part of membership and linked to conventions of practice, even though they can also cause conflicts within them. Furthermore, infrastructures are built on some installed bases. This refers to the fact that they do not suddenly come into existence but are grounded on some existing substructures, and oftentimes inherit their strengths and limitations.

In what follows, this chapter will first provide background information on the case community and its special domain. It will then describe the main phases in the development of the IS, and the parallel processes of shaping community knowledge and generating trust in the formation of social capital. Having described the evolution of these processes and the role of the IS in them, this chapter will end with a discussion of the results and conclusions.

The Case Community and Its Domain

The case community is that of Karelian bear dog (KBD) enthusiasts affiliated with the Finnish Spitz Club (FSC). The community breeds dogs for big game (elks, bears) hunting using an IT-based IS. The community has over two thousand members from a variety of educational and social backgrounds. Many of them live in rural areas, where the activities of the hunting dog hobby can take place more easily. Membership is based on ownership of a KBD, but participation in the community's events is voluntary, and in that sense the members' degree of activity is varying. The community also has some active members abroad—in Sweden, France, and Canada, for instance. The actual dog-breeding activity involves three interacting groups, which overlap to a degree. The first is composed of members in the FSC organization, the KBD division, and breeding consultants. This group is responsible for providing practical guidance to the members. The consultants are among the prime end users of the IT-based IS, and they serve other members almost on a daily basis by working at home during their spare time. The second group consists of the breeders (owners of dams), who put the consultants' breeding recommendations into practice, if they agree with them. The third group is made up of dog buyers along with the owners of sires and other dogs not yet used for breeding. Usually, then, everyone concerned is a KBD owner. All breeding services are free of charge. The community uses several forms of interaction to maintain communication, including phones, mail, and e-mail. Events that bring the community together include trial matches while the general meetings of the FSC, Breeding Days, and so on enable even larger gatherings of enthusiasts (Karelian Bear Dog Division 2001).

From the viewpoint of the members, the KBD's most important characteristic, often called its "hunting instinct" or "hunting ability," is problematic in more ways than one (Syrjänen, 2002, field notes). It has been said that "all dogs have this instinct," but not all dogs use it for hunting so as to get food for the pack—the original purpose of this feature (ibid.). It is not clear why some dogs turn out to be good hunters while others are not capable of hunting any game, or as some people say, some dogs are just "not interested" in game at all (ibid.). What is the distinction between these dogs? This so-called mystery has fascinated hunting dog people since the KBD was domesticated. Like a mental feature, this instinct is invisible to the human eye, making it somewhat difficult to understand how it can be deliberately bred at all. For instance, how do you choose the best pair of dogs to breed or pick out a new puppy, which should have this desirable hunting instinct?

The KBD is one of the original dog breeds in Finland, and its breed organization is the FSC, which is affiliated with the Finnish Kennel Club (FKC). To guide the breeding activity, the FKC collaborated with the FSC to approve rules for elk hunting trials in 1943, and the first trial under these rules was held in 1945 (Kilkki 1987; Perttola 1998). The current rules were introduced in 1991. An elk hunting trial is a field event held in a natural environment during the hunting season, but the actual hunting event is imitated, enabling the dog to test its skills. During the five-hour trial, the dog is free (it may have a radio locator in his/her collar) and trained judges observe every detail of its behavior. The results are assessed in cooperation with an experienced chief judge (Finnish Kennel Club 1991), and after being double-checked by experienced trial records controllers, the results are entered in the database.

Up until the late 1980s, before the new IT-based IS was set up, a manual system was in use. This system contained information on about 20,000 dogs, 5,000 dog shows, and 4,000 hunting trial results, stored on paper in various forms (Finnish Spitz Club 1987; Perttola 1998). The FSC started to develop its new database IS in 1987 in order to facilitate its annual publishing work (chiefly, books presenting the results of hunting trials, dog shows, etc.). The KBD community introduced the new IS around 1990 to be used in breeding counseling, and such functions as registration management and pedigree browsing were continuously in use. Then an application called "Breed" was designed to calculate degrees of inbreeding (a tool for determining how closely any two dogs are related). The system was extended by the mid-1990s with a tool called "Male Recommendation," which automated the process of selecting a male dog by means of predetermined quality conditions. Many of the more recently designed functions attempt to address the development of the practice and provide statistical information, including a tool called "Breeding Points" (1996) and another labeled as "Breeder Collation" (1997). As the former is used to judge dogs' inheritance ability and the latter what kinds of dogs the KBD breeders have produced, they are both concerned with the hunting instinct. Web site applications (since 1998) form the newest parts of the current IS, which today is a kind of practitioner-made decision-support system (see figure 2.1). It consists of three databases and hundreds of different functions for information processes, and contains information on about 26,000 dogs, 14,000 shows and 13,000 trial results, health scanning records, and so on. Around 900 new KBDs are registered each year. All the development work was carried out voluntarily and without pay by FSC hunting dog practitioners. The outputs of the system, utilized to provide recommendations for breeding as well as statistics, are published in annuals or club magazine articles, or appear on the Web site, for instance.

Evolution Phases of the Case with Commentary

The First Episode: The IS and Its Cultural Background

KBD breeding started in 1936 with the goal of "creating a sturdy dog that barks at big game" (Finnish Kennel Club 1996). The basic stock of dogs originated from an area of Finland known as Karelia, where they were used for different types of game hunting. In accordance with that purpose and "its legendary background" in Karelia, "where game was plentiful" (Perttola 1998, 72, 21), the dogs were referred to as "all-round game dogs," and later as "traditional bear dogs," to preserve the view that the KBD is "a valuable part of Finnish cultural history" (Karelian Bear Dog Division 2001). Soon after breeding was started, the Winter War of 1939 broke out, and Finland lost precisely that part of Karelia where the stock dog population had been collected. This nearly destroyed the breed as "the eastern border area ended in the destruction of various kinds of dogs" (Perttola 1998, 33), forcing breeders to continue their work with only a handful of dogs inland. The war broke out again after short period of peace, and since the Finnish front remained for years in eastern Karelia, a number of black and white or greyish dogs called bear dogs were collected, but "if all the Bear Dogs that existed in the 1930s in Finland had been involved in the formation of the breed, our chances would have been much better than they were in reality" (ibid., 54).

As to the IS of the community, in those days there was some knowledge on appearance breeding, but little knowledge available as far as the hunting ability of dogs was concerned. As a result, KBD breeders began conducting experiments intuitively on their own dog reserves (Perttola 1998, 103). The main method up until the late 1980s was known as "linear inbreeding," which maintained that "the best results are achieved with interrelated dogs" (Miettinen, Joenpolvi, and Sirviö 1987, 100). This was a commonly accepted view among dog experts at that time, and thus it was recommended by KBD breeding consultants as well. By the mid-1980s, the KBD population included thousands of dogs, and a certain body of top-level hunting dogs had been bred. Nevertheless, "the peak was a narrow one," and after fifty years of breeding attempts, breeders were still suffering from a lack of good breeding dogs (especially females) (Syrjänen 2002, audiotapes). Moreover, "public opinion was not so convinced that the KBD could be a good elk hunting dog on a larger scale" (ibid.). When the breeding consultant retired, no one seemed willing to take on this responsibility. Once a new consultant was brought onboard around 1986, he found that the manual IS—containing a rather large, incoherent collection of related information on paper—was limited to guidelines as to how his duty should be carried out. The practical experience that would have helped him was hidden "in the heads of old men," the earlier experts and breeders (ibid., field notes).

Commentary
The above account illustrates the breeding practice of the early days from the 1930s to the middle of the century, indicating how the activity was to some extent grounded in visions of a lost, "legendary" setting in Finland. This notion of the past included such vague elements as traditions, breeders' own experiments, and "inadequate stories of hunting and [old, undocumented] records of hunting trials" (Perttola 1998, 77). Since useful knowledge was seen as existing in the heads of old men, there was an obvious lack of more tangible, socially explicit information sources. In addition, many breeders preferred to work alone, limiting the scope of shared practical knowledge. To further complicate matters, some breeders seemed more willing to sit on their knowledge than share it as, for instance, certain dogs were advertised as having been bred using the "hidden knowledge of old-timers" (Syrjänen 2002, field notes). Factors such as these, coupled with an unwillingness to take on the breeding counseling duty, made for weak or even missing links in knowledge sharing, thereby also indicating a lack of trust and social capital. The ideal breeding goal of "creating a sturdy dog which barks at big game," typifies the general nature of the guidelines for achieving hunting abilities among KBDs. Breeders were forced to develop the KBD breeding activity mainly based on their own insights. Furthermore, "breeders who did not much care about a hunting dog were continually joining the breed" (Perttola 1998, 79). As a result, "without knowledge of its background and without studying it, the KBD was accepted here as a dog that hunts big game . . . in practice an elk-hunter . . . but after a while some hunters began to make complaints" (75). The inbreeding method

was also accepted without evidence of its utility for KBD breeding. Yet there were few alternatives. Considering that the stock breeding population of KBDs was rather small and that new dogs were not available, linear inbreeding was the best available method to breed any dogs under such circumstances.

It is a curious fact, however, that even in the late 1980s when there already was a lot of information on paper, such information could not support the breeding practice to the extent one might expect. One reason was that the manual IS was incapable of processing source data sufficiently, resulting in a large untapped stock of information. Breeders bred "what was on display," and they ended up developing dogs according to that attitude. Some good hunting dogs may thus have been taken out of consideration—many bear dogs were already disqualified from the group of breeding dogs in the early years (Perttola 1998)—especially if their appearance didn't match the quality defined in the standard (ibid., 52, 71). In addition, many breeders themselves might have shared the above attitude (cf. Miettinen, Joenpolvi, and Sirviö 1987).

The Second Episode: The New IS and One-to-One Relationships
Given this situation in the late 1980s, the new consultant decided to use his common sense: "If we want to have elk-hunting dogs, then breeding dogs should bark at elks and their external appearance is less important" (Syrjänen 2002, audiotapes). He started to make his own tools, such as hierarchically ordered charts, to which he added further information obtained from several sources. This system functioned sufficiently well for a while, but it was hard to maintain, and the limitations of the manual tools became more apparent as information accumulated over time. Although he found some good breeding combinations, "the results were not as good as I and everybody else had expected" (ibid.). In 1987, the FSC started to develop a computerized database IS; its "main purpose was to help annual publishing work" so that "people could get more objective information of how things really are," one informant explained (ibid.). The KBD community acquired a new IS around 1990. Soon after its introduction, the breeding consultant rated it as a useful tool for planning new dog pairs and offering recommendations to breeders. He began to more closely study the pedigrees of the dogs as well as their trial results

(i.e., records of dogs' behavior made by trained judges during elk-hunting tests)—an undertaking facilitated by the fact that "all necessary information was in one place and each dog could be compared against a larger population" (ibid.). As expected, the statistics derived from this big stock of information supported the notion that the dogs who had been successful in the hunting trials by passing the test also had more successful offspring than other dogs. This visible, objective evidence revealed by the IS also enabled the consultant to convince many of his customers, especially new breeders, of the worth of this study, especially because the breeding practice conformed to their commonsense idea of heredity. The outcome was that breeders obtained more and better dogs than before, and an increasing number of interested customers turned to the consultant for advice.

Commentary
Through the integration of the data offered by the new IT-based system, the IS put breeding counseling on a more solid footing by linking it to the concrete setting of breeding and hunting practices. Breeding and hunting trial results could more easily be processed together. By confirming the role of hunting trials as a relevant test tool for achieving good breeding results, the people concerned seemed to switch from using subjective, personal, and "in those days—someplace" data to the factual, codified, and more "here-and-now" information that was revealed as a whole by the new IS. That option was previously unavailable since the old system could only display small bits of integrated information at a time in the form of manually produced annuals, and the like. Moreover, under the old system, the top results of breeding work were better recognized in public while lesser quality results were often overlooked. Thus, the IS made the everyday quality of the breeding practice more visible as a whole, allowing measures to be taken to develop it and thereby create new knowledge by integrating data. In that sense, the form of the source files and their suitability for processing as well as the IS's capacity to process data seem to be crucial. The new system and reduced need to process data enabled the utilization of the extensive stock of information amassed earlier. As a result, an increasing number of members became more interested in using the information made avail-

able by the new system and mediated by the consultant during one-to-one counseling events. Facts such as these show that the IS assumed the role of an information medium by transforming information on the breeding practice, and as such, facilitated the creation and exchange of meanings that serve sense making and argumentation (cf. Hirschheim, Klein, and Lyytinen 1995, 11), first to its actual users, and then through them to other members. Information became a communal resource, and public awareness of its existence and use increased. While reviving the dealings of the members, the system was supporting the emergence of social capital, too.

The notion that this new information was "more objective" than that provided by the manual system was based on the fact that it was derived from a big stock of information and that the quantity of source items could be cited exactly. This could be a difficult task with "a sackful of papers," as one informant described earlier forms of breeding data (Syrjänen 2002, audiotapes). Given that the new IS enhanced the credibility of the information, this also implies that a nonhuman artifact is seen as more capable of processing a large quantity of information than a human being (e.g., Latour 1992). In spite of the obvious improvement in processing data, however, the objective evidence also denotes the best feature of the new IS: it made visible the value of the socially explicit system of hunting trials as a way to inspire trust and confidence. Official rules, trained judges, the assessment of results, and so forth made the codified test results in the database more reliable than the subjective stories of individual members, and they thus offered a good basis from which to make comparisons. As one informant put it, "On the phone, it sounds that certain dogs bark so well" in hunting and the off-the-record tests (Syrjänen 2002, field note). Yet this information is not comparable as such to anything, albeit "it certainly has to be true" (ibid.). Callers were usually advised to test the dogs in hunting trials. Obviously those dogs are good hunting dogs, but are they good breeding dogs, too? It is difficult to say without objective, comparable information. In order to develop better breeding practices, the information used should be comparable (cf. Miettinen, Joenpolvi, and Sirviö 1987) against some relevant, codified base or standards such as the hunting trial system. Actual hunting dealings are changeable under multiform circumstances, which are difficult to represent comprehensively enough since all hunters are not qualified as trial judges. Hence, actual hunting dealings don't necessary meet the needs of comparison, although they certainly serve as a relevant feedback channel.

The Third Episode: The New IS and One-to-Many Relationships

Some time after the introduction of the new IS, the consultant became aware of a thorny problem: some breeders who had used good male hunting dogs as sires for years and had managed to breed a number of good dogs were no longer successful. At first, when the Breed application (i.e., the software for determining how closely any two dogs are related) was adopted, "no one believed that it could be useful for bear dogs," as the consultant thought (Syrjänen 2002, audiotapes). Most breeders put their trust in the linear inbreeding method to produce good dogs. While the health risks of the method were known—at least to some degree—they were not seen as causing visible harm as "the KBD has been lucky, the breed is still healthy" (Miettinen, Joenpolvi, and Sirviö 1987, 100); in short, there was no reason to break the long, tried-and-tested tradition. Thus the new tool was considered useless at first. After using it, the consultant discovered something surprising, and it inspired him to more deeply explore how inbred the dogs were. A representative sample of a thorough analysis of the results of hunting trials compared with inbreeding degrees confirmed his observation: dogs with a high degree of inbreeding were less successful than those with a lower degree. The consultant also noticed that inbreeding had caused nearly unavoidable "bottlenecks" in the planning of new pairs of dogs: "Some very popular male dogs could be found in the pedigrees of almost every dog" (Syrjänen 2002, audiotapes). Hence, when the second version of the tool was introduced around 1993, it had already shown itself to be capable of supporting the breeders, who were challenged to cooperate rather than work on their own. The consultant's notion reopened the health risk question in regard to the inbreeding method, but above all it made the new, significant problem—could inbreeding work against hunting ability?—regarding its continuous usage visible and allowed it to be taken into account by the Breed application. The Breed outputs served as arguments against too tight inbreeding, and they were presented at

FSC meetings and published in the club's magazines as well as repeated in counseling events where the new tool was used constantly. As these arguments focused directly on the KBD's most important feature—its hunting ability—they were met with great response. Most breeders began to take the alternative seriously and gave up the inbreeding method. This was helped by the availability of objective, up-to-date statistics that could be used to support breeding attempts. As a corollary, communal meetings, annual reports, and magazines became more popular. The number of good hunting dogs gradually increased, and as such, decreasing the inbreeding degree of the population could be managed more easily and consciously than before.

Commentary
This episode confirms that the information infrastructures of a community need to be maintained; otherwise, the community can be blinded by the limitations of its own view (cf. Brown and Duguid 1998). This episode also provides convincing evidence of the importance of concrete, measurable elements such as those produced by the Breed software. As the consultant denoted, before introducing this new feature—inbreeding degree—hardly anybody believed that inbreeding could cause some serious defects as they trusted its opposite influences on dogs. Besides, calculating inbreeding degree manually was not so well-liked even among those who knew the formula. Now a computer could do it in seconds, and a seven-generation pedigree reduced to a single count could be utilized statistically. The information contradiction revealed with the help of the new tool enabled the consultant to expose the predominant breeding practice, thereby allowing its consequences to be the evaluated by the whole community. For this task—and to convince the many respected breeders—the consultant needed "hard data" to make his rationale clear, especially since the inbreeding method had been used for decades. Breeders would not have shifted their perspective or even reversed their opinion entirely without a powerful argument, which would have been hard to find as well as present without the new system. According to Dorothy Leonard and Susaan Straus (1997), individuals often have strong thinking-style preferences in regard to particular types of information that is hardwired into their brains and reinforced over years of

practice. Changing this kind of ossification will hardly be possible without relevant information addressed to a core interest and backed up by analysis (Leonard and Sensiper 1998). In this case, such a situation is easy to understand in light of the community's traditional breeding culture. Thus, this episode affirms the importance of taking cultural aspects (Virkkunen and Kuutti 2000) into consideration when attempting to foster the formation of trust and social capital in collective settings.

The fact that information produced by the IS gained popularity among members also meant that the IS supported the formation of collective conversation, action, and interaction space. This, in turn, facilitated the creation of new knowledge (Boland and Tenkasi 1995), especially shared tacit knowledge, which according to Nahapiet and Ghoshal (1998) is a crucial source of social capital. The information produced by the IS brought members together "from far away" because it provided relevant, good topics for discussion, particularly as these facts were gleaned from actual hunting trials, which most of the persons concerned had actually attended. Based on their own experience, they could estimate the value of the representation of reality that was revealed by the IS and discussed in the consultant's presentations. As these representations were linked through signs to known signifying events, the people concerned were able to talk about the common practice by using more consistent concepts that were created within the shared practice, which could also be suitably represented in the computer system. Thus, the shared breeding language began to unfold because the indexical nature of the signs[3] used (cf. Giddens 1994) was seen as meaningful and functional in practice. As a result, these events formed a good mechanism for sustaining conversations in order to generate trust, which according to Walter Powell (1996), manifests itself in a more natural way through voluntary involvement than through obligations and norms. All of these factors promoted collaboration, which in this case, turned out to be more productive than the isolated actions of individual members.

The Final Episode: The IS and Many-to-Many Relationships
By the mid-1990s, the KBD community had acquired more consultants with hunting experience; the consultants had also functioned in hunting

trials both as chief judges and competitors with their own dogs. At the same time, several useful tools, such as Male Recommendation, were added to the IS. But even these additions failed to shed light on the question of why all litters of good hunting dogs with a low degree of inbreeding did not reach the same quality standard. In 1996, the consultants started to look for an answer on the basis of the idea that "the younger the dog is, the less it has learned" (Syrjänen 2002, field notes). This resulted in a renewed emphasis on the importance of a dog's dam—a topic that had heated up some earlier debates, too. The idea was implemented by introducing a tool called Breeding Points, which enabled the consultants and breeders to test which combinations of dogs provided the best background for a litter. When planning breeding pairs, they always tried to come up with an optimal pair because "the mass involved can be scaled up only by using dogs with a higher than average hunting instinct intensity" (ibid.). Today, the community has even more tools, including one designed to support the selection of a new dog, referred to as the Breeder Collation, which provides feedback not only for the breeders on their efforts but also for the consultants on the success of their work. Another newer feature is the pedigree function on the Web site, which contains a wealth of relevant details including dogs' degrees of inbreeding. This information enables its users to conclude what kind of dogs each breeder has produced, how successful the dogs have been in hunting trials and dog shows, what sort of offspring they have, and so on. Many members use both the Internet and personal counseling when acquiring new dogs or planning new pairs of dogs (e.g., Kousa 2001). This is also illustrated by the simple cgi-based (common gateway interface) pedigree function, which was used 3,172 times between 1 January and 14 January 2002 (Syrjänen 2002, field notes). A new proposal, the so-called target directive for breeding, was presented at FSC meetings and published in the club's magazines in 1999. It defined a new vision, and included measurable, concrete items along with requirements for breeding dogs. The outcome of the whole breeding system is that KBD breeders are capable of producing litters in which several dogs have good "hunting instincts," and can take part in hunting trials at a younger age than earlier dogs (cf. Seluska and Tiensuu 2000, 2001). The success rate has also been affirmed by outsiders such as the FKC (Lindholm 2001).

Commentary

The community's members can access the information produced by the system through its outputs via several channels: counseling, communal meetings, club magazines, annuals, and the Internet, along with attending actual hunting trial events—in which many of them participate eagerly—as well as the consultants do. Through this dense network of peers, members have a fair chance of sharing experiences, which turns information into practicable knowledge by giving value and meaning to it. This situation is supported by the system, which has produced several useful tools including the new knowledge-based vision, as well as plans and strategies for achieving it. The practitioners themselves have designed these tools via joint action, but the tools' realization requires even more cooperation. This is the reason why the community's infrastructure has been extended to include its virtual existence on the Internet, where anyone can scan the quality of the KBD practice. Openness and public access to information relates to the fulfillment of the principle "people can get more objective information about how the things really are," which was the original goal in introducing the current IS. This episode demonstrates that several members have also adopted the IS as an integral part of their practice, highlighting the value of cooperation and trust as well as the belief that the application of relevant knowledge will promote success. It seems that the earlier period (compared with the first episode), which can be characterized as an era of hidden knowledge, served as a lesson. The current situation gives the impression that quite a number of community members put their trust in socially explicit knowledge, which is reputed to be the most advantageous form of knowledge (Boisot 1995). This generates trust because relevancy is formed through the socially explicit system of the hunting trials with the aid of the knowledge that has been produced. This and the availability of information evidently support interaction so that both the consultants and their customers can share rather similar conceptions of reliability and relevance. In counseling, consultants and their customers seem to be parties in a joint action, described by one informant in the following manner: "The computer mills information gathered from the stock," which is then given to a breeder or dog owner as "food for thought" (Syrjänen 2002, field notes). Having considered the alternatives offered by the consultant, the breeder in the first place and later the dog owner then agree on a combination of dogs to be bred.

Summary
The joint responsibility along with the strategies, plans, and common goals of the community can be considered as reflections of trust and social capital, which have ensued from information and knowledge sharing within the common practice. Counseling participants seem to share an identical notion of the relevant and necessary aspects of the practice they are trying to improve. The situation is moving from the materialization of expertise towards trust when a certain number of breeders reveal their faith in the breeding counseling by using it frequently. Slogans such as "recommended by the consultant" or the like also have been used at times when advertising a new litter or participating on a discussion site (Finnish Spitz Club 2001–2002). Interestingly, a similar trend—transitioning from expertise to trust—has been identified in the case of Silicon Valley (Cohen and Fields 2000). Miettinen (2001), as noted earlier, refers to this kind of trust as activity-based trust. As the KBD community shows, activity-based trust is grounded in concrete, shared activities. This approach emphasizes the social aspects of the activities, but perhaps more important, the role of knowledge must be considered a major incentive for interaction in the first place. The KBD community is keen to know how the dogs to which they are deeply devoted conduct themselves in the wild—as a hunting dog for big game hunting, which is, after all, the fundamental reason why this breed of dogs exists (Karelian Bear Dog Division 2001). This attitude, which can be traced back to the lost, legendary Karelia, is expressed by a traditional poem: "It fed the family, gave drink to the tribe, supported the forefathers" (Perttola 1998, 21). The rebuilt alliance with the KBD's legendary background emerged when the focus of attention moved from the dog itself to the social system of humans and its influence on dogs, especially on the selection of dogs to breed. The prevailing understanding of the KBD is framed by elements including both human-made settings and the original wild environment, which is where the dogs have proved their raison d'être. This kind of mutual understanding can be seen as referring to *sense making* as defined by Rudy Hirschheim, Heinz K. Klein,

and Kalle Lyytinen (1995, 37) when a group has arrived at socially shared meanings and thus increased its social capital.

Finally, the episodes described above verify the relationship between the intellectual and social capital of the community so that useful and relevant knowledge functions as an incentive in social interaction. In the case of the KBD community, most interaction events can be considered as knowledge-intensive actions. Examples include breeding counseling, communal meetings with presentations, and hunting trials where official rules are applied, data are collected, and the data are later codified into the IS to be used as a resource in counseling. These activities, which are either directly or indirectly connected to the community's IS, are also productive actions that yield concrete results. According to Engeström and Ahonen: "The more concretely the work community sees the infrastructure as an object of productive activity, the more likely the infrastructure is to contribute effectively to the formation of social capital" (2000, 6–7), and the case of the KBD community testifies to that. The case also corroborates that the tangible origin of the knowledge made visible by the IS made the situation more profitable than the earlier one, not because the knowledge of the time would had been considered as less believable per se but because the community lacked appropriate equipment to manage and process its source data. The relevance of the new knowledge sprung from its accessibility and publicity, which contributed to its truthfulness: it illustrated a wider and more integrated slice of reality accurately enough to assist in decision making.

The Role of IS as a Material Artifact in the Rooting of Social Capital

According to Susan Leigh Star and Karen Ruhleder (1996), infrastructures are basically relational concepts. As such, they appear and should always be seen in relation to organized practices as part of social systems in organizations. In this sense, IT-based ISs play an ambiguous role in that they are both the medium for information transmission and the infrastructure for the representation of information. As these systems cross over the boundaries of organizations and social systems, they can function as entries and tools for change (Star 1995). The change is not always for the better (e.g., Hales 1995; Star 1999), but to a certain degree it is also a matter of how we perceive "the artifacts that make up large sections of our social ties," as Bruno Latour (1992, 254) states in emphasizing the role of mundane artifacts in our society.

The case of the KBD community demonstrates the capacity of ISs to support the development of social ties, particularly as regards changes in trust in the formation of social capital. The most evident change in this case is how the IS aided the formation of an infrastructure for social interaction (cf. figure 2.1), through collective conversation, action, and interaction space, which can be seen to facilitate the creation of new knowledge and thus the formation of social capital (Boland and Tenkasi 1995; Nahapiet and Ghoshal 1998). The sequence in this instance was from one-to-one relations via one-to-many to many-to-many relationships through the network of peers in information-sharing and knowledge-shaping processes. A similar sequence from a meaningful object to social capital is reported by Nitin Nohria (1992), who argues that the development of social capital requires a "focus," and is usually a spin-off of other activities. The aspiration to know "how things really are" seems to form the basis of the social system of the KBD community, and the use of the word "really" indicates that not all knowledge is considered to be equally reliable and relevant. Richard McDermott (1999) writes that our ideas are meaningful only in relation to the beliefs of our community. These beliefs define what knowledge is relevant and are closely connected with the success of collaboration. Collaboration, in turn, is backed up with related communication through common language and shared concepts, as the KBD case shows. According to Juhani Iivari and Henry Linger (1999), shared relevant and explicit knowledge frames the common knowledge space of collaboration. This kind of shared knowledge is seen as the basis from which social interaction and alignment ensue (Berger and Luckman 1966).

Hence, socially explicit knowledge also highlights aspects of power in collectives, and the influences of that power on social interaction and alignment. In the KBD community, the IS supported the diffusion of relevant knowledge, which was the incentive for increasing interaction. This development, in turn, was instrumental in shaping trust and social capital among community members. In addition, by making "hidden knowledge" public, the IS has extended the ownership of social capital and social power in the community. According to Ronald S. Burt (1992), social capital is usually owned jointly, and individuals are not capable of having private mastery over it. Thus, the members who were thought to possess special hidden knowledge or who were powerful among the community because of their reputation for achieving top results in the past cannot be seen as having social capital as such. Rather, they possessed individual intellectual capital in Bourdieu's sense, or even financial capital in a business sense, which often attracts people as "capital finds its way to capital" (cf. Bourdieu 1998, 17); reproduction was therefore circulating. The reasons why members in the KBD community were involved in such reproduction are diverse. One obvious factor was the old IS, which could not display the quality of the breeding practice as a whole. Another reason was the lack of alternatives. As one informant conceded, "We knew of no better at that time" (Syrjänen 2002, audiotapes).

The perceived lack of possible alternatives may also be a matter of social alignment due to various features of information infrastructures. Some feature may facilitate one person's activity even as it acts as a barrier to another person (Star 1999). Such features as limitations to information access or its one-sided or fragmented service also reflect the social norms, values, and moral aspects of a collective. The importance of social norms such as cooperation, the disclosure of information (Starbuck 1992), a willingness to value and respond to diversity, or an openness to criticism and feedback (Leonard-Barton 1995), for instance, all influence social alignment, but this influence may also be negative. Our view is that it is relational and depends on the chosen point of view. In this case, the old IS was a hindrance that prevented members from seeing the quality of the breeding work as a whole. This was an advantage for those breeders who could sell dogs based on earlier top results. While the new IS supported the work of consultants and other breeders aligned with new ideas, this didn't always please everybody. As one informant maintained, "It had to be irritating for those with poor results to be told this statistic," and such a situation was likely unpleasant for all stakeholders including the consultant (Syrjänen 2002, audiotapes). This observation is corroborated by the fact that some of the breeders withdrew from actual counseling. On the other hand, after they became accustomed to the new tactic, "many breeders were longing for feedback" (ibid.). The current situation is also seen on the Web discussion site (Finnish Spitz Club 2001–2002) and through calls to stakeholders if statistics or updatings are late regardless of the cause of the delay. It shows that a preference to access-related information is important as well. Hence, in sustaining the formation of social capital, we should also ask *whose* social capital we are focusing on, as we probably cannot amend everything.

The acceptance of new ideas and common principles are often tied up with moral and cultural issues. Because they are embedded in conventions of practice, moral and cultural concerns may be invisible, thus making the situation difficult to change. In this case, the IS acts as a kind of gatekeeper that permits everybody to do anything they might want to, but whether positive or negative, the results will be public at some point in time in the form of dog information. This allows everyone to evaluate how they perceive the object of your work, the respect you show to the common goals, the principles you value, and so forth, As Latour describes it, "The sum of morality does not remain stable, but increases enormously with the population of nonhumans" (1992, 232). This holds true for the KBD community, as indicated by the members who regard breeding of closely related dogs as objectionable today. Many of them seem to understand that the situation was different earlier, due to the obviously unsatisfactory breeding conditions (cf. the first episode), but that the inbreeding method is now usually no longer necessary and may even be rather risky (e.g., Finnish Spitz Club 2001–2002). This epoch-making shift in the members' mode of thinking can largely be credited to the Breed tool of the IS, which in the words of one informant, "opened up the eyes of many breeders" including the consultant (Syrjänen 2002, audiotapes). It has been one of the most utilized functions since its introduction, and can therefore be seen as facilitating the alignment of community members to the new way of thinking, which is entirely opposed to the old way. Thus, when breeders were challenged to cooperate rather than work on their own, the IS proved itself capable of supporting their efforts.

Concurrently, given the IS's potential to organize the community's members' cooperation, this case highlights the significance of the IS as a

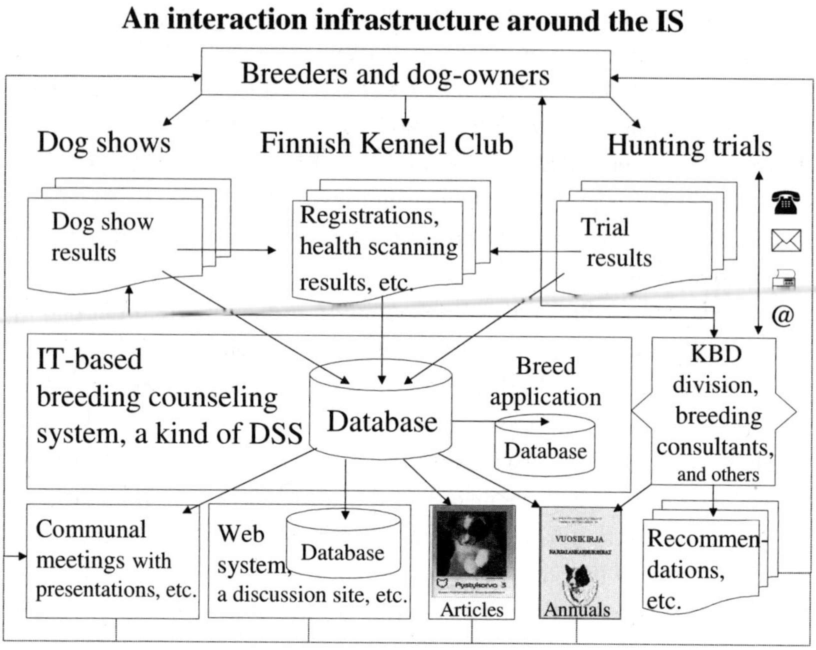

An interaction infrastructure around the IS

Figure 2.1
An infrastructure for social interaction in the KBD community.

model for the community and its basic practices. It follows, then, that the IS can be viewed as a social artifact, and as such, a technical implementation of social systems (cf. Hirschheim, Klein, and Lyytinen 1995, 11). The IS is a material link here between different social systems (see figure 2.1), yet it is also an artificial matrix where cooperation can be both designed and investigated through its results. This cooperation is arranged via the counseling, which connects breeders, dog owners, and dog buyers together. A setup of this kind can be viewed as a system comprising microcommunities of knowledge (or knowing). Such systems have been described by George von Krogh, Kazuo Ichijo, and Ikujiro Nonaka (2000), who emphasize their potential in knowledge creation because participants can, for instance, share their knowledge and as well as naturalize their common values and goals. Thus, these systems have great value for tailoring social capital.

In the present case, this kind of a system of microcommunities forms a dense network of peers, which obviously supports the diffusion and percolation of information along with knowledge sharing, thereby facilitating the formation of social capital. Community members are not only shaped by these collective processes but assume an active role in them as they can both evaluate and produce information by themselves, too. Hence, information itself can be seen as a social artifact with a social life and culture of its own within the social collective that created it, as suggested by John Seeley Brown and Paul Duguid (2000). Star (1999) maintains that every imaginable form of variation in practice, culture, and norms can be inscribed into information infrastructures. Consequently, these infrastructures are the repositories of the social system, practice, and culture that generated them, and are therefore linked to their material structures and tangible roots. The KBD community can even present concrete, living embodiments of its practice, culture, and social system—the dogs. The dogs carry in their genes the consequences of breeding practices, which have been codified into the community's information infrastructure as a testimony of the beliefs, values, norms, and attitudes held within the social systems that created them.

To summarize, ISs like that of the KBD with tangible roots possess great potential for analyzing social capital, organizational advantage, and the development of community knowledge and intellectual capital in collective settings.

Conclusions

The work presented in this chapter concentrated on examining how the IS of the community functions as a material artifact in which the social capital of the community can be rooted. We focused on the role of IT in processes and interactions among participants where the social capital manifests itself in generating as well as maintaining the trust, acceptance, and alignment necessary for successful cooperation, highlighting that these cannot be separated from their context, domain, and culture.

The study of our microcosm community shows that trust, acceptance, and alignment were generated and maintained through daily interactions among community members, and that the interactions were facilitated by the IS. The study also proves that IT—in this case, initially a relatively simple bookkeeping system that grew in complexity and sophistication during the years—became an integral part of these interactions to the degree that it eventually constituted the infrastructure vital for the day-to-day functioning, self-determination, and development of the community. The community adopted a certain direction based on the influence of a group of members with a concrete vision. The trust invested in this group resulted in their vision becoming generally accepted, and quite a number of community members aligned themselves and successful cooperation with this vision. This was to a certain extent influenced by the fact that the emerging IS made the rationale behind the actions and interactions of the key persons more visible and the activity as a whole more transparent, and thus more accountable and trustworthy. Besides, by revealing the everyday quality of the practice in its entirety, the IS allowed measures to be taken to develop the breeding practice, new knowledge, and the IS itself.

The significant role of the IS in facilitating the emergence of trust and social capital can be encapsulated in one intertwined spiral of effects as follows. The IS could display in an integrated way that relevant knowledge had its tangible roots in a reality that was firmly linked to concrete objects, events, artifacts, and people through social interaction processes. All that created opportunities for trust to evolve. The IS supported communication by unifying concepts in such a way that a shared language could unfold and thus conversations could be more fruitful. That, in turn, served as a precondition for the formation of shared explicit knowledge, which then helped the members understand and accept new ideas. As a result, this knowledge formed the basis from which social interaction and alignment ensued. We can conclude that the IS contributed to the formation of a common knowledge space for collaboration, and through that, the growth of social capital in the community.

On this basis, the study demonstrates that social capital and intellectual capital are interdependent properties of the community. This relationship can be considered as causal in that knowledge (intellectual capital) functions as an incentive for social interaction (which creates social capital). Hence, if necessary, it is possible to address social capital by mirroring the relationship between social capital and intellectual capital through the development of community knowledge. Social activities that contribute to the development of knowledge and the formation of social capital should be investigated against the backdrop of the whole meaningful system within which they occur. An appropriate unit for such analysis is a community of individuals who are involved in activities that realize or should realize a common set of issues, interests, or needs. With concrete results, such knowledge-in-action processes form the basis for addressing features of collaboration, the development of knowledge, and the formation of social capital.

Interestingly, the role of the IS in the KBD community strangely resembles the ideal of the "total information system" that was popular in management IS research in the 1970s (Davis and Olson 1985, for instance). Total ISs were to mirror the functioning of an organization in such a way that everything was visible, accountable, and manageable. Although the concept failed to catch on, the system described in this paper seems to be rather successful. This begs the question, What is the difference? The answer may be found in the fact that total ISs were aimed toward managers at the exclusion of others, while the KBD system—made by and for the practitioners themselves—is accessible to all. Not everyone uses it directly, but in principle, information used by key members is available to anyone through several channels. Thus, the rationale behind key members' decisions as well others' acceptance of them can be accounted for, and vice versa.

Finally, ISs used to process information with a view to generating new knowledge can be seen as highly social artifacts, when viewed as reduced agents of the social system of practices within the culture that generated them. As a result, they are linked to tangible material structures and roots. It then follows that social capital can be understood as firmly rooted in and practically inseparable from certain types of tangible material structures and artifacts—including the materiality of (human) beings as bodily actors. ISs can be discerned among structures and artifacts that can be utilized in analyzing social capital, organizational advantage, and the development of community knowledge along with intellectual capital in collective settings.

Notes

1. Many participants were surprised to see Putnam—a traditional sociologist who believes in statistics and society-wide theoretical models—as the keynote speaker at a conference that has usually been considered as the home field of microsociologists, who strongly oppose the research paradigm Putnam represents.
2. Fukuyama (1999) is clearly aware of the looseness of this kind of definition, and he criticizes it as based on manifestations of social capital rather than on social capital itself. He defines social capital as "an instantiated informal norm that promotes co-operation between two or more individuals" (3). When Fukuyama's examples of such norms range from reciprocity among friends to religions like Christianity, the current authors have difficulties in seeing much improvement in the analytic accuracy of this suggestion.
3. A sign is composed in the first instance of expression, such as words or symbols, and the content that is seen as completing the meaning of the expression (Hjemslev 1961). A sign always requires a context, which links the expression to a significant event (Manning and Cullum-Swan 1998, 252).

References

Berger, P. L., and T. Luckman. 1966. The social constructions of reality. London: Routledge.

Boisot, M. 1995. Information space: A framework for learning in organizations, institutions, and culture. London: Routledge.

Boland, R. J., and R. V. Tenkasi. 1995. Perspective making and perspective taking in communities of knowing. *Organization Science* 6:350–372.

Bourdieu, P. 1986. The forms of capital. In *Handbook of theory and research for the sociology of education*, edited by J. G. Richardson. New York: Greenwood Press.

Bourdieu, P. 1998. *Practical reason: On the theory of action*. Oxford, U.K.: Polity Press.

Brown, J. S., and P. Duguid. 1998. Organizing knowledge. *California Management Review* 40, no 3 (spring): 90–111.

Brown, J. S., and P. Duguid. 2000. *The social life of information*. Boston: Harvard Business School Press.

Burt, R. S. 1992. *Structural holes: The social structure of competition*. Cambridge: Harvard University Press.

Cohen, S. S., and G. Fields. 2000. Social capital and capital gains in Silicon Valley. In *Knowledge and social capital*, edited by E. L. Lesser. Woburn, MA: Butterworth-Heinemann.

Davis, G. B., and M. H. Olson. 1985. *Management information systems: Conceptual foundations, structure, and development*. New York: McGraw-Hill.

Engeström, Y., and H. Ahonen. 2001. On the materiality of social capital: An activity-theoretical exploration. In *Activity theory and social capital*, edited by Y. Engeström. Helsinki: Helsinki University Press.

Finnish Kennel Club. 1991. Rules and instructions for elk hunting trials. Valkeala: Valkealan Paino Ky.

Finnish Kennel Club. 1996. Official breed standard: Karelian bear dog.

Finnish Spitz Club. 1987. *The Finnish Spitz Club, 1938–1987: Fifty years*. Oulainen: Pyhäjokiseudun kirjapaino Oy.

Finnish Spitz Club. 2001–2002. Karelian bear dog's web site, <http://poro.koillismaa.fi/spj/>.

Fukuyama, F. 1999. Social capital and civil society. Paper presented at the IMF Institute Conference on Second Generation Reforms, <http://www.imf.org/ external/pubs/ft/wp/2000/wp0074.pdf>.

Giddens, A. 1994. Risk, trust, reflexivity. In *Reflexive modernization: Politics, traditions, and aesthetics in the modern social order*, edited by U. Beck, A. Giddens, and S. Lach. Cambridge, U.K.: Polity Press.

Hales, M. 1995. Information systems strategy, a cultural borderland, some monstrous behaviour. In *The cultures of computing*, edited by S. L. Star. Padstow, Cornwall: Blackwell Publishers.

Hirschheim, R., H. K. Klein, and K. Lyytinen 1995. Information systems development and data modelling: Conceptual and philosophical foundations. Cambridge: Cambridge University Press.

Hjemslev, L. 1961. Prolegomena to a theory of language. Rev. ed. Madison: University of Wisconsin Press.

Iivari. J., and H. Linger. 1999. Knowledge work as collaborative work: A situated theory view. Proceedings of the thirty second annual Hawaii International Conference on System Science, January 1999, Maui, Hawaii.

Karelian Bear Dog Division. 2001. Karelian bear dog—a Finnish breed for big game hunting: *Pystykorva* 1B (breeding issue):20–36.

Kilkki P. 1987. Review of the Finnish Spitz Club's activities in 1938–1967. In *The Finnish Spitz Club, 1938–1987: Fifty years*. edited by J. Simolinna. Oulainen: Pyhäjokiseudun kirjapaino Oy.

Kousa, J. 2001. Some thoughts of a member of the FSC. *Pystykorva* 4:51.

Krogh G. von, K. Ichijo., I. Nonaka. 2000. From managing to enabling knowledge. In *Enabling knowledge creation. How to Unlock the Mystery of Tacit Knowledge and Release the Power of Innovation*, edited by G. von Krogh, K. Ichijo, and I. Nonaka. New York: Oxford University Press.

Kuutti, K. 2001. ECSCW'01 Tutorial on community knowledge. Paper presented at the 2001 European Computer Supported Cooperative Work conference.

Latour, B. 1992. Where are the missing masses? The sociology of a few mundane artifacts. In *Shaping technology—building society: Studying sociotechnical change*, edited by W. E. Bijker and J. Law. Cambridge: MIT Press.

Leonard, D., and S. Sensiper. 1998. The role of tacit knowledge in group innovation. *California Management Review* 40:112–132.

Leonard, D., and S. Straus. 1997. Putting your company's whole brain to work. *Harvard Business Review* 75, no. 4:110–121.

Leonard-Barton, D. 1995. *Wellsprings of knowledge: Building and sustaining the source of innovation*. Boston: Harvard Business School Press.

Lesser, E. L. 2000. Leveraging social capital in organizations. In *Knowledge and social capital: Foundations and applications*, edited by E. L. Lesser. Woburn, MA: Butterworth-Heinemann.

Lesser, E. L., M. A. Fontaine, and J. A. Slusher. 2000. Preface; Communities: people, places, and things. In *Knowledge and communities*, edited by E. L. Lesser, M. A. Fontaine, and J. A. Slusher. Woburn, MA: Butterworth-Heinemann.

Lindholm, J. 2001. Karelian bear dogs achieve success and top results at a much younger age than ever. *Koiramme—våra hundar*, Finnish Kennel Club, 105, no. 12:b10–b14.

Manning, P., and B. Cullum-Swan. 1998. Narrative, content, and semiotic analysis. In *Collecting and Interpreting Qualitative Material*, edited by N. K. Denzin and Y. Lincoln. Thousand Oaks, CA: SAGE Publications.

McDermott, R. 1999. Why information technology inspired, but cannot deliver knowledge management. *California Management Review* 41, no. 4:103–117.

Miettinen, R. 2001. Social capital and innovations. In *Activity Theory and Social Capital*, edited by Y. Engeström. Helsinki: Helsinki University Press.

Miettinen, V., M. Joenpolvi, and U. Sirviö. 1987. Karelian bear dog: development as a hunting dog. In *The Finnish Spitz Club, 1938–1987: Fifty years*, edited by J. Simolinna. Oulainen: Pyhäjokiseudun kirjapaino Oy.

Nahapiet, J., and S. Ghoshal. 1998. Social capital, intellectual capital, and the organizational advantage. *Academy of Management Review* 23:242–266.

Nohria, N. 1992. Information and search in the creation of new business ventures. In *Networks and organizations: Structure, form, and action*, edited by N. Nohria and R. G. Eccles. Boston: Harvard Business School Press.

Perttola, J. 1998. The story of the Karelian bear dog. Finnish Spitz Club. Kotka: Painokotka Oy.

Powell, W. W. 1996. Trust-based form of governance. In *Trust in organizations: Frontiers of theory and research*, edited by R. M. Kramer and T. R. Tyler. Thousand Oaks, CA: Sage Publications.

Putnam, R. D. 1993. The prosperous community: Social capital and public life. *American Prospect* 4, no. 13:35–42.

Seluska, H., and S. Tiensuu. 2000. Karelian bear dogs' annual. Finnish Spitz Club. Saarijärvi: Gummerrus Kirjapaino Oy.

Seluska, H., and S. Tiensuu. 2001. Karelian bear dogs' annual. Finnish Spitz Club. Saarijärvi: Gummerrus Kirjapaino Oy.

Spender, J-C. 1996. Making knowledge the basis of a dynamic theory of the firm. *Strategic Management Journal* 17(S2):45–62

Star, S. L. 1995. Introduction. In *The Cultures of Computing*, edited by S. L. Star. Padstow, Cornwall: Blackwell Publishers.

Star, S. L. 1999. The ethnography of infrastructure. *American Behavioral Scientist* 43:377–391.

Star, S. L., and K. Ruhleder. 1996. Steps toward an ecology of infrastructure: Design and access for large information spaces. *Information System Research* 7:111–134.

Starbuck, W. H. 1992. Learning by knowledge intensive firms. *Journal of Management Studies* 29:713–740.

Syrjänen, A-L. 2002. Recordings of thematic interviews with the key informants in the FSC case 2000–2002 (audiotapes, field notes). Translated from Finnish by A-L. Syrjänen.

Virkkunen, J. and K. Kuutti. 2000. Understanding organizational learning by focusing on "activity systems": *Accounting, management, and information technologies* 10(2000):291–319.

3

The Effects of Dispersed Virtual Communities on Face-to-Face Social Capital

Anita Blanchard

How do virtual communities affect face-to-face communities? Researchers and activists have been asking this question since virtual communities first became popular. Virtual communities have been frequently conjectured to increase, decrease, augment, supplement, or otherwise change the ways in which people interact with their neighbors in their face-to-face (FtF) communities (Rheingold 1993; Wellman et al. 2001). Despite this interest and the vastly different potential outcomes, there are only a few empirical studies that examine how participation in virtual communities actually affects participation in FtF communities.

This chapter will add to that empirical knowledge. In particular, it will explore how virtual communities do and do not increase social capital in FtF communities. First, this chapter will review the relevant research on social capital and virtual communities. Then, a virtual community will be examined to determine how its members participate in their FtF communities. The chapter will end with a discussion on how we can better understand and promote the relationship between virtual communities and FtF social capital.

Social Capital

Although the study of social capital is popular, social capital does not have one universally recognized definition (Pruijt 2002). Some researchers consider it to be membership in groups and social networks (e.g., Belliveau, O'Reilly, and Wade 1996). Other researchers consider it an umbrella term encompassing a variety of other well-studied concepts including social cohesion, social exchange, and social support (Adler and

Kwon 2002; Timms, Ferlander, and Timms 2001). In an extensive review of the research on social capital, Adler and Kwon argue that, essentially, social capital is goodwill between group members. More concretely, social capital is any aspect of a group that facilitates the group's interactions, benefiting both the group and its individual members (Coleman 1990; Putnam 1995).

What creates social capital? As with the lack of agreement on the definition, there is a lack of agreement on where social capital comes from. The networks of relationships between group members are often considered important in creating social capital. Some researchers focus on networks of relationships such as market relationships in which there is the exchange of products and services as well as hierarchical relationships, which include obedience to authority (e.g., Adler and Kwon 2002). It is the networks of social relationships between people, however, that most researchers converge on as essential in creating social capital (Adler and Kwon 2002; Paxton 2002; Nahapiet and Ghoshal 1998; Putnam 1995; Coleman 1990).

Adler and Kwon also contend that to understand what creates social capital, researchers cannot ignore the substance of what is transferred through these networks. There is not widespread agreement on what the significant substance or content is, though. Many researchers have focused on the norms of appropriate behavior (Nahapiet and Ghoshal 1998; Putnam 1995; Coleman 1988) and information about the trustworthiness of individuals (Paxton 2002; Nahapiet and Ghoshal 1998; Putnam 1995; Coleman 1988). Other researchers have concentrated on obligations to and shared histories of the group (Nahapiet and Ghoshal 1998).

For this study, I have chosen to examine the most stable and agreed on components of what creates social capital: the network of social relationships, the norms of behavior, and trust among members. Exploring the network of social relationships will allow us to assess who is interacting with whom within the virtual community, and if and how those interactions move to FtF interactions. As a primary structural component in the creation of social capital (Nahapiet and Ghoshal 1998), these interactions (or lack thereof) will allow us to see how virtual interactions do and do not translate into FtF interactions.

By scrutinizing the norms of behavior and trust among members, we can assess how two important relational components of social capital (Nahapiet and Ghoshal 1998) move through the virtual and FtF networks (or not) and determine their effect of FtF social capital. Norms of behavior and trust among members are closely linked in the creation of social capital (Putnam 1995). That is, as groups establish prosocial norms such as helping each other, the level of trust that other members will be helpful and otherwise "good" increases. The question of whether norms of behavior and trust among members will move between virtual to FtF interactions is, therefore, crucial.

Virtual Communities

Virtual communities are groups of people who interact primarily through computer-mediated communication and who identify with and have developed feelings of belonging and attachment to each other. These subjective feelings are known as a "sense of community" (McMillan and Chavis 1986). They are an essential part of virtual communities and may separate virtual communities from mere virtual groups (Jones 1997; Koh and Kim 2001; Blanchard and Markus in press).

Virtual communities can either be *place based*, in which the electronic group is centered around some geographic locale, or *dispersed*, in which the electronic group is not (Blanchard and Horan 1998). For example, a place-based virtual community could be an e-mail discussion group (i.e., a list server) for a neighborhood association, a bulletin board sponsored by an elementary school's parent-teacher association, or a business organization's on-line computer help group. Examples of a dispersed virtual community include a bulletin board for a popular television show, a list server of owners of Volkswagen buses, or a list server for human resource professionals.

Whether a virtual community is place based or dispersed may have different effects on the social capital in an FtF community. A place-based virtual community may increase the density of an FtF community's network of social relationships (Blanchard and Horan 1998). That is, in a place-based virtual community, it is possible that the network of people with whom one interacts in the virtual community may overlap with the people one interacts with in an FtF community. This overlap is not likely to be complete; there will be some people in the virtual community network who are not in the FtF network and vice versa. This makes the FtF community's network of social relationships more dense because more people are now connected. Information about norms and trustworthiness is believed to flow more easily within this more densely connected network, thereby increasing social capital (Putnam 1995; Coleman 1988).

Dispersed virtual communities, on the other hand, may decrease the social capital in an FtF community because they may decrease the density of the social networks of relationships in the FtF community (Blanchard and Horan 1998). That is, the networks of relationships in the dispersed virtual communities are not likely to overlap at all with the FtF social networks. The networks of relationships may decrease social capital because information about the trustworthiness of members and norms of behavior cannot move between the virtual and FtF community networks.

Nonetheless, dispersed virtual communities may be quite beneficial to their members. Dispersed virtual communities may in fact have an active social capital process within them. Still, because they do not foster the FtF community's network of relationships, they are not likely to increase social capital within the FtF community. These dispersed virtual communities may even decrease an FtF community's social network if members prefer to spend more time in their virtual communities because of their greater socioemotional rewards.

Current research offers intriguing challenges to these arguments. Tonn, Zambrano, and Moore (2001) report that place-based community networks *do not* increase the social capital in their FtF communities. In this study, however, community networks were essentially Web pages of local information. They did little to foster interactions among community members. Therefore, these networks did not increase the density of the members' network of relationships. Because previous researchers have proposed that dense networks are so important in developing social capital (e.g., Putnam 1995; Coleman 1988), it is not surprising that this community network did not increase FtF social capital.

Pruijt (2002) examined a dispersed virtual community. This lively virtual community developed around computer programming. In one critical incident, members of the virtual community found and publicized a serious error in a popular software program. The attention given this issue within the virtual community spread outside the virtual community and induced a change in computer hardware that benefited nearly everyone. Although this virtual community clearly benefited society at large through drawing attention to and solving a problem, it is less clear that the virtual community increased (or affected at all) the social capital in any specific FtF community.

Finally, Wellman, Witte, and Hampton (2001) analyzed the effects of the Internet on social capital. They surveyed visitors to the National Geographic Society Web site to determine if Internet use increases, decreases, or supplements social capital as defined as network capital (interactions with others), participatory capital (activity in civic and political groups), and community capital (feelings of community). Their results show that Internet activity supplements interactions with others, increases activity in civic and political groups, and is associated with a sense of community with the general on-line community and kin.

Although Wellman, Witte, and Hampton (2001) define social capital differently than some other researchers, their study suggests that dispersed virtual community participation may not decrease some forms of social capital. In particular, their finding that increased Internet use was associated with an increase in civic group participation is significant because it may show an increase in activity in FtF networks of relationships. Wellman, Witte, and Hampton note the problems of causality in this correlational finding, and assert that the true relationship between Internet use and civic group participation may be a positive feedback loop.

What, then, are the effects of virtual communities on FtF communities' social capital? Although it has been argued that dispersed virtual communities can decrease FtF social capital and place-based virtual communities can increase it, this is not necessarily the case. One problem with the current research on this topic is that it is not clear that these studies examine virtual communities in which members have developed a sense of community with each other or whether these studies examine virtual groups in which members merely interact on-line. This distinction is important. Members may be more likely to spend more of their free time in a virtual community than in a virtual group because of their higher level of emotional involvement. Thus, participation in a virtual community may more likely take the members away from participation in an FtF community than in a virtual group.

Second, the studies do not look at the processes by which virtual communities affect social capital in a *particular* FtF community. Although Wellman and colleagues show that civic group participation is positively associated with Internet use, their work does not develop a deeper understanding of how this participation affects the networks, norms, and trust of a particular FtF community.

This chapter will address these issues by asking the research question, "How does participation in a dispersed virtual community affect social capital in the members' FtF community?" To answer this question, the study will empirically examine a dispersed virtual community and its members' participation in FtF social capital using the social capital model of networks of social relationships along with the information about norms and trust that travel through these networks.

Method

The virtual community explored here is Multiple Sport Newsgroup (MSN), a newsgroup for people interested in learning about and training for multiple sports (e.g., triathlons).[1] MSN was established in the early 1990s as an offshoot of a single-sport newsgroup when members wanted to be able to focus on the unique needs of multiple-sports athletes.

MSN was chosen because it is an active newsgroup.[2] In 1995, MSN had an estimated seventeen thousand daily readers (Atkinson 1995). At the time of the study, an average of one hundred messages was posted daily. Membership in MSN appears to be stable; members refer to each other in their posts and discuss past events. These interactions display shared knowledge of group history and particular members. MSN members also reveal a level of friendship in their interactions that suggests the presence of a sense of community among at least some members. The analysis presented in this chapter is part of a larger project that looks at MSN as a virtual community.

MSN was examined using naturalistic inquiry (Lincoln and Guba 1985). Data were collected using participant observation and member interviews. During the seven months of participant observation, two main sources of data were used. First, messages posted to the group were downloaded and stored. Second, after interacting with the newsgroup, the researcher recorded impressions of the conversations, members, and software characteristics using the guidelines of naturalistic inquiry.

Semistructured interviews were conducted with ten members. The advantage of a semistructured interview in this research is that although the researcher began the discussion with a series of topics to cover, interviewees had the ability to bring up key issues that the researcher had not considered or anticipated.

Members who were interviewed came from around the United States, and they lived in primarily metropolitan and suburban areas. Most members interviewed did not have careers or particular expertise related to multiple sports. The one exception was a member who worked for a sports equipment company.

Three different types of members were interviewed: leaders, "regular" participants, and lurkers. *Leaders* are members who the researcher identified as being influential in the group, who identified themselves as leaders, and who other participants identified as leaders.[3] *Participants* are members who posted messages to the group but were not considered leaders by themselves or others. *Lurkers* are members who read the messages but rarely if ever post messages to the group. After being identified, leaders and participants were recruited via private e-mails. Lurkers were recruited through a post to the entire group. Interviews were conducted over the telephone for about one hour. Interviews were tape-recorded with the interviewees' permission and professionally transcribed.

Data were collected and analyzed using Miles and Huberman's (1992) strategies for iterative data analyses. Data were first collected through the participant observations. Interviews were used to verify the initial analysis of the participant observations and to gather new data. The interview data were then verified through additional participant observations and interviews. Data collection continued for both the observations and interviews until no new insights were gained from analysis of the observations or interviews.

The quality of the data and analyses was assessed using Lincoln and Guba's (1985) checklist of trustworthiness. Strategies of prolonged engagement, triangulation of data and methods, negative case analysis, thick description, an audit trail, and an outside review of data and analysis were used to ensure that the methodology and its analyses met the criteria of qualitative research trustworthiness.

Results

Overview of MSN

As mentioned above, MSN is a virtual community for people interested in and training for multiple sports events, which usually consist of some combination of swimming, biking, and running. Some MSN members are active multiple-sports athletes while others are single-sport athletes interested in finding out more information about multiple sports. Vendors specializing in multiple-sports equipment comprise a sizable portion of the group. Although MSN members live all over the world, most are in North America.

MSN members exchange a recognizable pattern of messages consisting of asking for and providing help (e.g., what is a good multiple-sports bicycle), sharing personal experiences (e.g., injury recovery), financial exchanges (e.g., buying and selling specialized equipment), and discussions about relevant multiple-sports issues. For the most part, messages are higher in informational than socioemotional exchange. Yet messages are often funny or contain some humorous element. MSN considers itself "family friendly," and as such, cursing and flaming (i.e., extremely hostile messages) are not appropriate.

From the outside, MSN appeared to be a virtual community in which its members shared a sense of community with each other. This perception was validated during the member interviews. The majority of members interviewed spontaneously described MSN as a community and needed little, if any, further questioning to explain what "community" meant to them. These members felt that MSN was a community because they could recognize other members and even identify the personality of some members. They believed MSN had an active informational (e.g., how to swim in open bodies of water) and socioemotional (e.g., how everyone is afraid of swimming in open bodies of water) support system. It was the informational exchange, however, that was more important to them as members.

The more active MSN members reported that they had developed close, personal relationships with others in the group. All members (i.e., those in these relationships and those who observed them) felt these relationships are crucial to the community. The most active MSN members even expressed strong feelings of attachment and obligation to the group. Because of the observations of MSN, the shared interpretation of the members, and the location of members around the world, MSN can be considered a dispersed virtual community.[4]

Social Capital within MSN

MSN creates and exchanges a great deal of social capital within the group. The primary function of MSN is the exchange of help for and support in training for multiple-sports events. Although the exchange of socioemotional support is important, members were observed exchanging more informational support, and they reported that they valued this type of exchange more. Informational support includes advice about injuries, experience with particular events, and general questions about training. Socioemotional support usually comes as a by-product of the informational support and is rarely a focus of members' interactions.

Participation in the social capital process varied by the type of members in MSN. The more active members—the leaders and participants—contributed more to the creation of social capital. The leaders, in particular, were observed as being quite active in answering questions and providing support.

Yet members of all types (including lurkers) benefited from this public exchange of support. It is worth noting that the number of lurkers greatly outweighs the number of participants and leaders in this virtual community. Of the 17,000 estimated daily readers of this group cited earlier (Atkinson 1995), approximately 250 people posted messages during the observation period. That means that the vast majority of MSN members are lurkers—members who read the group's messages, but do not publicly contribute to the conversations and discussions. Although lurkers tend to be regarded negatively by researchers and virtual community practitioners, publicly lurking in these groups is most likely the way that most people participate in and benefit from them.

It is important to note the difference between publicly lurking and privately participating. Several members of MSN reported that they tend to receive more correspondence from other members privately (i.e., through their personal e-mail) than publicly (i.e., through posts to the group). Although most of these private e-mails were from friends or other known members in the group, many were also from unknown others, presumably lurkers. Thus, lurkers may actually be members who prefer to be active participants behind the scenes than in the public part of the group.

Social Capital outside of MSN

Now we can turn our attention to how participation in MSN affects social capital in the members' FtF communities. This section will focus on how MSN affects the social networks of relationships, norms of behavior, and information about the trustworthiness of people in members' FtF communities.

Networks In considering the issue of networks, the question is whether participation in MSN increases or decreases the networks of people with whom members interact on an FtF basis. If the MSN and FtF networks overlap, then it is possible that the networks will become more dense, thereby increasing social capital. If the networks do not overlap, then the networks will become less dense, decreasing social capital.

MSN members' on-line networks did and did not overlap with their FtF networks. Through the participant observations of conversations, it was clear that some members interacted with other members FtF because they would talk about the interactions in their messages. Other members appeared not to know people in their FtF community because they would post messages about the loneliness of training for multiple-sports events.

In interviews, some members reported that they interacted FtF with other MSN members. Some members trained for multiple-sports events with others MSN members from their FtF community. Other members reported meeting MSN members at races to "carbo load" beforehand, complete the races with each other, or meet afterward to "commiserate" about their performances.⁵

Others, conversely, were explicit about *not* interacting with MSN members FtF; that is, they purposefully avoided MSN members in their FtF interactions. One member reported that she appreciated the anonymity of interacting on-line versus FtF. This member was not a typical multiple-sports athlete, and she felt that when she met MSN members FtF, she "quite lost [her] credibility" because her appearance did not mesh with others' expectations of her. She did, however, train and interact with other athletes who were not involved in MSN in her community. These people had already evaluated her credibility and so she did not have to risk being negatively evaluated. Other members, particularly lurkers, reported that they did not feel enough like a "real" multiple-sports athlete to meet with the MSN members, even at events where they knew MSN members would be participating.

There are differences in the types of people who chose to interact FtF with others from MSN and the places in which members chose to interact. MSN leaders were the most likely to interact with other members at races, but *not in their FtF community*. For them, it may have been socially rewarding to meet the other members at an MSN-focused special event such as a race. MSN members considered certain leaders of the group to be famous and held them in high regard. Members would even discuss meeting the leaders in their posts to the group. During an interview, one active participant described observing a leader meet other members of the group:

Somebody like [this leader] is tremendously enthusiastic and revels in [the fame]. Shamelessly, shamelessly revels in it.... She's mother [MSN].... It's funny to watch people approach her.... It's amazing to me how much people know you from what you write. That amazes me. It's fun to watch people approach [her] because they can recognize her because she has a [Web] site.... [She] gets those [gatherings at races] together usually.

This participant goes on to depict how meeting at races allowed members to attach faces to names:

[A] lot of people who are active in [MSN] go to Ironman Canada and there is a fair amount of socialization amongst the [MSN] people.... And it's fun to put the faces of the names in writing and so a lot of us that have done Canada have gotten together a couple of years.... But there is a huge bonding that goes on and it brings the group of people, this core group of people closer together.

Thus, this participant explains that meeting at races allows leaders to reap the rewards of their hard work and establish deeper bonds among members. Although leaders interacted with others at races, they were unlikely to interact with other MSN members on a day-to-day basis in their FtF communities. Two of the leaders interviewed commented that they did not have time to interact with others in their FtF communities. As multiple-sports athletes, they remarked that they were not sitting in front of their computers in their spare time; they were exercising.[6] Yet they did not engage in their FtF community with others, either.

MSN participants (i.e., not the leaders or lurkers) reported a good deal of FtF interactions in their communities with both MSN members and other regular members of their community. In messages posted to the group, these were also the most common type of member to comment on some sort of ongoing FtF interaction with other MSN members. In the interviews, some participants reported being active in a variety of sports-minded civic groups including running FtF multiple-sports clubs and organizing athletic training events with others. These members also expressed more attachment to their FtF sporting community than to MSN. One member who was active both in MSN and his FtF sports club said:

> I certainly don't feel about any of these people on the newsgroup the way I feel about the people I race with and train with. Although I get the feeling that other people in the newsgroup are closer to each other than I am to them.... I don't really think of the people in [MSN] as my friends.... [In comparison to my FtF multiple-sports club], well, obviously the [sports] club is real. The [sports] club is active. It's something that I go out and do. You know, I do rides, swims, races with my friends in the club. But [MSN] is much broader. And I'm getting a lot wider spectrum of opinions and views. Although it's all cerebral. And it doesn't take up a lot of my time. But it gives ... but I get a lot out of it in a very short amount of time.... [MSN is] just talking about, or getting information or reading about the sport. It's not doing it. So from that point of view it's much less satisfying and much less gratifying than going for a workout. And given the option, I would never choose to read [MSN] over doing a workout or participating with somebody in real life. But it is a quick substitute in a way, a quick way to extend my interest in the sport of triathlon. (Laughter) Look at the Power Bars against the wall.[7] That's my job in the club: to find sponsor relations. I'll take them and pass them out at the meetings and the workouts.

This participant clearly preferred his FtF sports club, although he still felt he benefits from what he describes as the small amount of time spent reading MSN. The end of the quote makes reference to his role in the FtF sports club of finding corporate sponsors. Obviously, he is actively engaged in what would be considered FtF social capital. All the participants interviewed reported active FtF interactions with MSN and other regular athletes. Therefore, based on the interviews and the messages collected in the observations, participants appear to be the most active of the member types in their FtF communities.

The lurkers did not report any FtF interactions with other MSN members either at races or in their regular FtF communities.[8] For some, they did not express an interest. Others did not feel they had the opportunity. It is not clear whether this group participated in other nonsports-related civic groups. Due to the structure of the interviews, none of the members were asked to describe their nonsporting FtF activities. *All* the leaders and participants, however, reported interacting with others either at races or in their FtF communities; *none* of the lurkers did. Even with the low number of people interviewed, this sort of discrepancy in reported experience between member types (Miles and Huberman 1992) is interesting. It suggests the possibility that lurkers may lurk both on-line and FtF.

Norms In considering norms and FtF social capital, the issue is how appropriate behavior norms of the virtual community affect the FtF communities. We can expect from knowledge about groups that an active virtual community would have strong norms about behavior in the group (McMillan and Chavis 1986). This was certainly true in MSN. In MSN, though, norms about appropriate behavior also extended beyond on-line group behavior. Some of the most talked about norms in MSN dealt explicitly with behavior in FtF interactions, especially at races. Generally, these norms had to do with being a good citizen at the races. For example, during multiple-sports events, many athletes consume prepackaged food specifically made for endurance sports. In one active message thread, members discussed the inappropriateness of the littering that occurs at races when participants eat and then throw the food wrappers on the ground.

One of the strongest norms of the group was that of not "drafting" in races. Drafting occurs when bicyclists ride closely behind each other and one rider is carried along in the draft of the bicycle ahead of him or her. In some multiple-sports events, drafting is prohibited. Nonetheless, some competitors continue to draft during draft-prohibited events.

This norm was a dividing point for MSN members. Good MSN members did not draft. During interviews, several members discussed that they did not watch multiple-sports events in which drafting was legal because it did not represent a "real" multiple-sports event. During the participant observations, conversations about the evils of drafting were common, and several prominent members were observed including messages against drafting in their permanent signature files when they posted messages. In one message thread, members also discussed options for what to do if they saw people drafting during the race. Ironically, this particular message thread turned into an on-line discussion of proper FtF sanctioning behavior of inappropriate FtF behavior in races.

Thus, MSN had explicit and strong norms about appropriate behavior interacting with others at races. Adhering to these prosocial norms could even lead to the detriment of one's individual performance in a multiple-sports event; not drafting slows cyclists down. While it is interesting that MSN had norms about how to behave at races, no norms emerged as being appropriate for behavior in members' FtF communities outside of races.

In fact, local FtF norms may be more influential on members than MSN's norms. In one incident during the observational period, a disgruntled member posted a complaint about a vendor who was also an MSN member. The complaint sparked a great deal of controversy in the group: not all members felt it was appropriate to use the group to force vendors into a "trial by Internet." In an interview, an MSN member who was a friend of the complainer commented on the incident. He explained that the person decided to post her concerns about the vendor because her FtF friends encouraged her to do so. She did this knowing that MSN would disapprove, but that her local friends approved her actions.

This became an example of using the virtual community to sanction behavior that occurred in an FtF community. Many members felt the behavior was inappropriate. Nevertheless, the local norms and desires were more important than the virtual community's norms.

Trust In social capital terms, trust means that members of the group can expect that the help they provide to the community will be returned to them. Members are considered trustworthy if they help others.

In MSN, members were observed asking for and providing help to others. This was the most common form of communication within the group. Leaders and participants also reported that receiving as well as providing help were significant reasons for participating in the group. One leader even said she felt obligated to participate because of the help she had received from others.

Lurkers, however, are different. They do not necessarily provide public help but do benefit from it. One lurker admitted that he just waited until someone else posted a question that he had. He could benefit from the answer without having to ask the question. Therefore, lurkers in MSN could trust that, eventually, they would benefit from being a member of the group.

Lurkers may help privately, though. Two lurkers reported offering help by responding to public posts privately through personal e-mails. Of course, not all lurkers can be counted on to supply assistance. The lurker mentioned in the previous paragraph was explicit that he did not respond publicly or privately to any MSN messages. Lurkers can certainly benefit from the group and even count on the helpfulness of the group without always having to provide help in return. Their trustworthiness can only be assessed by the people who they privately help. Hence, it is not likely to travel through the group's social networks.

Does trust of being aided extend to FtF interactions? In MSN, it does occasionally, but not very often. One MSN leader reported that she caught up with an MSN member at the end of an Ironman race. This member helped her run the last four miles of the marathon even though he could have "run circles" around her at that point.[9] The support she felt from that FtF interaction was central to why she valued MSN. Therefore, members did exchange some support FtF, although it was within the group where most help could be obtained.

In MSN, trust and individual trustworthiness took on additional meanings. Trust also referred to the belief that people were actually real and that they were not interacting as personas. This sort of trust in MSN overlapped with their FtF communities in several ways. As mentioned previously, members often discussed their FtF interactions with others in their messages. After races, members would post race reports about their experiences and would frequently include interactions with other MSN members at the race. As a result, MSN members believed that other members were real. The discussions of real-world interactions helped members to trust each other and contributed to their sense of membership in a community of real people.

Members were also aware of their FtF friends when they posted online. They reported being careful about what they posted to increase their own level of trustworthiness in MSN, but more important, *with their FtF network*. One member who had recently increased his level of activity reported:

> I think more about what I post in [MSN] than [other groups]. Because I meet these people. Something I really might want to say, I might be a little bit more carefree with my words on some of the other newsgroups. Sometimes I've logged on to, offhand, Howard Stern and half the people in there are just like personas anyway. They make up things about themselves and stories and stuff so it's a totally different atmosphere. I know I'll never meet any of those people and I wouldn't care if I did anyways. But the people in [MSN] are some of my close friends. So I do think I think more about what I post there because of that.

Because he may meet other members FtF and because some of them are his close friends, this member takes care to post messages that will reflect better on himself and make him more trustworthy. Unlike the networks of relationships and the norms that we can see moving from the virtual group to FtF interactions, trust has a stronger influence when it moves from the FtF interactions to the virtual group.

Discussion

The goal of this chapter was to explore how participating in a virtual community affects social capital in FtF communities. Specifically, it looked at how MSN participation affected members' social network of relationships, the norms of appropriate behavior, and trust in their FtF communities.

Participation in MSN did not have a universally positive or negative effect on FtF social capital. Members who were active in MSN either as leaders or participants were active in FtF communities either at races or with local sporting groups. Lurkers, who were the majority of the group, did not appear to be active in their FtF sporting communities. Even though lurkers may have been active in creating and exchanging social capital with other FtF groups, it was not revealed in the interviews.

There was a relationship between MSN member type and where members were most likely to engage in FtF interactions. Leaders were much more likely to engage in FtF interactions primarily at races. Participants were much more likely to be involved in their local community.

Does this have a differential effect on FtF social capital? Quite possibly, yes. Although the social capital created at races is good for the multiple-sports community, the social capital created through interacting with one's FtF sporting community is more likely to help one's community at large. As observed in this study, despite being a dispersed virtual community, the social networks between the FtF and virtual communities overlapped for some members. The resulting increased density of the networks meant that information flowed through these networks from the virtual community to the FtF one and from the FtF community to the virtual one.

The same is not true for MSN members who interacted with others at the races only. Although the networks of interactions at the races clearly moved from on-line to FtF, it is not apparent that they overlapped with any other permanent social networks. Given this, members may have left the races feeling a more real connection with MSN, but their interactions had no lasting effect on any FtF community.

Why was there a difference in where leaders and participants interacted with others on an FtF basis? One clue may come from the fame that leaders had in MSN. The FtF interactions at races may have been an opportunity for them to cash in on all the hard work they put into the group. For participants who were active in their FtF communities, it was clear that their FtF community is where they put the most effort and found the most rewards. The virtual community simply enhanced their sporting experiences.

In this study, the decision was made to focus primarily on the norms of behavior and trust as the substance that flowed through the social networks of interactions. This case demonstrates that norms of appropriate behavior (e.g., how to behave at races) and information about the trustworthiness of members do flow between the virtual community and FtF communities in ways that positively affect both communities. Future research should concentrate on identifying additional sources of content that flow through these networks and affect FtF social capital. Although norms and trust are clearly important, they are not exhaustive as far as the content exchanged in the networks that could be considered significant(see Adler and Kwon 2002; Nahapiet and Ghoshal 1998). Moreover, there might be content that is unique to the overlap of FtF and virtual networks of interaction.

A few comments must also be made about the appropriateness of generalizing from MSN to other virtual communities. Because case study results can only be generalized to other similar cases, it is imperative on the researcher to provide information to potential users of the research (Lincoln and Guba 1985). As an athletic virtual community, MSN is uncommon because many of its members presumably engage in strenuous physical activities within their FtF communities. Still, the FtF sporting community is a prototypical example of a civic group as described by Putnam (1995). Most appropriately, then, the results of MSN may be generalized to dispersed virtual communities that correspond to or overlap with FtF civic groups within a community.

The lurkers in this case present some interesting opportunities for future research. It cannot be presumed that the lurkers interviewed in this study are representative of all the lurkers in MSN or lurkers in general. Indeed, this research points out that we know very little about lurkers in the research despite their being the most common type of virtual community member. This study presents two possibilities of how lurkers participate in virtual communities and social capital. First, lurkers may participate behind the scenes by communicating with virtual community members through private e-mails. Second, lurkers may not participate at all—not in their virtual communities and not in their FtF ones. This distinction is key. More research needs to be conducted to understand how and why lurkers participate in virtual communities, and how that relates to social capital in FtF communities.

Conclusion

What can this study tell researchers and activists about the effects of displaced virtual communities on FtF social capital? First, this chapter has demonstrated that dispersed virtual communities are not completely placeless. Although this virtual community was not associated with any specific geographic location, members in the virtual community did interact in meaningful ways with others on an FtF basis. Activists may be able to take advantage of popular, well-established virtual communities to connect people in FtF communities around particular topics. With the growing use and acceptance of the Internet, people's global, virtual villages are likely to overlap with their local, FtF ones.

Additionally, when the virtual community members were active in their FtF communities, it was the FtF not the virtual community that came out ahead. Members were less attached and obligated to the virtual community as compared to their FtF one. The strength of the FtF over the virtual community also included one of the two relational forms of social capital examined: the norms of behavior. Local, FtF norms took precedence in governing people's behavior over the group's norms.

The help exchanged within the virtual community, however, appeared to be more beneficial to more people than the help exchanged in the FtF communities. That is, MSN members were able to receive answers to a wider range of questions and hear from a larger number of experiences than is easily possible in FtF interactions. This is most likely to be the reason that people participate in the virtual community. Virtual communities that are explicitly place based (e.g., neighborhood associations) should pay close attention to providing opportunities for people to exchange support and assistance so that they can develop the attachment and loyalty seen in this virtual community in their members.

Community researchers and activists may not need to fear dispersed virtual communities. As our empirical studies of the interactions between virtual and FtF communities grow, we are likely to find a complicated relationship. Although challenging, it will allow us a better understanding of why people participate in these communities so that we can encourage the development of social capital for the betterment of all.

Notes

1. MSN is a pseudonym.
2. The newsgroup was initially identified when the researcher was training for her first multiple-sports race.
3. Contrary to the researcher's initial expectations, leaders were not necessarily the best athletes in the group.
4. For a more detailed analysis and discussion of MSN's sense of community, see Blanchard and Markus (in press).
5. Carbo load means to eat a large, carbohydrate-filled meal before a long sports event.
6. For even the shortest multiple-sports events, training times can exceed ten hours per week.
7. Power Bars are a type of food that endurance athletes eat.
8. By definition, it is impossible to observe lurkers in the newsgroup.
9. An Ironman race consists of a 3.8-kilometer (2.4-mile) swim, a 180-kilometer (112-mile) bike, and a 42.2-kilometer (26.2-mile) run.

References

Adler, P. S., and S. Kwon. 2002. Social capital: Prospects for a new concept. *Academy of Management Review* 27, no. 1:17–40.

Atkinson, K. 1995. *Usenet info center: Newsgroup info center*, <http://metalab.unc.edu/usenet-i/groups-html/[multi.sport.newsgroup].html>.

Belliveau, M. A., C. A. O'Reilly III, and J. B. Wade. 1996. Social capital at the top: Effects of social similarity and status on CEO compensation. *Academy of Management Journal* 39:1568–1593.

Blanchard, A. L., and T. Horan. 1998. Virtual communities and social capital. *Social Science Computer Review* 16:293–307.

Blanchard, A. L., and M. L. Markus. In press. The experienced "sense" of a virtual community: Characteristics and processes. *Data Base for Advances in Information Systems* 34.

Coleman, J. S. 1988. Social capital in the creation of human capital. *American Journal of Sociology* 92:1287–1335.

Coleman, J. S. 1990. *Foundations of social theory*. Cambridge: Harvard University Press.

Jones, Q. 1997. Virtual communities, virtual settlements, and cyber-archaeology: A theoretical outline. *Journal of Computer-Mediated Communication* 3, <http://www.ascusc.org/jcmc/vol3/issue3/jones.html>.

Koh, J., and Y. Kim. 2001. Sense of virtual community: Determinants and the moderating role of the virtual community origin. Proceedings of the 2001 International Conference on Information Systems, New Orleans, Louisiana.

Lincoln, Y. S., and E. G. Guba. 1985. *Naturalistic Inquiry*. Newbury Park, Calif.: Sage.

McMillan, D. W., and D. M. Chavis. 1986. Sense of community: A definition and theory. *Journal of Community Psychology* 14:6–23.

Miles, M. B., and A. M. Huberman. 1994. *Qualitative data analysis*. 2d ed. Thousand Oaks, Calif.: Sage.

Nahapiet, J., and S. Ghoshal. 1998. Social capital, intellectual capital, and the organizational advantage. *Academy of Management Review* 23:242–266.

Paxton, P. 2002. Social capital and democracy: An interdependent relationship. *American Sociological Review* 67:254–277.

Pruijt, H. 2002. Social capital and the equalizing potential of the Internet. *Social Science Computer Review* 20:109–115.

Putnam, R. D. 1995. Bowling alone: America's declining social capital. *Journal of Democracy* 6:65–78.

Rheingold, H. 1993. *The virtual community: Homesteading on the electronic frontier*. Reading, Mass.: Addison-Wesley.

Timms, D., S. Frelander, and L. Timms. 2001. Building communities: On-line education and social capital. In *Learning without limits: Developing the next generation of education*, edited by A. Szucs, E. Wagner, and C. Holmberg. Proceedings of the EDEN tenth anniversary conference held, Stockholm, Sweden.

Tonn, B. E., P. Zambrano, and S. Moore. 2001. Community networks or networked communities? *Social Science Computer Review* 19:201–212.

Wellman, B., A. Q. Haase, J. Witte, and K. Hampton. 2001. Does the Internet increase, decrease, or supplement social capital? Social networks, participation, and community commitment. *American Behavioral Scientist* 45:437–456.

4
Find What Binds: Building Social Capital in an Iranian NGO Community System

Markus Rohde

This chapter presents a project aimed at supporting the community building of Iranian NGOs. Such support for Iranian civil society has to be seen in the context of the democratic transition within state and society. In this project, an integrated organization and technology development (OTD) approach is used for the participatory design of a community system that supports the NGO community-building process. The project took place in a politically isolated Islamic state that is characterized by a fast-growing civil society. Other dynamics at play in the community-building process are the ongoing, rapid growth of a technological infrastructure and the exploding number of Internet accounts in Iran. Concerning the introduction of a technological community system, development nearly starts from scratch. These conditions make the building of an e-community of Iranian NGOs a real challenge.

Furthermore, the war between some of the Western democratic states and Iraq influences the political stability of the whole region while also affecting most Middle Eastern countries. Without being able to discuss all the implications of the political background, this chapter sees the ongoing process of Iranian NGO-community-building as a step in the development of a lively Iranian civil society.

This chapter introduces the concept of civil society and specific information about state-of-the-art Iranian NGO networking. It then presents the theoretical framework for this project along with a case study. This chapter then goes on to describe the action research approach taken here as well as its measures and achievements. Next, the results of the process and a first evaluation are presented. The last section discusses the implications of the empirical findings for the theoretical approach and further research.

Iranian Civil Society

Civil Society: Definition and Discussion of a Concept

In this study, I consider NGOs to be the actors in civil society. I thus use the terms *NGOs* and *civil society organizations* (CSOs) synonymously. Nevertheless, there is no formal and universal definition of the concept of either *civil society* or *NGOs*. According to the broad definition offered by the Food and Agriculture Organization (FAO) of the United Nations, NGOs are "all not-for-profit actors who are not governmental or intergovernmental." Definitions such as this one do not differentiate between "good" and "evil" organizations, and therefore enfold not primarily socially oriented organizations like, for example, the U.S.-based National Rifle Association or even criminal syndicates, insurgent militias, and terrorist formations. Thus, the "dark side of civil society" is included (Roth 2003).

Therefore, most democracy theorists refer to a normative concept of civil society and NGOs in which tolerance, fairness, and civic engagement for public goods are the most important characteristics. NGOs should be focused on communicative action and the exclusion of illegitimate forms of physical violence (cf. Lauth and Merkel 1997). In their situation analysis of Iranian NGOs, Iranian experts (NGO activists and political scientists) define an NGO as "an independent, non-government, non-profit, voluntary association of a group of citizens, rallying around a common community cause, and accountable to the clients they profess to serve" (Namazi 2000, 18). Yet even the normative approaches and more limited definitions of civil society include a wide range of different organizations, social movements, networks, pressure groups, and others.

Most approaches to structure this quite fuzzy area of civil society are based on the positioning of NGOs between the market and state sectors. NGOs represent intermediary organizations in a "third sector" or a "value driven sector" (Fowler 1992, 22). In contrast to national policies and particularistic interests, NGOs are advocacy organizations that promote general public goods. Categorizations of NGOs refer to the origin of the organizations in a global civil society, and they differ between the North (European and United States) and South (developing countries) NGOs. Others are referring to different organizational structures and the degree of institutionalization (networks of autonomous movements versus professional and hierarchical interest or pressure groups) or thematic issues (environment, development, women's rights/gender, peace, health, etc.) (cf. Rohde and Klein 1996, 7).

Concerning international NGO networks and the discourse of a "global civil society" (cf. Anheier, Glasius, and Kaldor 2001), recent developments within CSOs show a trend toward the establishment of umbrella organizations and professional, well-organized associations. As far as the trends of institutionalization and internationalization go, two problematic effects have to be taken into consideration: the risk of corporatism and co-optation. Corporatism means that the highly institutionalized international CSOs might take on filter functions, therefore equalizing the specific characteristics, issues, and activities of different NGOs. So the autonomy of single NGOs is at risk. Co-optation is a threat that results from narrow relationships with governmental organizations such as new forms of cooperation, partnership, and coalition. Co-optation might cause a loss of autonomy and specific principles such as participation, or social and cultural embedding. Moreover, additional management requirements (e.g., resource allocation, public relations, international campaigning, etc.) might increase the pressure to adapt to the logic of governmental organizations. NGOs might lose their specific characteristics by adapting themselves to traditional organizations, along with their structures and management problems (cf. Rohde and Klein 1996, 7).

Concerning the role of civil society in a global world, Benjamin Barber (1996) identifies two processes as the main threats to democracy: economic globalization, mass media entertainment, and commerce ("McWorld"), on the one hand, and ethnic/religious closures and fundamentalism ("Jihad"), on the other hand. These opposing trends of globalism and tribalism mark a development that might lead to a "clash of civilizations" (cf. Huntington 1998). Barber (1996, 294) expects solutions for these conflicting trends to come from transnational cooperation among NGOs in a global civil society rather than from international conventions and UN resolutions.

Civil society plays a crucial role in the process of democratic transition. A comparative analysis of civil society development in several transition countries reveals that this role is most important during the liberalization stage, at the start of the transition process, and the democratic consolidation stage (cf. Rohde and Klein 1997, 3). Since Iranian society is currently in the liberalization stage, we expect CSOs to play a prominent part in this process of transition. Therefore, the project presented was supposed to help strengthen Iranian civil society.

Background on Iranian Civil Society

In this section, some demographic data are presented to describe the background on the development of Iranian civil society. The state of the art of (national and international) networking of Iranian NGOs will be sketched, and some preconditions for successful community-building will also be discussed.

The Islamic Republic of Iran has a total population of sixty-six million people (as of 2001), of which 49 percent are female and 51 percent male. Only 39 of the populace lives in rural regions, or conversely, 61 percent of the population resides in urban regions. Population growth is at 1.7 percent, and 33 percent of the population is under fourteen years of age. The unemployment rate was 14 percent in 1999. Iran accommodates two million refugees (Namazi 2000, 13). In higher education, women account for more than 35 percent of the students (as of 1996; cf. Anheier, Glasius, and Kaldor 2001, 277).

The international NGO Human Rights Watch reports human rights violations in Iran such as a lack of freedom of expression and association, disappearances and extrajudicial executions, arbitrary detentions, torture, and discrimination against minorities (Anheier, Glasius, and Kaldor 2001, 264). At the same time, Iranian society is characterized by a fast-growing NGO and CSO sector: In 2000, there were 1,500 to 2,000 traditional community-based CSOs, more than 5,000 women-run cooperatives, 1,500 modern NGOs, and about 3,000 easy credit funds (cf. Namazi 2000). Traditional community-based organizations have a long history in Iran. They are sustained by community funds, focus on the most pressing needs of the people, and have survived pressures from both the monarchy and Islamic government. The growing number of so-called modern NGOs in Iran concentrate on gender issues, youth and children interests, health and population matters, sustainable development, and environmental protection. All of these organizations are building the basis for Iranian civil society.

State-of-the-Art National NGO Networking

In 1997 and 2001, two national conferences for Iranian NGOs took place in the cities Busher and Mashad, respectively. At the Mashad conference in September 2001, the more than 120 participants included NGOs staff, scientists, government delegates, and UN representatives (cf. Hamyaran NGO Resource Center 2001). The meeting was organized by the Hamyaran NGO Resource Center in Tehran, the center, officially registered as an NGO with the Iranian government in March 2001, is coordinating the ongoing networking process of Iranian NGOs. The next national NGO conference is planned for December 2003.

These conferences mark a turning point for NGO networking in Iran, which is already fairly advanced in many areas. National women-run NGO networks, environment and development NGO networks, youth organizations, and health/population umbrella organizations already exist. The conferences started a process of networking these networks.

In early 2000, the Ford Foundation and the Iranian Population Council funded a study to analyze two areas of conflict for Iranian NGOs: the differences between rural NGOs and urban organizations, which are located centrally in Tehran and the provincial capitals; and the significant distinctions between the so-called new NGOs and the traditional community-based organizations, which are mainly focused on relief work (cf. Namazi 2000). The study found that communication and cooperation between these different types of Iranian NGOs are quite weak.

According to the aforementioned situation analysis of Iranian NGOs, the recent networking process is largely shaped by the integration of these different NGOs, which are segmented by origin, style of work, and organization. The Hamyaran NGO Resource Center reports that the emerging NGO movement tends to neglect the significant role of traditional organizations, even though they have strong community ties, a deep knowledge of civil engagement, and many experiences from which the new, modern NGOs can learn and benefit (Namazi 2000). This lack of cooperation certainly poses a challenge for NGO networking in Iran.

Another challenge is the relationship between NGOs and governmental organizations. Over the past few years, the reform-oriented Iranian government and President Mohammad Khatami have officially fostered Iranian civil society, NGOs, and a networking process. Nevertheless, Iranian NGOs still face a lot of difficulties: "Khatami and his Government declare progressive policies everyday. Old legal and procedural forms, however, are still in place. Even worse is the negative attitude of senior officials in the executive, judical and legislative branches that need to be overcome" (Namazi 2000, 7). Given the necessity of cooperating with governmental organizations, NGOs must carefully take risks of cooptation into account.

The State of International NGO Networking in Iran

The statistics on international tourism reported in the *Yearbook of Global Civil Society 2001* for Iran serve as an indicator for international civil networking: From 1988 to 1998, inbound tourism increased about 5.5 percent—a large figure when compared internationally, but fairly low as an absolute number (1,008,000 in 1998). Outbound tourism increased about 2.2 percent, but again, that is quite low in absolute numbers (1,354,000) (Anheier, Glasius, and Kaldor 2002, 245).

In terms of international NGOs (INGOs) in Iran, the yearbook reports only three (in 1999) and nine (in 2000) first-level or national-level secretariats (ibid., 284). The number of Iranian leaders of international governmental organizations and INGOs decreased from five in 1996 to two in 2000 (ibid., 294).

These statistics indicate that in comparison to the international average, Iranian civil society is characterized by a low degree of international organization. This is due to the far-reaching political isolation of the Iranian state and the restrictions imposed by the Iranian regime. Thus, the project presented had to take this international networking of Iranian NGOs into consideration.

Challenges for the Technological Support of NGO Networks

As the *Yearbook of Global Civil Society 2001* documented for the years 1999 and 2000, only 11.2 percent of the Iranian population had access to a telephone, only 3.2 percent to personal computers, and a mere 0.2 percent to the Internet (ibid., 256). While the illiteracy rate decreased significantly (by about 12 percent) during the 1990s, it was still at 25 percent in 1998 (ibid., 277). Such statistics might underestimate the actual development in Iran. According to the UN Development Program (UNDP), the youth literacy rate was at 93.7 percent in 1999.

Although the number of Internet accounts and providers is growing rapidly, especially in the urban areas of Iran, the lack of Internet accounts for private households as well as NGOs and the still-high illiteracy rate are key barriers in terms of networking for the national NGO community in the rural areas of Iran.

Among other barriers to the development of civil society, the Hamyaran NGO Resource Center stresses internal management and resource problems. Iranian NGOs face problems "such as lack of respect for professionalism, and for open and participatory management systems, weak technical and financial capabilities, and insufficient affinity with the community. Information and data is scarce and unreliable. The culture of transparency, accountability and experience sharing is very weak" (Namazi 2000, 7). Such a culture is another big obstacle, then, for the project of building an Iranian NGO community system.

Theoretical Framework

In light of all these challenges, the so-called modern Iranian NGOs defined as one of their central requirements the introduction of a communication system to foster cooperation as well as the exchange of information and experience. In the ongoing process of social networking, this technological support is looked on as an appropriate means to improve transparency and participation.

Given the requirements of Iranian CSOs, one useful approach is integrated organization and technology development (OTD; Wulf and Rohde 1995), which was introduced to combine the interwoven processes of socio-organizational and technological developments. In the intellectual tradition of sociotechnical systems approaches and action research methodologies, the OTD approach was designed to introduce technical systems, especially groupware, into organizations. This approach does not focus on CSOs, however.

So I augmented my framework by considering the theoretical approaches of "social capital" (cf. Bourdieu 1983; Coleman 1988; Cohen and Prusak 2001; Putnam 1993) and "communities of practice" (CoP; Lave and Wenger 1991; Wenger 1998). In my interventions during the project, I tried to make these approaches fruitful for the practical support of community-building processes.

In the following section, these three approaches are briefly described to sketch the theoretical background of my practical work and the framework for my action research.

Integrated Organization and Technology Development
In my project, I focus on both organizational change and technological development, which are closely linked together. Organizational structures and practices are the context and background for the development and usage of IT, and they define the requirements for technological development. On the other hand, technological developments influence cooperation, practice, and therefore organizational processes. Thus, I integrate the concepts of evolutionary organization development with user-centered software engineering in the OTD approach (cf. Wulf and Rohde 1995). OTD was developed in the tradition of the sociotechnical systems approach, which focuses on the relation between nonhuman and human systems (Emery and Trist 1960; Ropohl 1999). From this perspective, processes of organizational change can be characterized by interwoven technical and social developments.

Organization development can be understood as an initiated, long-term, organization-wide process of change in the behavior, attitudes, and abilities of its members as well as its structures and processes (cf., e.g., French and Bell 1990; Pieper 1988; Wulf et al. 1999). It is performed in an evolutionary, cyclic process that involves collecting data about the organization and its problems, presenting and discussing these data within the organization, planning interventions to overcome the problems, and performing the intervention within the organization. With this iteration of data collection, feedback, intervention, and new data collection I refer to the tradition of action research methodology.

This evolutionary, cyclic approach was transferred to software engineering processes. Thus, software development takes place as a process of cooperation among software developers and users. According to the iterative cycle described above, the appropriation and use of a system and its evaluation are important aspects for the redesign of the system. In order to keep pace with environmental changes, the approach assigns an iterative development process that establishes a revision as soon as the system's functions do not match the requirements of the users (cf. Floyd, Reisin, and Schmidt 1989).

The OTD process, which tries to integrate both concepts of organization development and iterative software engineering is characterized by its concentration on the parallel development of organizational and technical systems, the management of (existing) conflicts by discursive and negotiative means, qualification and training measures, and the immediate participation of the organization members affected (cf. Rohde and Wulf 1995).

Regarding this case study of Iranian NGOs, rather than focusing on a traditional organization, I explore a network of organizations (or even a network itself) that exhibits the basic characteristics of a "virtual organization" (cf. Davidow and Malone 1993; Nohria and Berkley 1994; Strausak 1998; Travica 1997). I assume that networks of NGOs have more in common with virtual organizations than with traditional ones. The virtual Iranian NGO community can be characterized by:

• absence of traditional organization structures and principles which imply a formal definition of internal order;
• transboundary amalgamation of organizations or enterprises;
• temporal instability of the virtual unit which affiliates, changes fluidly, expands or reduces itself, and disappears after achieving its purpose;
• non-simultaneity of collaborative processes and acceleration of organizational development;
• spatial distribution, and;
• modern ICT (information and communication technology) as a precondition for the existence of the (virtual) organization. (Rohde,

Rittenbruch and Wulf 2001, 3; cf. Mambrey, Pipek, and Rohde 2003).

The introduction of a community system to support NGO networking by technical means will probably emphasize this virtual character. Nevertheless, there is no reason to assume that the integration of organizational and technological developments might not make sense for virtual networks or communities. In this context, the application of OTD to virtual organizations and networks means a further development of the approach in terms of innovative types of organizations and new fields of applications. But since the OTD approach was refined mainly for traditional working groups and organizations, it has to be evaluated to determine if it can usefully be applied to virtual organizations and networks.

In the case of the Iranian NGO network, one is dealing with an emerging "community" of CSOs. Thus, to cope with the specific requirements of communities, I have turned to the theoretical approach of CoP.

Communities of Practice

The CoP approach integrates identity theory, theories of practice, and theories of situated experience (Wenger 1998, 12). In their research work on situated learning, Jean Lave and Etienne Wenger (1991) focus on common daily practice, active membership, and in-group awareness. The most important inclusion mechanisms concerning these communities are processes of collective learning along with the production of shared meaning and collective identity.

Lave and Wenger analyzed processes of learning in organizational units. Their findings characterize processes of learning as engagement in the social practice of groups and networks. For CoP, rather than organizational structures, (mostly informal) working and cooperative relationships are constitutive: "These practices are thus the property of a kind of community created over time by the sustained pursuit of a shared enterprise" (Wenger 1998, 45).

For individuals, learning is situated in processes of social participation in these CoP. Individual learning in a CoP is mainly based on "legitimate peripheral participation" (Lave and Wenger 1991). During the participation process, an individual might enter the community as a newcomer at the periphery and, over time, gain a more centered position through a "cognitive apprenticeship." Therefore, this acquisition process means an intensified inclusion into the social practice of the community. The CoP themselves can be seen as "shared histories of learning" (Wenger 1998, 86).

The development of a common practice integrates the negotiation of meaning among the participating members as well as the mutual engagement in joint enterprises and a shared repertoire of activities, symbols, and artifacts. This practice of a community is inseparable from issues of (individual and social) identity, which is mainly determined by a negotiated experience of oneself in terms of participation in a community and the learning process concerning one's membership in a CoP (Wenger 1998, 145ff.).

These processes of identification are the subject of sociopsychological theories on social identity (cf. Tajfel 1978, 1982) or social categorization (cf. Turner et al. 1987). Both approaches postulate that people tend to categorize themselves as a "group" if the salience of perceived differences among these individuals is minor relative to the perceived differences to other individuals. Thus, perceived similarities among different persons in terms of attitudes, beliefs, norms, and values, a common task or shared history, a shared perception of threats or common enemies, and so on, are significant conditions for social identification and group cohesion. Although Lave and Wenger do not refer to these sociopsychological theories and therefore to the specific conditions under which identity-building takes place, processes of identification are pivotal for their CoP approach.

In this sense, the CoP approach combines the "two sides of the medal" of community participation: the social practice of the community as a collective phenomenon, and the (social) identification of its members as an individual one.

Nevertheless, neither OTD nor CoP are focused on NGOs or CSOs. OTD was developed to study processes of introduction of technical systems mainly in business organizations, small and midsize enterprises, and public administrations. It deals with organizational changes and technological developments regardless of the profit or nonprofit character of the organization. CoP have been analyzed mainly in the professional daily practice of profit organizations. I assume that the main findings of both approaches are true for nonprofit organizations as well. Since I focus on nonprofit-oriented CSOs, however, the social capital approach provides another perspective on community building.

The Social Capital Approach

The paradigm of social capital gained prominence as an approach to explore societal and political networking processes. Over the last few years, the social capital approach has been increasingly adapted to analyze cooperation in (NGO) networks as well as collaboration in companies and working groups. Yet the concept of social capital is not defined universally, and is used by various authors in different ways.

Social capital is defined by Robert D. Putnam as the sum of networks and social contacts, trust (respectively trustworthiness), and reciprocity relations that a person owns. Putnam highlights the importance of voluntary associations and organizations for the creation of social capital, and comes to the remarkable conclusion that "social capital makes democracy work" (1993, 2000). Still, Putnam's definition focuses predominantly on an individual perspective of social capital. This leads to the critical objection that Putnam's approach—like, for example, a communitarian perspective (cf. Etzioni 1993)—concentrates too much on personal contacts, face-to-face networks, and neighborhood relationships (cf. Hellmann 2002, 50).

To deal with the social capital of collective actors (like groups or organizations) and more distributed networking processes, the concept has to refer to the network contacts, *generalized* trust, and *norms of* reciprocity that the collective subject has established. To analyze the social capital of societal communities, Francis Fukuyama (2000) defines social capital as an indicator of mutual trust in societies. It is based on self-generated ethical conventions and mutual commitments concerning activities and behavior, and it results in solidarity. James Coleman (1988) introduces social capital as a resource available to actors, not by focusing on individuals but on social actors—for example, interrelated individuals in groups and networks. Therefore, in his understanding, social capital is a public good.

Pierre Bourdieu defines social capital as the actual and potential resources that are based on ownership of sustainable networks, (institutionalized) relationships, and mutual respect (cf. Bourdieu 1983). After analyzing the relation of social capital and economic, symbolic, and cultural capital, he describes social capital as the (individual and social) reputation that is needed to enter the "good society" and political sphere. In Bourdien's view, social capital is a mechanism of political inclusion/exclusion.

To adapt the concept to collaboration processes in companies Don Cohen and Laurence Prusak conclude: "Social capital consists of the stock of active connections among people: the trust, mutual understanding, and shared values and behavior that bind the members of human networks and communities and make cooperative action possible.... Its characteristic elements and indicators include high levels of trust, robust personal networks and vibrant communities, shared understandings, and a sense of equitable participation in a joint enterprise—all things that draw individuals together into a group" (2001, 4). The authors refer to the concept of social capital mainly to study and support information and knowledge management within companies, departments, and working groups.

Concerning processes of gaining and fostering social capital, the approach assumes that it is accumulating when it is used (productively), but if not, it is decreasing. In this sense, social capital tends to be self-reinforcing and cumulative. People gain connections and trust by successful cooperation, and these achievements of networks and trust support good cooperation in the future. To gain and foster social capital, Cohen and Prusak suggest the following (organizational) investments in trust-building processes: social capital can be gained by being trustworthy, by being open and encouraging openness, and by trusting others (2001, 45f.).

In the case of Iranian NGO networks, which generally cannot be characterized by traditional organizational structures or a common practice that has been established for a long time, I therefore assume that social capital and mutual trust are basic concepts for the community-building process.

The Iranian NGO Community Project

In 2002, the International Institute for Socio-Informatics (IISI) concluded a contract with the Department for International Cooperation of the German foundation Friedrich-Ebert-Stiftung for a research and development project that attempted to support community building and networking among Iranian NGOs. The Friedrich-Ebert-Foundation (FEF), founded in 1925, is a political legacy of Germany's first democratically elected president, Friedrich Ebert. The foundation is committed to the ideas and basic values of social democracy, and aims at political and social education in the spirit of democracy and pluralism, while also contributing to international understanding and cooperation. According to its mission, the FEF sees its activities in developing countries as a contribution to:

- promoting peace and understanding between peoples and inside the partner countries,
- supporting the democratization of the state and society and strengthening the civil society,
- improving general political, economic and social conditions,
- reinforcing free trade unions,
- developing independent media structures,
- facilitating regional and worldwide cooperation between states and different interest groups and
- gaining recognition for human rights. (<http://www.fes.de/intro_en.html>)

In defining Iranian civil society, I look to the understanding of the Iranian NGOs themselves (Namazi 2000). As such, urban and rural NGOs are included as well as grassroots movements and institutionalized organizations, or the so-called new NGOs and traditional community-based CSOs. Although in the project I cooperate closely with these new or modern NGOs, no Iranian NGOs should be excluded.

On the other hand, to cope with the previously described risks of co-optation and corporatism, I focus here exclusively on the requirements of NGOs, rather than on their relationship to governmental institutions. This exclusion of governmental actors (especially from the cooperation system) should improve mechanisms of identity building and foster the building of trust within the Iranian NGO network. This is not done in order to avoid a dialogue with Iranian governmental organizations but to create a space of exclusive internal interaction within the community first. Moreover, I treat all organizations as equals, disregarding their organizational structure or degree of institutionalization, so as not to exclude any NGO from communication and cooperation.

Besides national NGO networking, I also take international networking and cooperation into account, both to overcome political isolation as well as to enable Iranian NGOs to adequately respond to global causes and problems. Given these assumptions, strengthening Iranian civil society contributes to the ongoing process of transition in Iran.

The Iranian NGO community project was developed on the basis of a situation analysis of Iranian civil society that was conducted by scientists with the Hamyaran NGO Resource Center in Tehran and a prestudy carried out by IISI that aimed at the analysis of the specific requirements of Iranian NGOs. This requirement analysis was realized in spring 2002 by a detailed questionnaire, which was answered by Iranian scientists and NGO experts. This questionnaire included questions on the number and type of Iranian NGOs involved, the ICT (Information and communication technology) infrastructure and Internet connectivity, computer and language skills, requirements concerning technical support and trainings, the state of socio-organizational networking, and constraints and limitations concerning the networking process. Based on this requirement analysis, the project plan was devised.

During 2002, the IISI provided the cooperation platform Basic Support for Cooperative Work (BSCW) to Iranian NGOs in order to support their networking process by technical means. The BSCW system was developed by the Fraunhofer Institute for Applied Information Technology (FIT). The IISI organized the introduction of the BSCW to the Iranian NGO network, consulted an integrated process of OTD (cf. Wulf and Rohde 1995) referring to the Iranian NGO community project, and realized a train-the-trainer program for members of Iranian NGOs that was focused not only on technical trainings but community building, cooperation trainings, and project development as well.

This project began in March 2002 and ended with a delegation visit of leading Iranian NGO members and civil society experts to Bonn, Hamburg, Berlin, and Brussels in December 2002.

Project Approach
As was mentioned above, the project Iranian NGO Community System (NGO–CS) followed the OTD approach, in which socio-organizational and technological networking is looked on as interdependent and combined in a participatory process (cf. Wulf and Rohde 1995).

The process of socio-organizational networking among Iranian NGOs was already up and running several years before the project started. Thus, the planning of the Iranian NGO–CS did not take place in a social or organizational vacuum but had to cope with the requirements that had been previously formulated during the ongoing organizational NGO networking. To support the Iranian NGO networking by technical means, the introduction of the BSCW and the participatory design of the NGO–CS had to be embedded in the long-lasting networking process. Additionally, the effects of the NGO–CS and its usage for the socio-organizational process had to be evaluated, and cycles of technological (re)design and management of organizational change/development had to be provided.

According to these preconditions, the following strategy for the project was chosen:

• Requirement analysis: Based on process documents written by the Iranian NGOs and a prestudy involving Iranian NGO experts, the requirements of the Iranian CSOs in terms of technological support were analyzed.

• System introduction: During a first expert visit to Tehran, the technical platform BSCW was introduced to a group of leading Iranian NGO members.

• System design: Together with Iranian NGO practitioners, a structure for the Iran NGO–CS was developed, the BSCW was adapted to the NGOs' needs, and the initial batch of content was loaded into the system. This system was meant to support the already-running and ongoing socio-organizational process of NGO networking. Thus, a computer-induced process of organizational development was not intended. Instead, the introduction of the technical system was meant to serve as a catalyst for an already-ongoing process. A special focus was to bring together people and organizations in the different Iranian provinces regardless of time and location.

• Qualification measures and trainings: In several training measures, Iranian NGO members were trained to use the system and train other Iranian civil society practitioners by themselves. Additional measures were taken to strengthen social capital and CoP.

• Socio-organizational interventions: In several meetings and workshops, the establishment of a common practice was supported by initiating collaborative projects. Tools for proposal writing and project management were offered. Furthermore, adequate methods had to be used to foster trust building and social capital within the Iranian NGO network. According to the theoretical approaches presented above, the basic presumption is that the participatory development and the common work in collaborative projects should support trust-building within the Iranian NGOs, strengthen the network relationships, and thus foster social capital. These shared experiences with collaborative tasks should also enable the establishment of a common practice as a vital characteristic of the new Iranian NGO CoP.

• Evaluation and redesign: Each measure and intervention was evaluated with questionnaires and expert interviews. The system's usage was also evaluated by means of anonymous log files. During two follow-up visits to Iran, additional trainings and workshop meetings were conducted, and the system was redesigned for further use. These two visits had been planned in advance due to the cyclic approach of the project. Moreover, during the second and third visit, the process was broadened, and as a result, trainings and workshops in Iranian provinces outside Tehran were conducted.

• International networking: Due to the far-reaching political isolation of the Iranian state and civil society, the project also attempted to support the international networking of Iranian NGOs by fostering the international exchange of experiences, mutual visits, and common research projects. Especially if the global character of many problems—for example, environmental pollution or development issues—is taken into account, this international NGO networking is necessary.

The aim at the beginning of the project was that in the long run, a total of about 500 NGOs with around 25,000 members in all Iranian provinces should be able to benefit from the project's outcome by accessing the system, receiving training, and getting involved in the ongoing (re)design process. Therefore, the project (whose financial support by FEF was limited to eight months during the first phase) focused on a sustainable development and consulting approach, which required Iranian NGO practitioners to take ownership of the process themselves (after the project's end). This participatory approach is based on the belief that the Iranian NGO members are the real experts of their own NGO networking.

Thus, the project was set up in close cooperation with Iranian partners: the Hamyaran NGO Resource Center and, later, the newly founded Iranian Civil Society Organizations Resource Center, both located in Tehran.

Based on this approach, the IISI provided expertise during the first stage of the project for the development of the cooperation between Iranian networks of women-run NGOs, health NGOs, school building philanthropists, youth NGOs, and environmental NGOs. Most of these cooperation partners were one-issue networks or umbrella organizations. As such, the project was directed toward a "networking of networks."

First, a defined group of facilitators within the NGO network was enabled to (tele)cooperate with each other via the NGO–CS. During the next step, these facilitators had to be trained in order to work as trainers themselves for their colleagues and other NGO members. These the trainers then conducted workshops in which Iranian NGO members were trained in the theoretical basics of the development of socio-organizational structures that enable project development, expertise sharing (cf. Ackerman, Pipek, and Wulf 2002), and community learning.

Technical Support
According to the results of the prestudy—conducted in spring 2002 by the IISI—one central demand of Iranian NGO practitioners was the tech-

nological support for the dissemination of information. Iranian experts expressed their need for an enabling technology for cooperative projects and the exchange of relevant data. The desired technology had to be easy to use even with old computer systems and low data transfer speed. The system also had to be accessible to Iranian NGO members only for internal communication and collaboration. Based on these requirements, the BSCW system (cf. Appelt 1999) seemed to be the most appropriate fit for the Iranian NGO community system.

The BSCW is a Web-based groupware system with extensive functionality to support cooperation. One only needs an Internet account to access the system with a standard Web browser. No installation of additional software is required (cf. Koch and Appelt 1998; Appelt 1999). In 2000, the system was utilized by more than 80,000 users worldwide; in March 2003, there were more than 100,000 users.

The use of the BSCW is free of charge for not-for-profit purposes and scientific research projects at universities. Each user of the system gets ten megabytes of Web space for free. It would only be necessary to buy a server license if there were specific requirements for server hosting and system administration (security aspects, system performance, etc.) in Iran. For the Iranian NGO–CS, the system had to be free of charge and able to be used with any Internet-enabled computer. The BSCW avoids advanced graphic features. Thus, even with a low data transfer speed (fifty-eight kilobytes per second), the performance of the system is quite comfortable.

In the project, the BSCW system was used to develop structures for networked communication and the dissemination of national as well as international information among the Iranian NGO community. On the technical basis of this BSCW system and with the help of the IISI experts, the Iranian users set up an NGO–CS. The NGO–CS requires Internet connectivity, and provides workspaces for different kinds of working groups, NGO networks, topics/issues, and so on, which are to be set up by the users themselves. In this way, the IISI supported a process of participatory design in which members of Iranian NGOs defined their own spaces and activities, selected the members addressed, defined groups, invited new members, sent group mail, and started discussions on self-chosen topics. Moreover, the system supports Iranian NGO members

Figure 4.1
Snapshot of the Iranian NGO Community System.

through a version-management system for coauthoring, features that allow for the storage and exchange of most document types, and the ability to set up links to external Web sites. The system also provides a lot of awareness features to inform its users of activities and events within the NGO–CS (see figure 4.1).

Measures and Achievements of the Project

In May, July, and October–November 2002, two IISI experts visited different Iranian provinces three times. Several consulting services were provided to Iranian NGOs during these expert visits. Based on the project's approach and goals, a number of meetings, trainings, and workshops for Iranian NGO members were conducted.

The cooperation platform BSCW was introduced, and the structure of an Iranian NGO–CS was designed. This NGO–CS was set up as a closed system exclusively for invited members of Iranian NGOs. This restrictive access policy was chosen by the Iranian NGO members due to the aforementioned risks of co-optation by governmental organizations and corporatism in relation to international NGOs. As of December 2003, about 296 members from a variety of Iranian NGOs were registered in the system.

All of the trainings and interventions were based on the project's theoretical approach, and were in accordance with the assumptions introduced above. Several design and redesign cycles were planned, participation and sustainability were supported, and continuous evaluation measures were conducted. The following sections describe the different measures of the project in more detail.

Train-the-Trainer Measures
Sixteen Iranian NGO members in Tehran, another fourteen NGO members in Urumieh in the province of West-Azarbeijan, twenty persons in Shiraz, and another eighteen participants in Esfahan were trained as trainers for the BSCW system. Most of the participants had basic English-language and computer skills. More than one-third of the participants were women of all ages (figure 4.2).

The trainings were conducted in English, although some parts were supported by English-Farsi interpreters. Each training course lasted two full days, and contained six training sessions on the BSCW, six practical exercises in small groups with Internet access, and discussions on (tele)-cooperation and computer-mediated communication. Theoretical basics and scientific findings on collaborative computing were also offered, and in a reflection phase, needs for adaptation and tailoring of the system along with specific requirements for the system design were elaborated.

Support for Self-Learners and Self-Organized Trainings
Self-learners were supported through the translation of the training materials (mainly, Power Point slides and commented snapshots of the system's functions) into Farsi (figure 4.3). These self-learning materials can be downloaded from the system. Additionally, several tools for planning, organizing, and conducting trainings were provided to enable

Figure 4.2
Training participants in Urumieh.

participants to realize self-organized courses. Different institutions in Tehran began offering these trainings in fall 2002.

Design Workshops
By means of several workshops, the Iranian NGO–CS structure was developed, strategy and culture for the usage of the community system were discussed, the process of tailoring was started, and tasks for the system design were defined. Specific problems related to Internet connectivity of the (mainly) provincial NGOs were also addressed, and crucial aspects of the ongoing community-building process among Iranian NGOs were identified and scrutinized during these workshops.

Facilitators and a Working Group
During the second and third expert visits, advanced training courses and follow-up trainings were conducted for the sixteen participants of the first basic training. An Iranian working group (consisting of three NGO network managers) for the coordination of the process was set up in

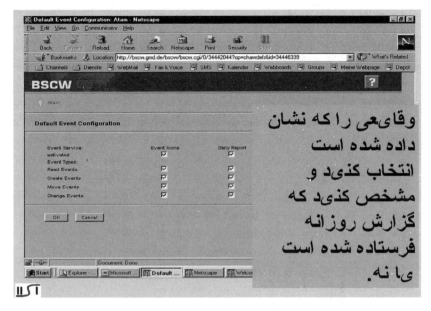

Figure 4.3
BSCW training slides, translated into Farsi.

Tehran, and fifteen facilitators were picked. In the provinces, a reasonable number of trained trainers were also prepared to take over, if necessary, the role of facilitator.

The working group and facilitators were established to promote and coordinate a sustainable, self-organized development process after the project's end. Neither group is a "closed shop," meaning that new members can join the groups and work as coordinators or facilitators of community building in the future.

Code of Ethics
Together with the facilitators and Tehran working group, the project coordinators drafted a proposal for a "code of ethics" for the Iranian NGO–CS, the proposal was then published on the system for discussion. This code of ethics deals with

- the criteria for membership;
- the nonhierarchical system's structure;

- the participatory introduction and adaptation process;
- the content structure;
- some rules for information and document management;
- roles and access rights;
- privacy matters; and
- some cultural aspects of cooperation and trust.

This code of ethics was developed as a "living document," to be discussed and modified if necessary. According to the OTD approach (Wulf and Rohde 1995), the project combined participatory system development with socio-organizational development processes. The mediated discussions and negotiations on the code of ethics concerning the community system's usage are an example for this integration of technological and organizational matters.

In particular, the statements regarding the criteria for membership and the system's structure as well as the rules for information management, access rights, and privacy were established to support the building of trust within the electronic community by technical means. According to the assumptions of the social capital approach (Bourdieu 1983; Coleman 1988; Putnam 1993; Cohen and Prusak 2001), when developing the code of ethics, we focused on the creation of far-reaching transparency and binding commitments concerning the usage of the system. Some transparency mechanisms were offered by the system's awareness functionality. The members offered some personal information about the NGO they belong to, their profession, voluntary engagement, and work. Additionally, some structured context information about uploaded documents was given by the author (type of document, file size, etc.). Well-known "netiquette" standards were integrated in the code of ethics. There were also several rules about morally and politically incorrect postings, personal attacks and insults, and the deletion of documents. The transparency as well as the commitments that were agreed on by the community members should contribute to the building of trust/trustworthiness and social capital within the Iranian NGO network.

The discussion on this code of ethics within the system ensured a process of negotiation between the community members about the common practice of system usage. Thus, the code of ethics should support the process of establishing a CoP (Lave and Wenger 1991).

Facilitating Joint Projects and a Common Practice

During the expert visits, tools for project proposals, project planning, and documentation were provided to Iranian NGO members, project development workshops were conducted; comprehensive cooperation projects for the Iranian NGO community were planned; and several practical projects were initiated. Most workshop participants complained about their lack of experience with a "culture of cooperation" in Iran (cf. Namazi 2000). To support such cultural experiences and foster the building of social capital (cf. Putnam 2000; Cohen and Prusak 2001), several cooperation projects were planned in order to establish CoP (Lave and Wenger 1991; Wenger 1998).

These projects were planned during the workshops using the tools provided for developing proposals and professional project planning. They deal with the production of an Iranian NGO database, publication of an electronic and paper-based newsletter, planning and conducting of training measures, design of a Web portal for Iranian NGOs, and translation of the system's user interface and training materials into Farsi. In each project, members of different NGOs will work together—via face-to-face meetings and via the community system—to fulfill the project tasks. The division of labor was discussed, and personal responsibilities were defined.

Besides offering training in project development and planning activities, tools and artifacts for common use were presented. By establishing these shared tools and common enterprises, the basics for the development of CoP were worked out together in participatory practical trainings. Within these CoP, the social construction of meaning, community learning, and collective processes of identification—that is, "social identity" (Tajfel 1982) or "collective identity" (Simon and Klandermans 2001)—should be enabled and supported. This process of social identification seems to be important for the Iranian NGOs, which lack of a tradition of collaboration and expertise sharing (cf. Namazi 2000).

NGO Delegation Visit to Germany

As discussed earlier, the international networking of NGOs is necessary to cope with the effects of globalization and threats of tribalism to democracy. Based on this assumption and the requirements of the Iranian NGOs, a delegation of Iranian NGOs visited Bonn, Hamburg, and

Berlin and the European Commission in Brussels in December 2002. Ten leading members of Iranian NGO networks, managers of resource centers, and civil society researchers participated in this NGO delegation. In Germany and Belgium, they met with German NGO networkers (mainly from umbrella organizations), scientists and civil society experts, politicians, European Union consultants, ministry representatives, and businesspeople to exchange experiences.

Further meetings in Iran and Germany, scientific exchange, and several cooperation projects between Iranian and European experts were agreed on and planned for 2003/2004.

Sustainability of the Process
According to the OTD's cyclic approach, further steps toward the organizational network development and technological adaptation of the community system are necessary in the future. Therefore, together with the working group and facilitators, Iranian NGO members as "experts of their own" will coordinate the following activities to support the sustainability of the process:

Members of three Iranian NGOs have been delegated to translate the system's user interface into Farsi. The German BSCW developers will support this translation.

An electronic and paper-based Farsi-language newsletter with information about the project, community system, and networking process is in the works. The Hamyaran NGO Resource Center in Tehran took responsibility for this newsletter, which will be distributed via the Iranian NGO network.

A Web site (IRANngoCS.net) is also being designed. Intended to serve as an NGO community portal, it will be accessible to the public, and offer information about the NGO community and the system. While this "entrance portal" will initially only allow access to the community system to Iranian NGO members, future plans include the addition of content for governmental organizations, potential donors, and an international public.

In cooperation with both resource centers in Tehran, the development of a database with information on all Iranian NGOs and their activities as well as civil society practices in Iran is planned. This Iranian NGO database should be accessible to all Iranian NGOs via the Iranian NGO community system.

During the Iranian delegation visit to Germany, and in meetings with German scientists, NGO networkers, and politicians, additional strategies for the national and international networking of Iranian NGOs were developed. These plans include mutual visits, common proposals and projects, scientific exchange, workshops and conferences on civil society topics, and the establishment of international fellowships for female scientists.

Evaluation of the Project

In what follows, some short-term results for social capital building will be presented. Even though some project activity results are already visible, the longer-term effects will have to be evaluated at a later time.

Methods

The project has been evaluated using a variety of different methods. Every official meeting, training, and workshop was evaluated by the participants. They rated the content and performance by means of a structured questionnaire. As well, an open format also allowed participants to suggest further activities. Additionally, several semistructured interviews with experts from the Iranian NGO community were conducted. All meetings with Iranian experts, facilitators, and the working group, along with the workshops training session discussions, were documented in written reports. The documentation of these measures (including appendixes) fills a total of 275 printed pages. Mail communication between the Iranian NGO community members and the IISI has been stored as well. The activities in the NGO–CS were also recorded in the log file, and the absolute frequencies of activities were registered anonymously (figure 4.4).

The first interventions took place nine months before this chapter was written. Since the last training in October–November 2002, four months have passed. Most of the intended effects like trust building or the establishment of a common practice cannot be expected to show up after such a short time; they need additional evaluation after a longer period. This

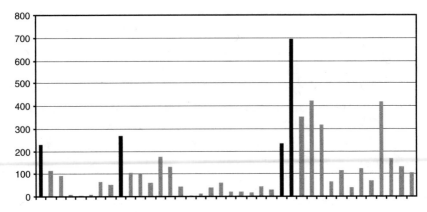

Figure 4.4
Statistics on the Iranian NGO community system's usage: absolute number of user activities per week, based on thirty-eight weeks of usage; each column represents a week; the dark columns are training weeks; the twenty-fifth (training) week was cut off at 700, but counts 1,105 activities in total.

short evaluation period, however, allows us to observe the first indicators of developments and hints at future ones.

Usage of the Community System

Concerning the usage of the Iranian NGO–CS, some data are already available. In December 2003, about 296 Iranian NGO practitioners in several Iranian provinces were working with the introduced NGO–CS. The activity statistics already reveal that usage of the system is slowly increasing (figure 4.4). The highest absolute numbers of activities were registered during the weeks of the BSCW trainings in Iran. This result is not surprising since in each training measure, fifteen to twenty members participated in practical exercises with the system. Outside these training weeks, the access rates are lower and they oscillate weekly.

The number and range of activities are expected to rise, as more NGO members in several Iranian provinces are registered in the community system and as more specific cooperative projects are established. The trainings took place in Tehran and three other Iranian provincial capitals (Urumieh, Shiraz, and Esfahan). Therefore, mainly urban NGO members joined the courses. These participants have been trained to work as trainers themselves to provide additional courses for Iranian NGO members. Such self-organized trainings should include participants from rural NGOs as well. The Iranian working group has agreed to support trainings in rural regions. In May 2003, the working group provided training materials in Farsi and supported the organization of trainings for rural NGOs. Several self-organized trainings of Iranian trainers have already been conducted in different provinces of Iran. According to reports from Iranian users, other trainings are already planned. Thus, it is hoped that the number of trained and experienced users will rise to a "critical mass," which might be needed to foster the regular use of and a living exchange within a virtual community.

For the time being, the system is mainly used to upload basic information about the work of NGOs and download training material. New members are also invited to join the system. In terms of the collaborative use of the system, only a few documents related to the management of projects are shared, and about six discussions on several topics (e.g., the code of ethics or the Iran NGO–CS Web site) have been started, but communication is low. To sum up the observations, the system is barely used for cooperation, active exchange, collaborative planning activities, and communicative interactions. As some of the registered and trained users reported in interviews, the management of their NGOs does not support the usage of the system. Because the management of some NGOs' is not convinced of the benefit for the NGOs' work, the trained activists are not allowed to use the NGO–CS during their working hours or conduct trainings for their colleagues. Furthermore, the low rate of interactive activities might be due to the aforementioned lack of cooperative culture and experiences in Iran, and the fact that most of the planned cooperative projects are not yet up and running.

Further Promotion
Nevertheless, besides the Iranian NGO–CS workspace structure, there are several additional workspaces under development in the BSCW system—for example, the new Iranian Civil Society Resource Center in Tehran is using the community system as a communication and cooperation platform for its own internal purposes.

All the materials for trainings, presentations, and checklists are provided in English and Farsi for both the trainers and self-learners. The trainings have been rated positively by the participants in the questionnaire survey. On a scale of one to six, the participants rated their overall satisfaction with the training course quite high; the outcome, atmosphere, environment, organization, and concept were all rated between five and six (mean score). Some of the participants, however, complained about the dominance of the English language in the trainings.

The expert visits aimed at the sustainability of the networking process of Iranian NGOs. The project and community system have been promoted in the Iranian provinces and among the broader public. After the third visit, about seventy NGO members in four Iranian provinces were trained with the community system and as trainers who can now conduct their own training courses in Farsi. As described earlier, a working group has developed a code of ethics as a living document for the Iranian NGO community, and this group now coordinates different projects, which are in various stages of development.

Sustainability and International Networking
The next steps are aimed at fostering a living culture of cooperation and CoP of Iranian NGO network practitioners. The questionnaire evaluation of the training measures as well as the meetings with the facilitators and Tehran working group underlines that there is a significant need for further trainings on issues such as communication and cooperation, project management, fund-raising and campaigning, and IT and new media in the Iranian NGO community. Additionally, Iranian NGOs have expressed a strong desire for future cooperation with German and international organizations.

In different questionnaires as well as interviews, training participants stressed that the physical meetings during these social events had positive effects on the networking itself. Besides the use of technical systems, it seems fruitful to initiate and intensify personal contacts by meeting physically. This observation has been made even in provincial capitals in which members of different NGOs work together without knowing each other personally. Indeed, the realization of interorganizational meetings, trainings, and collaborative projects is pivotal for Iranian NGO networking.

In terms of international networking, the Iranian delegation visit to Germany and the European Union headquarters marked the start of cooperation between Iranian and German/European NGOs. Several concrete project ideas have since been developed. Various proposals have already been written collaboratively by Iranian and German NGOs, and the NGO members have agreed to mutual follow-up visits in the near future. According to interviews with the Iranian delegation members, the initial visit was productive, forming the basis for further cooperation.

Conclusion

The project described above aimed to support community building and technical networking among Iranian NGOs.

With regard to the theoretical assumptions, I have to conclude that social capital building as well as the establishment of CoP within the Iranian NGO network are in their infancy. Both processes—trust building as well as social identification with a shared enterprise and common practice—need more time to show any substantive results. As Etienne Wenger states, "These practices are thus the property of a kind of community created *over time* by the sustained pursuit of a shared enterprise" (1998, 45). Social trust and common practices need a longer period of shared experiences to establish stable relationships. Moreover, the measurement of the effects of social processes is a methodological problem: Which criteria should be measured to evaluate social capital or social practice? An appropriate evaluation of trust, social capital, and identification with a common practice should therefore integrate qualitative and quantitative methods investigating communication and cooperation over a longer period.

In this case study, the trainings and workshops have been evaluated as networking events fostering social capital within the community. Furthermore, the individual ratings of the measures in the questionnaires have been quite positive. Both results could be due to the attraction that foreign experts might have for Iranians, who have been isolated internationally for a long time. Additionally, the fact that a free computer system was introduced and computer trainings were offered might have caused positive ratings. Especially for NGO practitioners who are volunteering in civil society activities, these computer systems and trainings provide a personal benefit.

On the other hand, the rate of activities in the community system is not high, especially in relation to the number of registered members. Regarding the establishment of CoP, I found initial indicators of a shared practice in several joint projects. Based on the on-line activities until May 2003, however, the common practice in the established projects is not pronounced. This might be due to the reported lack of cooperative culture in Iran (cf. Namazi 2000). Contrary to the traditional community-based charity organizations, the new Iranian NGO movement has not developed a tradition of collaboration and teamwork. As well, the establishment of mutual trust and trustworthiness will be more complicated and difficult in a society that faces violations of human rights by a restrictive regime.

Last but not least, the establishment of a common practice, the negotiation of meaning and social identification, and the building of trust and social capital are not processes that can be manipulated directly by training. The successful establishment of social capital and CoP is based mainly on (personal and collective) experiences, shared history, and common activities. Thus, the project presented here could only provide a sociotechnical infrastructure to enable these social processes and develop various preconditions. In the future, a more intensively used community system may support these social processes by technical means.

One shortcoming of this study is the time frame for evaluating the results. Due to the fact that the last sociotechnical interventions took place only some months ago, middle- and long-term effects did not have time to show up. Therefore, a reevaluation of the project's achievements is planned for late 2003.

According to the short-term effects evaluated in this chapter, training measures, official meetings, and workshops supported the socio-organizational networking of Iranian NGOs. The evaluation questionnaires and interviews showed that communication and interaction in physical meetings were seen as positive improvements for social networking. The introduction of the technical community system and related training courses were highly rated by the participants, but a quantitative evaluation of the user activities in the system shows that they do not play an important role in networking at this time. The main results of the project can be seen in the social effects that brought together Iranian NGO activists in different trainings and workshops. Technical support by the community system played a role in ensuring participation in these meetings, the exchange of experiences, and common project development. Nevertheless, the ongoing processes of participatory system introduction and system development as well as the self-organized trainings mark starting points for further socio-organizational and technical networking among Iranian NGOs.

Besides national NGO networking, the project aims to support international cooperation with Iranian NGOs as well. In several meetings, Iranian delegates exchanged experiences with German NGO practitioners and experts, politicians, and civil society researchers. The delegation visit to Germany marked a starting point for cooperation and international relationships between Iranian and German CSOs.

To guarantee a sustainable process of building an e-community, a living culture of cooperation and a CoP of Iranian NGOs are needed. The future success of the system is dependent on the engagement and activities of Iranian facilitators and supporters of the networking process.

Concerning the processes of community building and the establishment of a common cooperative culture and shared practice, I hope that the socio-organizational interventions will reveal middle- and long-term effects. The ongoing development of a vital civil society in Iran can play an important role in any future democratic transition. Furthermore, the establishment of stable and sustainable transnational cooperation could significantly contribute to overcoming the international isolation of Iranian civil society. Such international cooperation among civil societies is more imperative than ever given that official political relations are increasingly being strained due to the Iraq war and public classifications of Iran as part of an "axis of evil."

Acknowledgments

The author would like to thank the German foundation Friedrich-Ebert-Stiftung (the Department for International Cooperation—namely, Andrä Gärber), which funded the Iranian NGO community project. Furthermore, the project would not have been possible without the scientific work of Meryem Atam and the engaged cooperation of my Iranian partners, especially Hedieh Azad and Baquer Namazi from Hamyaran NGO Resource Center and Shorab Razzaghi from Iranian Civil Society Organizations Resource Center (ICSORC) in Tehran. Volker Wulf also provided valuable comments on a former version of this chapter. The BSCW developers at the Fraunhofer FIT—namely, Elke Hinrichs and Wolfgang Appelt—were extremely helpful concerning the adaptation of the system.

References

Ackerman, M., V. Pipek, and V. Wulf, eds. 2002. *Sharing expertise: Beyond knowledge management.* Cambridge: MIT Press.

Anheier, H., M. Glasius, and M. Kaldor, eds. 2001. *Global civil society 2001.* New York: Oxford University Press.

Appelt, W. 1999. WWW-based collaboration with the BSCW system. *Proceedings of SOFSEM '99: Springer Lecture Notes in Computer Science* 1725:66–78.

Barber, B. R. 1996. *Jihad vs. McWorld: How globalism and tribalism are reshaping the world.* New York: Ballantine Books.

Bourdieu, P. 1983. Forms of capital. In *Handbook of theory and research for the sociology of education*, edited by J. C. Richards. New York: Greenwood Press.

Cohen, D., and L. Prusak. 2001. *In good company: How social capital makes organizations work.* Boston: Harvard Business School Press.

Coleman, J. S. 1988. Social capital in the creation of human capital. *American Journal of Sociology* 94:95–120.

Davidow, W. H., and M. S. Malone. 1993. *The virtual corporation: Structuring and revitalizing the corporation for the twenty-first century.* New York: Harper Business.

Emery, F. E., and E. L. Trist. 1960. Socio-technical systems. In *vol. 2 of Management sciences, models and techniques.* London.

Etzioni, A. 1993. *The spirit of community: Rights, responsibilities, and the communitarian agenda*. New York: Touchstone.

Floyd, C., F.-M. Reisin, and G. Schmidt. 1989. STEPS to software development with users. In *ESEC '89: Second European software engineering conference, University of Warwick, Coventry*, edited by C. Ghezzi and J. A. McDermid. Heidelberg: Springer.

Food and Agriculture Organization (FAO) of the United Nations. NGOs: Definition of NGOs, <http://ngls.tad.ch/english/pubs/hb/hb4.html>, Jan. 2003.

Fowler, A. 1992. Building partnerships between northern and southern development: NGOs—Issues for the nineties. *Development* 1:16–23.

Friedrich-Ebert-Foundation (FEF). Mission, <http://www.fes.de/intro_en.html>, Jan. 2003.

French, W. L., and C. H. Bell. 1990. *Organisationsentwicklung*. 3rd ed. Bern: UTB.

Fukuyama, F. 2000. *Great disruption: Human nature and the reconstruction of social order*. New York: Simon and Schuster.

Hamyaran NGO Resource Center. 2001. *Mashad meeting (Busher+4: Challenges and approaches)*. Tehran: Hamyaran Center.

Hellmann, K.-U. 2002. Systemvertrauen und Sozialkapital im Vergleich. In *Demokratie und Sozialkapital: Die Rolle zivilgesellschaftlicher Akteure*, edited by AK "Soziale Bewegungen" der DVPW et al. Berlin: WZB papers.

Huntington, S. P. 1998. *The clash of civilizations and the remaking of world order*. New York: Simon and Schuster.

Koch, T., and W. Appelt. 1998. Beyond Web technology: Lessons learnt from BSCW. In *Proceedings of IEEE WET ICE Workshop '98*, Stanford University.

Lave, J., and E. Wenger. 1991. *Situated learning: Legitimate peripheral participation*. Cambridge: Cambridge University Press.

Lauth, H.-J., and W. Merkel. 1997. Zivilgesellschaft und Transformation: Ein Diskussionsbeitrag in revisionistischer Absicht. *Forschungsjournal Neue Soziale Bewegungen* 9, no. 2:12–34.

Mambrey, P., V. Pipek, and M. Rohde. 2003. *Wissen und Lernen in Virtuellen Organisationen*. Heidelberg: Physica-Verlag.

Namazi, B. 2000. *Iranian NGOs: Situation analysis*. Tehran: Hamyaran Center.

Nohria, N., and J. D. Berkley. 1994. The virtual organisation: Bureaucracy, technology, and the implosion of control. In *The post-bureaucratic organisation: New perspectives in organisational change*, edited by C. Hekscher and A. Donnellor. Thousand Oaks, Calif.: Sage.

Pieper, R. 1988. *Diskursive Organisationsentwicklung*. Berlin: de Gruyter.

Putnam, R. D. 1993. *Making democracy work. Civil traditions in modern Italy.* Princeton, N.J.: Princeton University Press.

Putnam, R. D. 2000. *Bowling alone: The collapse and revival of American community.* New York: Simon and Schuster.

Rohde, M., and A. Klein. 1996. Soziale Bewegungen und Nicht-Regierungsorganisationen. *Forschungsjournal Neue Soziale Bewegungen* 9, no. 2:3–14.

Rohde, M., and A. Klein. 1997. Transformation der Zivilgesellschaft. *Forschungsjournal Neue Soziale Bewegungen* 10, no. 1:3–11.

Rohde, M., M. Rittenbruch, and V. Wulf. 2001. *Auf dem Weg zur virtuellen Organisation: Fallstudien, Problembeschreibungen, Lösungskonzepte.* Heidelberg: Physica-Verlag.

Rohde, M., and V. Wulf. 1995. Introducing a telecooperative CAD-system: The concept of integrated organization and technology development. In *Proceedings of the HCI International '95*, 6th international conference on human-computer interaction, Yokohama. Amsterdam: Elsevier Science Publishers.

Ropohl, G. 1999. Philosophy of socio-technical systems. *Technè: Journal of the Society for Philosophy and Technology* 4, no. 3, <http://scholar.lib.vt.edu/ejournals/SPT/v4_n3pdf/ROPOHL.PDF>.

Roth, R. 2003. Die dunklen Seiten der Zivilgesellschaft: Grenzen einer zivilgesellschaftlichen Fundierung von Demokratie. Forschungsjournal *Neue Soziale Bewegungen* 16, no. 2. Stuttgart: Lucius and Lucius.

Simon, B., and B. Klandermans. 2001. Politicized collective identity: A social psychological analysis. *American Psychologist* 56, no. 4:319–331.

Strausak, N. 1998. Resumée of VoTalk. In *Organisational virtualness: Proceedings of the VoNet-workshop*, edited by P. Sieber and J. Griese. Bern: Simowa-Verlag.

Tajfel, H. 1978. *Differentiation between social groups: Studies in the social psychology of intergroup relations.* London: Academic Press.

Tajfel, H. 1982. *Social identity and intergroup relations.* Cambridge: Cambridge University Press.

Travica, B. 1997. The design of the virtual organisation: A research mode. In *Proceedings of the association for information systems '97*, Americas Conference, Indianapolis <http://hsb.baylor.edu/ramsower/ais.ac.97/papers/travica.htm>.

Turner, J. C., M. A. Hogg, P. J. Oakes, S. D. Reicher, and M. S. Wetherell. 1987. *Rediscovering the social group: A self-categorization theory.* Oxford: Blackwell.

United Nations Development Program (UNDP). Statistics on Iranian society, <http://www.undp.org/hdr2001/indicator/cty_f_IRN.html>, Jan. 2003.

Wenger, E. 1998. *Communities of practice: Learning, meaning, and identity.* Cambridge University Press.

Wulf, V., and M. Rohde. 1995. Towards an integrated organization and technology development. In *Designing interactive systems: Processes, practices, methods, and techniques*, edited by G. M. Olson and S. Shuon. New York: Association of Computing Machinery Publications.

Wulf, V., M. Krings, O. Stiemerling, G. Iacucci, P. Fuchs-Frohnhofen, J. Hinrichs, M. Maidhof, B. Nett, and R. Peters. 1999. Improving inter-organizational processes with integrated organization and technology development. *Journal of Universal Computer Science* 5, no. 6:339–365.

5
How Does the Internet Affect Social Capital?

Anabel Quan-Haase and Barry Wellman

Two trends intersect in this chapter. One is the dramatic increase in Internet use since the 1990s, affecting the way people live, work, and play in the developed world. Approximately 60 percent of North American adult households are on-line, with growing percentages in other countries (Howard, Rainie, and Jones 2002; Reddick, Boucher, and Groseillers 2000). For a large proportion of the population of Internet users, Internet access is a daily activity, with more than half of Internet users reporting having been on-line "yesterday" (Howard, Rainie, and Jones 2002).

The second trend is the emergence of *social capital* as a useful conceptual tool to examine the vitality of a neighborhood, city, or country (Putnam 1993, 1996, 2000). Although users of the notion sometimes lack conceptual clarity (Fischer 2001), thinking in terms of social capital allows researchers and policymakers to evaluate a number of core dimensions, such as public and private community, and civic engagement. There are two complementary uses of the social capital concept:

1. *Social contact*: Interpersonal communication patterns, including visits, encounters, phone calls, and social events.
2. *Civic engagement*: The degree to which people become involved in their community, both actively and passively, including such political and organizational activities as political rallies or book and sports clubs.

This chapter is about the intersection of these two trends: How the rise of Internet use affects social capital. We situate the discussion in an ongoing debate about the possible recent decline in North Americans' social capital. Robert Putnam uses a variety of survey data as evidence of declining civic and social participation (1996, 2000; see also Norris 2001). He argues that intertwined with this declining civic involvement is a decline in collective social activities, from family dinners to participating in clubs. Yet Claude Fischer (2001) claims there are two main problems with Putnam's interpretation. First, the decrease in social capital is not constant across all measures of social capital. Although most indicators of political involvement show a consistent decline, indicators of socializing and visiting are inconsistent. This inconsistency across measures questions the validity and reliability of the construct. The second problem is related to how to interpret the amount of decrease that is occurring. Putnam sees the decrease as substantial while Fischer maintains that it is often negligible or short-term.

The Putnam-Fischer debate is a continuation of a 150-year-long tradition in the social sciences to see if community is declining or flourishing since the Industrial Revolution (reviewed in Wellman and Leighton 1979; Wellman 1999). Analysts contrast contemporary community life with preindustrialized communities, composed mainly of locally based interactions in closely bounded, homogeneous groups. Although there were few opportunities for travel, people visited, provided social support, and were concerned with the well-being of their community. People in group-based societies deal principally with fellow members of the few groups to which they belong: at home, in the neighborhood, at work, or in voluntary organizations.

Has this traditional, pastoral community life been lost in modern times? One school of thought sees industrialization—accompanied by such other large-scale social changes as urbanization and bureaucratization—as the root cause of the decline, pointing to long work hours, regimented organization, urban sprawl that creates isolation, and a general lack of public spaces. Dora Costa and Matthew Kahn (2001) attribute the decline in entertaining at home to women's increasing work hours. Moreover, new modes of transportation and communication have emerged supporting distant interactions that remove people from their immediate vicinities, and ultimately, creating sparsely knit communities. With industrialization also came increased participation in more individualistic activities, such as watching television (Putnam 2000).

Counter to the community-lost view, advocates of the community-liberated stance argue that community life is not lost but has gone through radical transformations. Analysts in the 1960s began realizing that communities were flourishing outside of neighborhoods (Guest and Wierzbicki 1999; Wellman 2001; Wuthnow 1991, 1998). Their research shows that people continue to socialize, but that few immediate neighbors are known, and community has moved from local involvement to interactions with geographically dispersed friends and kin (Fischer 1992; Wellman 1979, 1999). Face-to-face visits are still the predominant means of communication, but the telephone also occupies a central role, particularly for distant communication (Wellman and Wortley 1990; Wellman and Tindall 1993).

The changes in how people socialize have created a need to develop new models for conceptualizing and, hence, measuring community. Considering that socializing occurs beyond the boundaries of the local neighborhood, useful approaches define community not in terms of locality but as social networks of interpersonal ties that provide sociability, support, information, a sense of belonging, and social identity (Wellman 2001; Wellman, Carrington, and Hall 1988). By examining people's social relationships, independent of narrowly defined boundaries based on location, researchers have discovered that many people live in long-distance communities (Wellman and Wortley 1990). Thus, this evidence suggests that industrialization did not destroy community, but instead helped transform its composition, practices, attitudes, and communication patterns.

These transformations in the expression of community are related to the development and use of technologies. Transportation technologies have been especially relevant for the development of unbounded, long-distance communities. The car, train, and plane have allowed people to mobilize easily and quickly from one place to another (Wellman 1999). Innovations in telecommunications such as the telegraph and telephone have also radically changed how people communicate. The telephone, especially, facilitated relationships among people who were geographically dispersed, and it allowed people who were located near each other to communicate conveniently and coordinate visits easily.

The latest technological innovation, the Internet, is affecting how people communicate, work, and use their leisure. The evidence suggests that the Internet has blended into the rhythms of everyday life and is used for a wide variety of purposes, such as surfing for information, playing on-line games, and chatting (Howard, Rainie, and Jones 2002; Quan-Haase and Wellman 2002). Moreover, a large proportion of people report using the Internet for making important life decisions (Howard, Rainie, and Jones 2002).

There are a number of ways in which the effects of the Internet on social capital can be conceptualized. In general, three different approaches can be identified:

1. *The Internet transforms social capital*: The Internet provides the means for inexpensive and convenient communication with far-flung communities of shared interest (Barlow et al. 1995; Wellman 2001). Coupled with the Internet's low costs and often asynchronous nature, this leads to a major transformation in social contact and civic involvement away from local and group-based solidarities, and toward more spatially dispersed and sparsely knit interest-based social networks.

2. *The Internet diminishes social capital*: The Internet through its entertainment and information capabilities draws people away from family and friends. Further, by facilitating global communication and involvement, it reduces interest in the local community and its politics (Nie 2001; Nie, Hillygus, and Erbring 2002).

3. *The Internet supplements social capital*: The Internet blends into people's lives. It is another means of communication to facilitate existing social relationships as well as to build patterns of civic engagement and socialization. People use the Internet to maintain existing social contacts by adding electronic contact to telephone and face-to-face contact. Further, they often continue their hobbies and political interests on-line. This suggests that the Internet helps increase existing patterns of social contact and civic involvement (Quan-Haase and Wellman 2002; Chen, Boase, and Wellman 2002).

We focus our discussion here principally on the relationship of Internet use to social contact. We draw from previous research done by our *NetLab*, especially data from "Survey 2000," hosted on the National

Geographic Society's Web site. Our discussion concentrates on the North American sample, which consists of 20,075 adults: 17,711 Americans (88 percent) and 2,364 Canadians (12 percent).[1] We also examine results from similar surveys: the Pew Internet and Everyday Life Project (Howard, Rainie, and Jones 2002), Projecte Internet Catalunya (Castells et al. 2003a,b), and other studies (mainly collected in Wellman and Haythornthwaite 2002; see also Kraut et al. 2002).

Does the Internet Transform Social Capital?

Many analysts see the Internet as stimulating positive change in people's lives because of its rapid diffusion to all strata of the population, diminishing costs for getting on-line, ease of use, and variety of information and communication tools (De Kerckhove 1997; Jones 1998; Lévy 1997). They foresee a digital revolution restoring a sense of community by connecting friends and kin near and far, providing information resources on a wide variety of topics, and engaging various groups in political and organizational participation. They hope that the digital realm will lead to new forms of community by offering a meeting space for people with common interests—one that overcomes the limitations of space and time (Hiltz and Turoff 1993; Baym 1997; Jones 1998; Wellman 2001). These analysts expect on-line communities to flourish because people would be able to choose communities of shared interests regardless of their physical location. The unique characteristics of digital, textual communication and its cue-reduced nature would have democratizing as well as equalizing effects by de-emphasizing the salience of such characteristics as race, age, and socioeconomic status (Sproull and Kiesler 1991). Electronic Frontier Foundation cofounder John Perry Barlow sums up this spirit nicely: "We are in the middle of the most transforming technological event since the capture of fire. I used to think that it was just the biggest thing since Gutenberg, but now I think you have to go back farther" (Barlow et al. 1995, 36).

Some evidence supports the community-multiplying nature of the Internet. Many Internet users participate in on-line communities, such as mailing list servers and newsgroups. A Pew study on on-line communities reports that 84 percent of U.S. Internet users have been members of

Table 5.1
Participation in online communities (Survey 2000)

	Mailing list servers and other group e-mails (percentage)	Newsgroups (percentage)
Never	28	57
Rarely	23	23
About monthly	6	5
About weekly	8	4
A few times a week	9	5
Daily	26	5
Total	100	99

an on-line community (Horrigan 2002). In the Survey 2000 study, 76 percent of North American users report having participated in an on-line community, such as newsgroups, mailing list servers, and other group e-mails. Within the population of members of on-line communities, 37 percent receive or send messages on a daily basis to list server discussion groups or "Usenet newsgroups." Forty-four percent of the sample reports participating in list servers at least once a week, while only 14 percent of the sample reports participating in newsgroups at least once a week (see table 5.1).

People seek out those who share similar interests, with mailing lists and newsgroups providing the means to connect on a regular basis. On-line communities have to do with a wide range of topics, with respondents reporting participation primarily in those related to their work (50 percent of respondents) or shared interest groups (50 percent), followed by sports fan clubs (31 percent) and television fan clubs (29 percent) (Horrigan 2002). For example, fans of soap operas discuss their favorite shows on-line, thereby creating a common understanding and reinterpretation of the events occurring on the shows (Baym 1997).

Such high levels of participation in on-line communities suggest that the Internet has become an alternative route to being involved in groups and pursuing interests. Therefore, Putnam's (2000) observed decline in organizational participation may not reflect actual disengagement from community but community becoming embedded in digital networks rather than in traditional, geographically bounded groups: in short, a movement of community participation from public spaces to cyberspace (see the related discussions in Lin 2001; Wellman 1999, 2001). Moreover, the positive relationship of the amount of time spent on the Internet with feelings of community on-line indicates that on-line participation may intensify reciprocity and trust (Quan-Haase and Wellman 2002). Similarly, the Pew study examining on-line communities shows that half of those who belong to on-line communities say that the Internet provides them with an alternative means to connect with people who share their interests (Horrigan 2002). The Internet, then, not only provides a new sphere of communication; it also helps in establishing new social relationships. These social relationships are often continued off-line, creating a mix of on-line and off-line interactions (e.g., Müller 1999; Rheingold 2000).

The Internet promises to create a global village consisting of sparsely knit communities by removing space constraints and allowing for far-flung interactions. This trend is enhanced by the large diffusion of e-mail as a communication technology. In Survey 2000, North American users report exchanging e-mails more than five times per week, with 68 percent checking their e-mail on a daily basis. Clearly, e-mail is a useful technology for communicating with friends and kin. Survey 2000 respondents use e-mail for 24 percent of their nearby contacts (within fifty kilometers) and 49 percent of their more distant contacts. This suggests that e-mail is especially useful for keeping in touch with those who are far away because of its low cost, which does not increase with distance. E-mail is also asynchronous, making it easy to contact people living in other time zones (Howard, Rainie, and Jones 2002; Quan-Haase and Wellman 2002). Yet the bulk of contact—e-mail and phone as well as face-to-face—remains relatively local.

Although dystopians fear that the Internet will lead people away from their local communities, the evidence reveals that the Internet also supports local community interests. For example, the Pew study examining on-line communities reports that 29 percent of members of on-line communities take part in a local community group via the Internet, which supplies them with information about local activities, issues, and debates. The study shows that such participation does more for fostering civic involvement than it does for social contact (Horrigan 2002). Evidence for the Internet fostering increased social contact, however, comes from an ethnographic study of a new residential area ("Netville") that was wired with high-speed Internet access. In Netville, people with access to high-speed Internet (and accompanying list server) socialized more frequently with their neighbors (Hampton and Wellman 1999, 2002). Those with access not only knew more neighbors locally but kept in contact with friends and kin at a distance via the Internet. Wired residents therefore became "glocalized": involved in both local and long-distance relationships. The Internet not only helped people to meet and exchange messages regarding the residential area; it was also used as a tool to organize and mobilize. Thus, in Netville, the Internet managed to combine far-flung connectivity with local interests.

Does the Internet Diminish Social Capital?

Not all Internet activity is social. Much is Web oriented, with people either seeking information or engaging in solitary recreations (Wellman et al. 2001). Moreover, social contact on-line can be immersive, drawing people away from face-to-face and phone contact. Indeed, when people with one telephone line use dial-up modems to get on the Internet, they cannot send or receive telephone calls. There is some empirical evidence for these suppositions. One longitudinal study found that as newcomers used the Internet more, their social contact off-line decreased, and their depression and loneliness increased (Kraut et al. 1998). As users gained more experience, though, the Internet was associated with an increased number of weak on-line ties and a decreased number of stronger off-line ties (LaRose, Eastin, and Gregg 2001; Kraut et al. 2002).

Is local community more adversely affected? If the Internet allows for easy access to on-line communities that span the globe, what consequences does this have for family ties and local interactions? The high level of global connectivity may have a downside, especially for local interactions and family ties. Even those activities that are social can lead to domestic conflict. For example, Survey 2000 data show a positive association between the time a person has been on-line and the amount of e-mail he or she sends and receives (Quan-Haase and Wellman 2002;

see also Howard, Rainie, and Jones 2002; Kavanaugh and Patterson 2002). The data also reveal that people are maintaining far-flung as well as local relationships. Maintaining many far-reaching ties may result in less time for interactions with household members. Moreover, if people are spending more time on-line, public spaces become less relevant for interaction and socializing.

To date, such suppositions are more deductive than supported by evidence. Two informal studies done with Wellman's students (in 1999 and 2002) show a preponderance of local e-mails. But these are students, not a broadly representative sample. Further data is supplied by Survey 2000, in which daily e-mail users report that 58 percent of their contact with friends and 83 percent of their contact with kin are with those living within fifty kilometers: within a one-hour drive in many developed areas (Quan-Haase and Wellman 2002).

Is the Internet failing to support a "global village" (McLuhan 1962)? It depends on how you look at it. Although local connectivity remains high, it is still a lower percentage of contacts than was the case prior to the coming of the Internet (Wellman 1996). The Internet may be differentially fostering contact with acquaintances, thereby tilting the balance between such weak ties and stronger ones. Yet weak ties have their own value, in providing new information and access to disparate networks.

The Internet may compete for time with other activities in an inelastic twenty-four-hour day. There are discrepant findings about whether on-line time sinks do or do not pull people away from other interactions inside and outside the household (Nie and Erbring 2000; Nie, Hillygus, and Erbring 2002). The Internet can draw people's attention away from their immediate physical environment because when they are on-line, they pay less attention to their physical and social surroundings (Nie and Sackman 1970). As the number of activities performed on the Internet increases and the amount of time spent on these activities also increases, there is a risk of the Internet reducing time spent in face-to-face contact with family and friends. For example, some evidence from research on children's heavy involvement with on-line games shows that it can reduce family ties and children's socializing.

Some scholars see a parallel between the effects of television and the Internet (Putnam 2000; Steiner 1963). Both technologies draw people away from their immediate environments, potentially alienating them from social interactions and civic engagement. Broadcast television is not a good analogue to the socially interactive Internet, however, because it is much less individually immersive and engages viewers much more passively than the Internet.

The Internet Supplements Social Capital

What if the Internet has neither radically transformed the nature of community nor markedly diminished it? Evidence is accumulating showing that the Internet adds on to existing patterns of communication, that it is "used in a manner similar to other, more traditional technologies" (Flanagan and Metzger 2001, 153). The Internet is a crucial, but not dominant, means of communication for contact with friends and relatives. E-mail, chat, and other communication capabilities supplement social contact by helping people to organize meetings and social events as well as filling communication gaps (Wellman and Haythornthwaite 2002).

For instance, e-mail is an important medium to keep in touch with friends and relatives, but as the amount of e-mail sent and received increases, interactions and phone calls do not decrease (Howard, Rainie, and Jones 2002; Quan-Haase and Wellman 2002). E-mail appears to support existing social contact, yet it does not become a substitute for phone and face-to-face communication. Our Survey 2000 study shows that most contact is over the phone (41 percent), by e-mail (32 percent), and through face-to-face encounters (23 percent), with a small amount (4 percent) of postal letter writing and greeting cards (table 5.2). Those with low phone and face-to-face contact also e-mail less. Similarly, people who visit and phone frequently contact also e-mail frequently. Thus, the capabilities of the Internet add on to interactions with other media. The stronger the relationship, the more media are used, and the more types of information are exchanged (Haythornthwaite and Wellman 1998).

Nor does the way the Internet fits into people's lives always follow the e-mail-heavy North American model (Chen and Wellman 2004). In 2002, the Open University of Catalonia surveyed 3,005 adult residents

Table 5.2
Social contact with friends and kin, near and far

	Phone (days/year)	F2F (days/year)	E-mail (days/year)	Letters (days/year)
Friends near	126	92	118	9
Kin near	114	58	49	7
Friends far	25	10	85	8
Kin far	43	10	72	10

of this autonomous region of Spain, of whom 1,039 (35 percent) were using the Internet.[2] The study shows that Catalan networks are more local than their North American counterparts. Nearly two-thirds (64 percent) of Catalan network members live within the same municipality. These 13.5 local network members consist of an average of 0.8 parents (including those living in the same house), 4.5 other kin, 5.5 friends, and 2.7 neighbors. Personal encounters are the predominant mode of communication, especially among the great majority of network members who live within the same municipality or elsewhere in Catalonia. Telephoning is of secondary importance. The Internet is hardly ever used except to communicate with those few friends who live in other countries (Castells et al. 2003a,b). There is a contrast on the other side of the globe: The residents of Hong Kong use the Internet even more than North Americans for socializing (Chau et al. 2002).

In short, the Internet has joined the telephone and face-to-face contact as a main means of communication—one that can be more convenient and affordable.[3] Although face-to-face and telephone contact continue, they are complemented by the Internet's ease in connecting geographically dispersed people, institutions, and organizations bonded by shared interests.

In the North American population as a whole, the Internet also does not appear to have radically transformed civic involvement in voluntary organizations and politics, although more active groups use it extensively (Kavanaugh and Patterson 2002; Norris 2001; Quan-Haase and Wellman 2002). Survey 2000 shows that people who engage in political and organizational activities tend to use the Internet as much as those who are not engaged. There is no strong statistical association between Internet use and active participation. Yet subtler dynamics are at work. The Internet helps as well as supports the activities of organizations and individuals who are interested in obtaining national and international news. For those with access, it facilitates accessing news at a low cost. The Internet's possibilities may not have a widespread mobilizing effect, however. The hope that the Internet would be especially useful in encouraging many people to join political discussions has not yet been realized (Norris 2001), although the Howard Dean 2004 presidential campaign had heavy Internet participation.

Considerations in Internet Research

Not only is the Internet an evolving technology that constantly re-creates itself, it is also a social technology. There is no simple technological determinism with the Internet driving social trends. The Internet's development also resonates with and responds to social trends. Our analyses of the Internet and social capital reveal that there are a number of challenges that researchers need to take into consideration:

• *Rapid, unpredictable changes*: The Internet has chameleonlike properties and is thus constantly changing. The most prominent changes are the large increase in content, the increase in bandwidth, and the ubiquity of access. Another important change has been the commercialization of the Internet. Most large international companies offer and advertise their products on-line (Castells 1996). The composition of Internet users has also changed from predominantly young, white, North American males to a more diverse set of people.

• *Measurement*: The Internet leads to new forms of social capital that cannot be easily captured with existing forms of measurement. Thus, to assess the full impact of the Internet on social capital, researchers need to develop new forms of measurement that complement existing ones.

• *Effect direction*: Most research is aimed at identifying an effect, regardless of whether the effect is positive or negative. Yet our analyses show that in many cases, no directional effect is present because the Internet adds on to existing patterns of communication and engagement.

- *Target group*: Many of the changes associated with the Internet are specific to a particular group. For example, women seek health information on the Internet three times more frequently than men do (Howard, Rainie, and Jones 2002; Horrigan 2002). By contrast, men seek information on stock markets five times more frequently than females do (ibid.). Hence, the particulars of a group have to be examined to understand how they are appropriating the Internet and how the Internet fits into their everyday routines.
- *Uses of the Internet*: Not all uses of the Internet are social. Although e-mail is a common use, the Internet is also a widespread tool for seeking information. Moreover, not all uses of the Internet are predictable. The Internet may not affect social capital when it is used for one-to-one e-mail purposes, but it might affect it when used for other purposes such as virtual communities. Therefore, analyses will be different when applied to different uses of the Internet.
- *Changing uses*: Until now, there has been an implicit assumption that as the Internet grows up around the world, it will increasingly resemble the North American version. That is, e-mail will be a principal use, complemented by Web surfing. Yet with time and research, two things are becoming clear. First, Internet use varies around the world. For example, Catalans use e-mail less frequently than North Americans do, and Japanese and Europeans often use short message texting instead of e-mail. Second, Internet use is changing within countries. E-mail attachments of text, photos, audio, and even video are becoming more widespread. Wireless connectivity means that people can be reached anywhere, not just where their desktop computers are wired into the Internet.

Conclusions

The evidence we have gathered suggests that the Internet occupies an important place in everyday life, connecting friends and kin both near and far. In the short run, it is adding on to—rather than transforming or diminishing—social capital. Those who use the Internet the most continue to communicate by phone and through face-to-face encounters. Although the Internet helps to connect far-flung community, it also helps to connect local community.

We have shown that what makes the communication possibilities of the Internet unique is its capability to support many-to-many information exchanges among geographically dispersed people. On-line communities around a wide variety of topics flourish by allowing people to exchange ideas and provide social support (Wellman and Gulia 1999). The Internet has led to new communication forms, with users often utilizing the communication tools in unforeseen ways. For instance, the use of short text messages on mobile phones leads to increased social contact because it is often used to arrange face-to-face meetings with close friends (Katz and Aakhus 2002).

The evidence to date indicates that like the telephone (Fischer 1992), the Internet's effects on society will be significant, but evolutionary. While the Internet's effects on social capital may be less dramatic than the "transformationists" had dreamed of, the effects may be extensive in the long run. The unique features of the Internet will interact with existing social factors, creating new, often unexpected, behaviors and changes.

Therefore, an analysis of the impact of the Internet needs to consider that the Internet may be contributing to new forms of interaction and community that cannot be measured using standard indicators of social capital. The fact that people are not interacting in visible public spaces does not mean that they are isolated. They may be going on-line to create new worlds, using instant messaging to chat with old and new friends, visiting cyberspace communities, or playing multiuser games. The Internet makes it necessary to redefine our understanding of what social capital is. We believe that the Internet will intensify the interpersonal transformation from "door-to-door" to "place-to-place" and individualized "person-to-person" networks (Wellman 2001).

Acknowledgments

Jeffrey Boase, Wenhong Chen, Keith Hampton, Kristine Klement, Monica Prijatelj, and Uyen Quach gave fine research assistance on the projects reported here, and Quach also helped with the final preparation of this chapter. We would also like to thank the editors for their valuable comments on an earlier draft. The Social Science and Humanities

Research Council of Canada, Communications and Information Technology Ontario, and IBM's Institute of Knowledge Management provided financial support for our research.

Notes

1. For details on the study and previous publications, see Quan-Haase and Wellman 2002; Wellman et al. 2001. Survey 2000 is available at <http://survey2000.nationalgeographic.com>. Supplementary tables are available at <www.chass.utoronto.ca/~wellman/publications>. For other descriptions of these data, see Witte, Amoroso, and Howard 2000; Chmielewski and Wellman 1999.

2. Manuel Castells and Imma Tubella (Open University) led the entire study, with Barry Wellman doing analysis of this section in cooperation with Isabel Diaz de Isla. For details, see Castells et al. 2003a,b.

3. The Survey 2000 study and others (Wellman 1979; Wellman and Leighton 1979) show that other than ritual greeting cards, people rarely send letters through the traditional post anymore, even as the Internet itself boosts the sheer volume of written communication. It would be interesting to compare the effects of the Internet to that of the introduction of the telephone as a complement to and replacement for face-to-face and postal communication. For the beginnings of such analyses, see Fischer 1992; Wellman and Tindall 1993.

References

Barlow, J. P., S. Birkets, K. Kelly, and M. Slouka. 1995. What are we doing on-line? *Harper's* 291 (August): 35–46.

Baym, N. K. 1997. Interpreting soap operas and creating community: Inside an electronic fan culture. In *Culture of the Internet*, edited by S. Kiesler. Mahweh, N.J.: Lawrence Erlbaum.

Bijker, W. E., T. P. Hughes, and T. Pinch. 1999. Introduction to *The social construction of technological systems: New directions in the sociology and history of technology*, edited by W. E. Bijker, T. P. Hughes, and T. Pinch. Cambridge: MIT Press.

Castells, M. 1996. *The rise of the network society*. Oxford: Blackwell.

Castells, M., I. Tubella, T. Sancho, I. Diaz de Isla, and B. Wellman. 2003a. *La Societat Xarxa a Catalunya*. Barcelona: Random House Mondadori.

Castells, M., I. Tubella, T. Sancho, I. Diaz de Isla, and B. Wellman. 2003b. The network society in Catalonia: An empirical analysis, <http://www.uoc.edu/in3/pic/eng/index.html>.

Chau, P. Y. K., M. Cole, A. P. Massey, M. Montoya-Weiss, and R. M. O'Keefe. 2002. Cultural differences in the online behavior of consumers. *Communications of the ACM* 45, no. 10:138–143.

Chen, W., J. Boase, and B. Wellman. 2002. The global villagers: Comparing Internet users and uses around the world. In *The Internet in everyday life*, edited by B. Wellman and C. Haythornthwaite. Oxford: Blackwell.

Chen, W., and B. Wellman. 2004. Charting digital divides: Within and between countries. In *Transforming Enterprises*, edited by W. Dutton, B. Kahan, R. O'Callaghan, and A. Wyckoff. Cambridge: MIT Press.

Chmielewski, T., and B. Wellman. 1999. Tracking Geekus Unixus: An Explorers' Report from the National Geographic Website. *SIGGROUP Bulletin*, no. 20:26–28.

Costa, D. L., and M. E. Kahn. 2001. *Understanding the decline in social capital, 1952–1998*. Cambridge, Mass.: National Bureau for Economic Research.

De Kerckhove, D. 1997. *Connected intelligence: The arrival of the web society*. Toronto: Somerville House.

Fischer, C. S. 1992. *America calling: A social history of the telephone to 1940*. Berkeley: University of California Press.

Fischer, C. S. 2001. Bowling alone: What's the score? Paper presented at the American Sociological Association, 17–21 August, Anaheim, California.

Flanagan, A., and M. Metzger. 2001. Internet use in the contemporary media environment. *Human Computer Research* 27:153–181.

Guest, A. M., and S. K. Wierzbicki. 1999. Social ties at the neighborhood level: Two decades of GSS evidence. *Urban Affairs Review* 35, no. 1:92–111.

Hampton, K. N., and B. Wellman. 1999. Netville on-line and off-line. *American Behavioral Scientist* 43, no. 3:478–495.

Hampton, K. N., and B. Wellman. 2002. The not so global village of a cyber society: Contact and support beyond Netville. In *The Internet in everyday life*, edited by B. Wellman and C. Haythornthwaite. Oxford: Blackwell.

Haythornthwaite, C., and B. Wellman. 1998. Work, friendship, and media use for information exchange in a networked organization. *Journal of the American Society for Information Science* 49, no. 12:1101–1114.

Hiltz, S. R., and M. Turoff. 1993. *The network nation*. 2d ed. Cambridge: MIT Press.

Horrigan, J. B. 2002. *Online communities: Networks that nurture long-distance relationships and local ties*. Washington, D.C.: Pew Internet and American Life Study.

Howard, P., L. Rainie, and S. Jones. 2002. Days and nights on the Internet: The impact of a diffusing technology. In *The Internet in everyday life*, edited by B. Wellman and C. Haythornthwaite. Oxford: Blackwell.

Jones, S. 1998. *Cybersociety 2.0: Revisiting computer-mediated communication and community*. Thousand Oaks, Calif.: Sage.

Katz, J., and M. Aakhus, eds. 2002. *Perpetual contact: Mobile communication, private talk, public performance*. Cambridge: Cambridge University Press.

Kavanaugh, A. L., and S. J. Patterson. 2002. The impact of community computer networks on social and community involvement in Blacksburg. In *The Internet in everyday life*, edited by B. Wellman and C. Haythornthwaite. Oxford: Blackwell.

Kraut, R. E., M. Patterson, V. Lundmark, S. Kiesler, T. Mukhopadhyay, and W. Scherlis. 1998. Internet paradox: A social technology that reduces social involvement and psychological well-being? *American Psychologist* 53, no. 9:1017–1031.

Kraut, R. E., S. Kiesler, B. Boneva, J. Cummings, V. Helgeson, and A. Crawford. 2002. Internet paradox revisited. *Journal of Social Issues* 58, no. 1:49–74.

LaRose, R., M. S. Eastin, and J. Gregg. 2001. Reformulating the Internet paradox: Social cognitive explanations of Internet use and depression. *Journal of Online Behavior, 1, no. 2* (on-line). Retrieved 23 March 2001, <http://www.behavior.net/JOB/v1n2/paradox.html>.

Lévy, P. 1997. *Collective intelligence: Mankind's emerging world in cyberspace*. New York: Plenum.

Lin, N. 2001. *Social capital: A theory of social structure and action*. Cambridge: Cambridge University Press.

McLuhan, M. 1962. *The Gutenberg galaxy: The making of typographic man*. Toronto: University of Toronto Press.

Müller, C. 1999. Networks of "personal communities" and "group communities" in different online communication services. In *Proceedings of Exploring Cyber Society: Social, Political, Economic and Cultural Issues, vol. 2*, edited by J. Armitage and J. Roberts. Newcastle, UK: University of Northumbria Press.

Nie, N. H. 2001. Sociability, interpersonal relations, and the Internet: Reconciling conflicting findings. *American Behavioral Scientist* 45, no. 3:426–437.

Nie, N. H., and L. Erbring. 2000. *Internet and society: A preliminary report* (on-line). Palo Alto, Calif.: Stanford Institute for the Quantitative Study of Society, Stanford University. Retrieved 18 July 2001, <http://www.stanford.edu/group/siqss>.

Nie, N. H., D. S. Hillygus, and L. Erbring. 2002. Internet use, interpersonal relations, and sociability: A time diary study. In *Internet in everyday life*, edited by B. Wellman and C. Haythornthwaite. Oxford: Blackwell.

Nie, N. H., and H. Sackman. 1970. *The information utility and social choice*. Montvale, N.J.: American Federation of Information Processing Societies.

Norris, P. 2001. *Digital divide: Civic engagement, information poverty, and the Internet worldwide*. New York: Cambridge University Press.

Putnam, R. D. 1993. *Making democracy work: Civic traditions in modern Italy.* Princeton, N.J.: Princeton University Press.

Putnam, R. D. 1996. The strange disappearance of civic America. *American Prospect* 24:34–48.

Putnam, R. D. 2000. *Bowling alone: The collapse and revival of American community.* New York: Simon and Schuster.

Quan-Haase, A., and B. Wellman, with J. Witte, and K. Hampton. 2002. Capitalizing on the Internet: Social contact, civic engagement, and sense of community. In *Internet in everyday life,* edited by B. Wellman and C. Haythornthwaite. Oxford: Blackwell.

Reddick, A., C. Boucher, and M. Groseillers. 2000. *The dual digital divide: The information highway in Canada.* Ottawa: Public Interest Advocacy Centre.

Rheingold, H. 2000. *The virtual community: Homesteading on the electronic frontier.* Rev. ed. Cambridge: MIT Press.

Sproull, L. S., and S. B. Kiesler. 1991. *Connections: New ways of working in the networked organization.* Cambridge: MIT Press.

Steiner, G. A. 1963. *The people look at television: A study of audience attitudes.* New York: Knopf.

Wellman, B. 1979. The community question. *American Journal of Sociology* 84:1201–1231.

Wellman, B. 1996. Are personal communities local? A dumptarian reconsideration. *Social Networks* 18:347–354.

Wellman, B. 1999. The network community. In *Networks in the global village,* edited by B. Wellman. Boulder, Colo.: Westview.

Wellman, B. 2001. Physical place and cyberspace: The rise of personalized networks. *International Urban and Regional Research* 25, no. 2:227–252.

Wellman, B., P. Carrington, and A. Hall. 1988. Networks as personal communities. In *Social structures: A network approach,* edited by B. Wellman and S. D. Berkowitz. Cambridge: Cambridge University Press.

Wellman, B., and M. Gulia. 1999. Net surfers don't ride alone: Virtual communities as communities. In *Networks in the global village,* edited by B. Wellman. Boulder, Colo.: Westview.

Wellman, B., and C. Haythornthwaite, eds. 2002. *The Internet in everyday life.* Oxford: Blackwell.

Wellman, B., and B. Leighton. 1979. Networks, neighborhoods, and communities. *Urban Affairs Quarterly* 14:363–390.

Wellman, B., A. Quan-Haase, J. Witte, and K. Hampton. 2001. Does the Internet increase, decrease, or supplement social capital? Social networks, participation, and community commitment. *American Behavioral Scientist* 45, no. 3:437–456.

Wellman, B., and D. Tindall. 1993. Reach out and touch some bodies: How social networks connect telephone networks. *Progress in Communication Sciences* 12:63–93.

Wellman, B., and S. Wortley. 1990. Different strokes from different folks: Community ties and social support. *American Journal of Sociology* 96, no. 3:558–588.

Wuthnow, R. 1991. *Acts of compassion: Caring for others and helping ourselves.* Princeton, N.J.: Princeton University Press.

Wuthnow, R. 1998. Loose connections: Joining together in America's fragmented communities. Cambridge: Harvard University Press.

II
Social Capital in Knowledge Sharing

Part II covers five chapters that deal with organizational aspects of knowledge sharing both within as well as among organizations.

In chapter 6, Rob Cross and Stephen P. Borgatti employ social network analysis and qualitative research to explore the criteria that individuals use to establish a knowledge-sharing relationship with another individual. In other words, they study what relationship characteristics influence to whom people turn to for information. The authors focus on the following dimensions: the information seeker's awareness of a potential source's expertise; timely access to the information source; the degree of safety in the relationship; and the willingness of an information source to cognitively engage in problem solving. The authors find that these four relational dimensions predict who people turn to when faced with a new problem or opportunity. In a follow-up study, the authors have actually discovered that people's position within this "latent" or "potential" information-seeking network is a higher predictor of performance than their position within the current information flow network. Particularly in knowledge-based work where projects and patterns of interaction are shifting dynamically, this more latent view of a network seems to be both more stable and a more important asset in terms of social capital.

The next chapter also deals with knowledge-sharing individuals. Bart van den Hooff, Jan de Ridder, and Eline Aukema explore the individual characteristics of social capital, and come up with the concept of *eagerness to share*, which they compare to the more collectively based notion *willingness to share*. The literature on social capital seems to lean more toward the latter motivation to contribute to knowledge sharing, whereas according to the authors, the eagerness to share is perhaps even more significant for social capital development. The authors also propose that IT plays a particular role in this process, as it provides a means of communication that contributes to knowledge sharing in two ways: in terms of enhancing the efficiency of the process as well as in terms of influencing the degree of collectivism within a group of users. They conclude that IT has a number of characteristics that call for a new perspective on the role of social capital in knowledge-sharing processes.

In chapter 8, Marleen Huysman continues the discussion on knowledge sharing in organizations. She explicitly looks at the sociotechnical gap in the tradition of technologically supporting knowledge management. Most existing knowledge management tools fail the test of institutionalization. Based on case study research that has been published elsewhere, she argues that this failure is largely because these tools do not sufficiently address the social capital of a social network. In order to align the tools with these networks, it is crucial to address the knowledge-sharing needs of communities, the motivation to share knowledge, and the need to embed the technology within the existing social networks. Huysman contends that the fallacies of so-called first-generation knowledge management can be circumvented in the cases where social capital analysis forms part of the design requirements of knowledge management tools. An analysis of the networks' social capital as part of the design and introduction process of knowledge management tools responds to these specific requirements, and will consequently increase the chance of adoption. Such an analysis looks at the structural opportunities to share knowledge, the relation-based motivation to share knowledge, and the cognitive ability to share knowledge. The chapter offers guidelines for social capital analysis, thereby contributing to the design of knowledge management tools.

In chapter 9, Charles Steinfield picks up the discussion on knowledge sharing, but looks instead at the interorganizational level. In particular, Steinfield focuses on the role of IT in supporting social capital development while transcending the boundaries of organizations. His emphasis is on so-called business-to-business (B2B) hubs. He argues that the primary weakness of most B2B hubs is the dominance of efficiency-related services and the relative lack of services for relationship building, which is of central importance to the formation of stable business relationships. Conversely, evidence from successful geographically defined business clusters suggests that location and proximity facilitate the formation of social capital as well as interfirm relationships in ways that do not require interorganizational ISs. Thus, interorganizational systems (like B2B hubs) are likely to be underused in geographically defined clusters. It is concluded that successful collaborative e-commerce in geographic business clusters must recognize as well as complement the rich communications and preexisting relationships that have served to enhance trust and cooperative behavior, rather than attempt being a substitute for such communication and relationships.

Part II ends with a critical account of the optimistic voices that characterize the literature on social capital and knowledge management within organizations. In their chapter, Mike Bresnen, Linda Edelman, Sue Newell, and Harry Scarbrough question the contention that the accumulation of social capital has a positive and proportionate effect on the performance of projects in organizations. Their argument is based on data collected from three research projects in which project-based learning played a major role. Their findings indicate that while social capital has many beneficial effects with respect to information access and retrieval, there are also a number of less beneficial aspects, which are underexplored in the current empirical literature.

6

The Ties That Share: Relational Characteristics That Facilitate Information Seeking

Rob Cross and Stephen P. Borgatti

Despite all the technology we have and the huge investment we make as a firm into it, people are the only way I get information that matters to me.... Learning how to use the constellation of people around you requires understanding what they can and will do for you. In part this means knowing what they are good at and can be relied on for, but just as importantly, it means knowing to what degree you can trust someone or how to get them to respond to you in a timely fashion.
—informant

In knowledge-intensive work such as professional services or software development, we can expect that those better able to find information and solve problems should be better performers. In part, people solve problems by relying on their own memory and by searching impersonal sources such as databases and file cabinets. They also rely heavily on other people, however (Granovetter 1973; Allen 1977, 182–228; Rogers 1995, 284–289; Lave and Wenger 1991, 27–42; Brown and Duguid 1991). We can thus anticipate that personal networks that facilitate effective information acquisition constitute an important form of social capital and contribute to the performance of those engaged in knowledge-intensive work. Yet while people are critical sources of information, aside from the construct of tie strength (Granovetter 1973), we know little about those characteristics of relationships that are learned and affect information seeking in field-based settings.

Traditionally, social network research in this realm has focused on the structural properties of information or communication networks, and paid less heed to the quality of relationships binding a network together (Monge and Contractor 2000; Adler and Kwon 2002). A potential limitation to this approach is its focus on current or past information flows. Mapping such interaction patterns misses those relationships not presently employed for informational purposes but that could be, when and if opportunities emerge requiring new expertise. This perspective is important because it can reflect an individual's, or even an entire network's, potential to recognize, assimilate, and take action on new problems or opportunities. This view is not necessarily revealed by analyzing current interaction patterns but rather requires one to map relational dimensions that precede information seeking.

What determines who goes to whom for information? We suggest that people develop relational awareness of others in organizations that affect information seeking. In part, this is sure to include awareness of other's knowledge and skills, as evidenced in studies of transactive memory (e.g., Wegner 1987; Moreland, Argote, and Krishnan 1996; Hollingshead, 1998; Rulke and Galaskiewicz 2000). A baseline condition for seeking information from someone is at least some perception (even if initially inaccurate) of their expertise as it relates to a given problem or opportunity. However, it is also likely, that social features of relationships are learned over time and determine who a given person seeks out for information in the face of new opportunities or problems.

When seeking information or knowledge held by other people, simply knowing where knowledge or expertise might reside is not enough. Getting information from someone requires his or her cooperation, which at some level is a function of the kind of relationship one has with that person (Lin 1982, 1988). Research on situated learning and communities of practice has richly demonstrated the importance of social interaction (from gossip to formal problem solving) in the development of knowledge (Lave and Wenger 1991; Brown and Duguid 1991; Orr 1996; Tyre and von Hippel 1997). This point is underappreciated in the distributed cognition and transactive memory literatures, perhaps because the bulk of this research has been done in regimented settings such as ship navigation or the laboratory.

This research seeks to contribute to the emerging relational perspective on learning by establishing characteristics of relationships that are learned and affect information seeking in field-based settings. Huber claims that an organization learns when "through its processing of information its range of potential behaviors has changed" (1991, 89). Along with social network analytic techniques, this relational perspective holds the potential to inform the modeling of organizational learning concerns, such as absorptive capacity, that to date have been underexplored from a social perspective (Zahra and George 2002). Our study employs a two-part methodology, consisting of a qualitative phase followed by a hypothesis-testing quantitative one. Next, we review the relevant literature that guided our empirical efforts and describe the results of the qualitative phase of our work. We then review findings from a quantitative study informed by our qualitative results, and conclude with a discussion of the implications for scholarship and practice.

Seeking Information from People

When considering information seeking as a dyadic phenomenon (person i seeks information from person j), there are three factors that researchers can examine: the attributes of the information seeker, the attributes of the relationship between the seeker and source, and the attributes of the source. The first and last of these factors have received the most scholarly attention. Examples of work on attributes of the seeker include research on gender differences in help seeking (Good, Dell, and Mintz 1989; Ibarra 1993), job categories (Shah 1998), and motivation (Lee 1997). Examples of work on attributes of the source include research on information seeking by newcomers (Morrison 1986), performance feedback (Ashford 1993), IT helping relationships (Rice, Collins-Jarvis, and Zydney-Walker 1999), and differentiating characteristics between information held by people and information stored in repositories such as file cabinets or databases (O'Reilly 1982).

Our interest lies in the relationship between the information seeker and source. There has been little research on the learned dimensions of relationships that affect information seeking, although there is work in other areas that has bearing. For example, we know that communication is likely to occur in homophilous relationships (McPherson, Smith-Lovin, and Cook 2001), and we have evidence of the importance of similarity (e.g., Wagner, Pfeffer, and D'Reilly 1984; Zenger and Lawrence 1989) and physical proximity (e.g., Allen 1977; Monge et al. 1985). We also know that information flow is affected by the strength of social ties. Granovetter (1973) argued that novel information is more likely to be obtained through weak ties than strong ones because weak ties are more likely to connect an information seeker with sources in disparate parts of a social network that are circulating information not known to the seeker. Subsequent research on the importance of weak ties has demonstrated that they can be instrumental in finding a job (Lin 1988) and the diffusion of ideas (Granovetter 1985; Rogers 1995). On the other hand, strong ties also have their uses (Krackhardt 1992). For example, Hansen (1999) found that strong ties are important for transferring complex knowledge across departmental boundaries in an organization. Similarly, Uzzi (1997) and others studying embeddedness (Granovetter 1985) have contended that strong ties facilitate the transfer of tacit knowledge.

Outside of tie strength, however, we have little empirical evidence regarding the learned relational characteristics that may affect information seeking. Beyond being aware of another's expertise, it is likely that social dimensions of relationships affect the extent to which another person is helpful and sought out for information. For instance, from the consistent findings relating physical proximity to the likelihood of collaboration, we can infer that access to other people's thinking is critical (Archea 1977). Of course in organizations, other, nonphysical barriers to access can exist as a product of either informal influence (Astley and Sachdeva 1984; Brass 1984; Burkhardt and Brass 1990) or relative position in a formal structure (Stevenson 1990; Stevenson and Gilly 1993). Thus, access might also be thought of as a relational variable not entirely based on physical propinquity.

Similarly, with the development of the concept of social capital (Hanifan 1920; Coleman 1988; Putnam 1995; Leenders and Gabbay 1999), there has been considerable interest in interpersonal trust and its contributions to group performance (Mayer, Davis, and Schoorman 1995; Kramer and Tyler 1996). This literature suggests that in organizations, trust may contribute to performance by enabling people to share valuable information with each other. From another perspective, trust and psychological safety have been associated with an ability to learn at both the individual (Argyris 1982) and group level of analysis

(Edmondson 1996, 1999). One might also therefore anticipate interpersonal trust or a perceived sense of safety to inform a model of information seeking.

In summary, while some research has bearing on those attributes of relationships that affect information seeking, the question has not been addressed directly. Consequently, we begin our empirical work with an exploratory, qualitative phase in which we use the results of previous research as sensitizing concepts (Strauss and Corbin 1990; Yin 1994).

Qualitative Phase: Methods

Research was conducted within the business consulting practice of one of the Big Five accounting firms. Consulting work done in this practice largely entailed process, organizational, and strategy projects. Typically, consultants were staffed on engagements, and spent the bulk of their time at client sites collecting information from interviews, third-party sources, archival data, and observation to identify crucial aspects of client problems. The output of their work entailed either crafting lengthy consulting reports or leading client teams through the various analysis and implementation phases of a project. As opposed to technology-based consulting, where efforts are focused on developing a specific application, the work of these consultants was much more ambiguous. This setting was deemed appropriate because knowledge seeking is central to consulting in defining important dimensions of problems facing a client, developing solutions, and convincing organizational members with diverse backgrounds and interests of the correctness of a given course of action. Within the consultancy, forty managers were interviewed from a wide cross-section of offices throughout the United States. We targeted this hierarchical level as these are the people who make the bulk of the decisions regarding problem definition and solution trajectory in consulting engagements.

An iterative process of data collection was employed in order to incorporate new elements as they emerged from interviews (Strauss and Corbin 1990). Specifically, the first ten interviews were conducted over a two-week period. Following each interview, the taped transcription and notes were reviewed to identify emerging themes as well as highlight those previously suggested in the literature. At the end of the first ten interviews, data collection was temporarily halted to review the results and begin to name the key dimensions of relationships that informants indicated were important to information-seeking behavior. A second group of ten interviews was then conducted using the same process as the first ten. Once these interviews were reviewed, the first twenty interviewees were recontacted by phone in order to ask them the additional questions that emerged during the course of these interviews. Finally, the last twenty interviews were conducted over a two-week period.

Interviews generally lasted between one and two hours, and followed a two-step process common in studies of individual networks (Scott 1990; Wasserman and Faust 1994). First, the composition of each respondent's network was determined using a name generator (Burt 1984; Marsden 1990). Specifically, respondents were asked to reflect back on a significant project they had been involved in over the last six months and write down the names of all the people they turned to for information or knowledge during that project. Respondents then selected the three people from their list that they considered most important. The remainder of the interview then focused on these three individuals. To help guard against memory errors, respondents were asked to ground recollections in specific behaviors, names, and dates, when possible (Dougherty 1992). Interviews were transcribed, coded, and assessed for interrater reliability using typical content analysis procedures (Lincoln and Guba 1985; Miles and Huberman 1994). A third party independently coded the transcripts with 93 percent agreement (inconsistencies were not considered as evidence in the analysis).

Qualitative Results

In the 120 relationships explored (3 for each of 40 managers), four key relational features emerged from the interviews: an awareness of another's relevant expertise; being able to gain timely access to that person; the willingness of a potential knowledge source to actively engage in problem solving; and a safe relationship to promote creativity.

Awareness

The managers we interviewed reported turning to 110 of the 120 contacts explored because they considered those people knowledgeable in relation to some aspect of the problem they were facing. Thus, these contacts provided a critical extension to the manager's own knowledge when she or he had at least a semiaccurate understanding of these people's expertise. Such relationships were valued for knowledge in two qualitatively different ways. First, people were often sought out for the specific knowledge they could contribute to some problem. These people were frequently skilled in technical domains and were looked to simply for the information that they were likely to be able to provide. Second, people were also sought out for their ability to help think through a problem. These people's expertise lay with either defining or refining complex problems and making salient important dimensions of such problem spaces.

However, the decision to ask another person for information on a given problem was often challenging, because it was difficult for informants to critically assess the quality of information received unless they had worked together before. Hence, while "weak link" relationships held potential to yield new, nonredundant information, they were also risky propositions. Our informants generally did not have a sufficient base of knowledge in the problem domain to assess the contributions of subject matter experts. Thus they frequently had to trust the source—which was difficult when they had little experience with that person.

Interestingly, of the 120 relationships, 78 involved cases where informants indicated that they would have initially preferred *not* to work with that specific contact if other options had been available in the required time period. These preferences were sometimes personal (e.g., not enjoying another person's style) and sometimes problem-domain related (e.g., not believing someone to be intelligent). Over time, they were altered or at least mitigated by favorable aspects of a person discovered through various interactions. Informants often described working together as the key mechanism by which they formed and modified their understanding of others' skills and abilities. For example, as one manager indicated,

I tend to form first impressions very quickly, so my first impression of him was that he was slow and dumb, and what I came to learn is that he is not dumb.

He is very bright and he is bright at all the things that will make you successful in this firm. So my opinion did change, but it took a lot of working together to do this.

Access

Informants also noted that information received from a source was only useful if it was received in a timely fashion. In 112 of the relationships explored, the accessibility and responsiveness of the source was considered critical to the effectiveness of the relationship for information seeking. Learning how to gain access to other people requires developing an understanding of a person's response style and preferred medium of contact. Several interviewees reported that this learning process was crucial to being able to utilize an important relationship. When others did not respond quickly, informants were often initially put off. Yet when they developed accurate expectations of how a potential information source would make themselves available, informants were better able to utilize that person's expertise. Often, nothing changed in the relationship to make it more effective—only the informants' understanding of how to make contact. For example, one manager said:

I have gotten less frustrated the more I have worked with him because I have realized that it is hard to get John to stick to a schedule. So now when I meet with John, I have a list of like ten things I need to get through. And we set up a meeting for 1:00, and I know that it is not going to happen until 2:30 or 3:00 or maybe not until the next day, but when I do see him, I have my list and I am ready, and we can run down it. It was important to learn to accept that rather than be frustrated by it.

While one would expect access to matter, less obvious was the importance of selecting the right strategy for gaining access to specific others. For instance, when asked how they would approach contacts for information or advice, interviewees were also asked to indicate whether they would make a personal appeal (e.g., "I need help") or an appeal based on benefits the other person would receive if they responded (e.g., "I have an opportunity for you"). By and large, personal appeals were perceived to result in more rapid responses. Estimated response times were much greater (and so the relationship itself was likely less functional) for the thirty-eight relationships where people indicated that they would probably employ a nonpersonal appeal. The average response time estimated for nonpersonal appeals was 32.8 hours, but it was only 18.1 hours for personal appeals (this is in an environment where people indicated that waiting more than a day or two for information often made it useless). Illustrations of personal and nonpersonal appeals include:

- Personal appeal: "If I had to get in touch with [the person]? I would give him a call directly and say that I was in a bind or needed some information quickly. This would get his attention a lot more quickly than an impersonal request."

- Nonpersonal appeal: "I would have to word it in such a way so that he would see what is in it for him," *or* "I would relate it to things I knew she was concerned with or tell her how it would be important from a client perspective."

Engagement

The willingness to be cognitively engaged in helping to solve an information seeker's problem is an attribute that emerged early in the interviews as a distinguishing feature of relationships that informants truly valued. The seventy-nine contacts who cognitively engaged with the information seekers were consistently described as particularly valuable in helping to create knowledge. Specifically, these people were valued for their willingness to actively inquire into and understand the respondent's problem, and then shape their own knowledge and experience to the problem in generating a solution. Such people were often contrasted to others who used verbal and intellectual skills as a defense to keep a person with a problem from consuming too much of their time. These less helpful people were frequently able to quickly craft an impressive-sounding answer. Without taking the time to listen and then actively shape a solution with the information seeker, however, they were often of little help in creating useful knowledge. As one manager put it,

> You know, I have been around people who give you a quick spiel because they think they are smart, and that by throwing some framework or angle up they can quickly wow you and get out of the hard work of solving a problem. [He], for all his other responsibilities and stature within the firm, is not like that. He helps you think about a problem. And by taking his time up front, he makes the solution much better and probably reduces the number of times I badger him after that.

People valued for their willingness to cognitively engage in problem solving were described in one of two qualitatively different ways. Some were depicted as being good at identifying and making salient to the respondent significant dimensions of a complex problem. Such people were important because they helped define and clarify a small number of key dimensions to what seemed an overly complex or ambiguous problem. They might also expand the perspective of the person seeking information by getting them to consider aspects of the problem not thought of before. One manager, for instance, recounted that

> the way I would describe [him] is that regardless of what the content issue is, [he] has a fantastic ability to frame issues and to challenge your thinking and help you look at things in a new way.... You will often get more value from him than if you go to someone who knows a ton of stuff, but will just spiel it out. So he is kind of a thinking partner, a framer.

Others were described as being good at seeing the consequences of a current course of action or plan. These people tended to excel at taking the perspective of third parties who would be affected by a proposed plan and making the respondent aware of their potential concerns. For example, a senior consultant indicated:

> [He] is very good at asking probing questions and exploring situations. He comes up with questions, you know, about a proposal or for a meeting with a prospective client, he comes up with questions that I do not even think of.... I just often do not see social or political consequences of actions I take, so I go to him to think these things through.

Safety

In the interviews, all of the managers indicated that they felt safe asking other people for information, claiming that they were willing to admit a lack of knowledge when necessary. In general, though, informants considered highly safe relationships penalty free and were willing to examine novel or ill-formed ideas in these interactions. In low-safety relations, this kind of latitude did not exist. Informants reported being more cautious in both admitting to their lack of knowledge as well as testing out ill-formed ideas. Thus, safer relationships offered certain advantages. They provided more learning value as informants said they were not afraid to explore new ideas. Second, informants indicated that they were more creative in safer relationships. An important feature of these relationships to them was that they were more willing to take risks with ill-formed ideas or opinions.

Summary
Qualitative results suggested that awareness of who knows what is not always sufficient for effective knowledge transfer in organizational settings. A person's awareness of the knowledge or expertise of contacts does dictate whether and for what kinds of problems a contact will be consulted. Still, our interviews elicited additional features of relations that underlie effective information sharing: access, safety, and engagement. Access was anticipated as important as it has been a basis for weak tie, structural hole, and social resource theories claiming advantage to derive from bridging relationships (Granovetter 1973; Lin 1988; Burt 1992).

A somewhat surprising finding lay with the limited role of safety. As would be anticipated from theories of learning, relationships considered highly safe seemed to provide certain benefits in terms of learning and creativity in problem solving (Argyris 1982; Edmondson 1996, 1999). Yet safety did not seem to emerge as an overriding concern in dictating who one would turn to for information. At least in the specific organization under study, people seemed to seek information or knowledge from others without an excessive concern for safety.

It was also interesting to observe the role that engagement in problem solving played because it highlights a frequently overlooked consideration in models of knowledge transfer. Recent discussions of knowledge transfer have fingered the limited absorptive capacity of the recipient as a block to effective transfer (e.g., Szulanski 1996; Simonin 1999). For example, Szulanski's (1996) study, while acknowledging the difficulty created by an "arduous" relationship, generally pointed to impediments to knowledge transfer from the perspective of the acquirer of knowledge rather than the interaction created by both the acquirer and sender. The idea of engagement elicited from informants suggests an alternative dimension—that our energies should be refocused to better understand the interaction between the sender and receiver of knowledge. Our findings suggest that a critical behavioral difference between effective and ineffective knowledge exchanges lay with a source's willingness to engage in problem solving in the interaction. This simple behavior of the knowledge provider first understanding the problem as experienced by the knowledge seeker and then shaping his or her knowledge to the evolving definition of the problem appeared critical to the receiver's ability to be able to take action on newly acquired knowledge.

Quantitative Phase: Methods

The objective of the quantitative phase was to assess the extent to which the relational dimensions emerging from the qualitative phase were in fact related to information-seeking behavior. Informants may bring up relational attributes that are salient, but that have little impact on their behavior. Since each of the four relational dimensions were thought by informants to have a facilitating affect on information seeking, we infer the simple model of information seeking shown in figure 6.1, where the dependent variable is information seeking between any two parties within the network. The model is stated at the dyadic level. That is, for any pair of persons, we claim that the degree to which one person goes to another for information is a function of perceptions on the four dimensions of awareness, access, engagement, and safety.

In order to test the model, we collected social network data from a group of thirty-eight members of a telecommunications practice in a global consulting organization. This is a separate organization from the

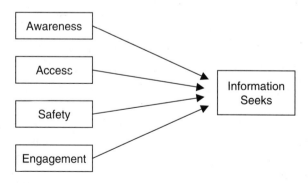

Figure 6.1
Model of information seeking inferred from respondents.

Table 6.1
Questionnaire items for relational variables

Variable name	Question
Awareness	I understand this person's knowledge and skills. This does not necessarily mean that I have these skills or am knowledgeable in these domains but that I understand what skills this person has and domains they are knowledgeable in.
Access	When I need information or advice, this person is generally accessible to me within a sufficient amount of time to help me solve my problem.
Safety	Please indicate the extent to which you feel personally comfortable asking this person for information or advice on work-related topics.
Engagement	If I ask this person for help, I can feel confident that they will actively engage in problem solving with me.
Information	Please indicate how often you have turned to this person for information or knowledge on work-related topics during the last three months.

one in which the qualitative data were collected. Nevertheless, the work and organizational structure (in terms of roles and hierarchical levels) were almost identical to the first consultancy. Table 6.1 gives the wording of the key survey items. All of the data were collected on a five-point scale. Since all of the variables are dyadic, the unit of observation is an ordered *pair* of persons. For each ordered pair, (i,j), we have measured the extent to which person i seeks information from person j (the dependent variable), as well as the degree to which i believes they know what j's expertise is, how accessible j is to i, how safe i feels in seeking information from j, and the extent to which i feels that j will engage in problem solving with i (the independent variables).

As is typical of social network research, a single sociometric question was employed to measure each theoretical variable. While some have faulted this practice (Rogers and Kincaid 1981), a review by Marsden (1990) suggests that these indexes are largely reliable when appropriate procedures are followed to help individuals report their network links accurately. Here, we both pretested the items as well as constructed question items that were highly specific and that elicited typical patterns of interaction rather than one-time events (Rogers and Kincaid 1981; Freeman, Romney, and Freeman 1987).

An important concern in collecting social network data is respondent accuracy. In particular, Bernard, Killworth, and Sailor (1982) have shown that respondents have difficulty recalling accurately who they did what with in a specified time period. This is not a concern in the case of our independent variables since they are explicitly respondent perceptions and not objective truths. It is a concern, however, for our dependent variable, which is meant to capture the respondent's behavior. To mitigate the accuracy problem, we asked not only "How often did you turn to X for information?" but also "How often has X turned to you for information?" Then, to measure the extent to which person i sought information from person j (i.e., construct the dependent variable), we took the smaller of two quantities: the amount i claimed to seek advice from j, and the amount that j indicated that he or she was sought out by i (Borgatti, Everett, and Freeman 1999).

We also collected demographic data, including each person's position in the organization (partner, manager, senior consultant, staff consultant) and tenure within the group (in months), both of which were used as control variables. In both cases, a dyadic difference variable was constructed from the individual-level data. That is, for each ordered pair of persons, we subtracted the second person's value (such as number of months in the organization) from the first person's.

The model was tested using bivariate correlation and multivariate regression. Network data do not satisfy the assumptions of statistical inference in classical regression since the observations are not independent. Consequently, special procedures, known as QAP and MRQAP (Baker and Hubert 1981; Krackhardt 1988; Borgatti, Everett, and Freeman 1999), were used to run the correlations and multiple regressions, respectively. QAP and MRQAP are identical to their nonnetwork counterparts with respect to parameter estimates, but use a randomization/permutation technique to construct significance tests. Significance levels for correlations and regressions were based on distributions generated from ten thousand random permutations.

Quantitative Results

Bivariate correlations among all variables are reflected in table 6.2. As shown in the table, the correlations among the independent and dependent variables are all high and significant, indicating strong initial support for the model. The table also reveals that the correlations among the independent variables are high as well. If no other evidence were available, we might be forced to entertain the suspicion that the four independent variables actually measure a single underlying construct. Yet our qualitative work makes it clear that respondents conceptualize these dimensions as distinct, and since our model is fundamentally perceptual, our working assumption is that the variables are distinct. We attribute their strong intercorrelations to a dynamic process in which favorable relational conditions encourage information seeking, and in turn, an information-seeking event between a given pair of persons results in changes to all four relational conditions for that pair in a similar direction (depending on how the interaction went).

We can, however, see if the theorized variables remain significant even when we control for all the others. Table 6.3 gives the results of regressing information seeking on all four independent variables simultaneously while also controlling for differences in tenure and hierarchical rank. In the table, model 1 is a baseline that contains only the two control variables. Neither is significant, and the proportion of variance accounted for is negligible. Model 2 adds the four variables obtained in the

Table 6.2
Correlation matrix

	Awareness	Access	Safety	Engage	Information	Hierarchy	Tenure
Awareness	1.00						
Access	.61	1.00					
Safety	.58	.71	1.00				
Engagement	.58	.75	.79	1.00			
Information	.58	.48	.43	.46	1.00		
Hierarchy	−.11	.02	−.05	−.00	−.04	1.00	
Tenure	−.05	.15	.05	.09	−.04	.64	1.00

Note: N = 1722

Table 6.3
Predicting information seeking

Variable	Model 1	Model 2
Hierarchy	−.02	.07***
Tenure	−.03	−.09***
Awareness		.42***
Access		.15***
Safety		−.01
Engagement		.12***
R-square	.00	.37
N	1722	1722

Notes: ***$p < 0.001$, **$p < 0.01$, *$p < 0.05$ (all significance based on 10,000 permutations)

qualitative study. Here we find that even when controlling for each other, all the independent variables are significant except for safety. Further, the percentage of variance accounted for is fairly large (37 percent), suggesting that the qualitative phase has in fact identified at least three key relational conditions relating to information seeking.

Discussion and Conclusion

This research has sought to define learned characteristics of relationships that affect information seeking in organizational settings. While existing research in the transactive knowledge tradition has focused on a person's perception of others' knowledge, our research suggests that simply knowing who knows what does not ensure that valuable knowledge can be drawn out of a network. Other factors such as access, engagement, and (perhaps) safety are also important. Our results suggest that as we move beyond the restricted settings in which the concepts of transactive memory (Wegner 1987; Moreland, Argote, and Krishnan 1996; Hollingshead 1998) and distributed cognition (Hutchins 1991, 1995; Weick and Roberts 1993) have been developed, we must begin to take into account other features of relationships that facilitate knowledge exchange in organizations. This is not to say that knowledge of what others know is not significant. Both the qualitative and quantitative analyses suggest it is perhaps the single most important variable in knowledge seeking. However, in field-based settings, we find that other learned relational characteristics are also key.

First, accessibility was shown important. Knowing that someone has information or knowledge you need does not translate to actually utilizing that knowledge if the person is not accessible. In fact, in today's time-constrained world, it is likely that access alone might be uniquely influential in whether and how others are tapped for information or knowledge. We rarely, if ever, make optimal decisions but rather satisfice (March and Simon 1958), and the extent to which we satisfice is a function of the ease with which solutions are located—as solutions are harder to find, the standards of search fall (Cohen, March, and Olsen 1972; Perrow 1986). Thus, both knowing who to turn to and having access to them are critical in pulling information from a network of contacts.

Another key factor is a source's willingness to engage in problem solving with the knowledge seeker. Qualitative findings suggested that this does not necessarily mean a lengthy problem-solving session but rather a simple shift in how people engage with others around problems. While engagement is perceived by respondents to be a quality or attribute of their contacts, in this study we choose to treat it as a quality of the relationship between the respondents and their contacts. One reason for this is empirical: respondents disagreed in their ratings of how willing specific individuals are to engage in respondents' problems. This probably reflects both perceptual differences (which are important in themselves as perceptions drive behavior) and differences in actual experiences they have had with the same contact. Another reason is theoretical: a contact's willingness to fully engage with specific others is likely to vary according to their relationship with them. This may be a function of many things, including the perceived significance of the seeker or social capital possessed by the seeker (Lin 1988).

Finally, we found that while safety was elicited from respondents in the qualitative phase, it did not play an independent role in predicting who turned to who for information in the quantitative phase of our study. A closer look at the qualitative data suggests that when respondents talked about safety, it may have been primarily with regard to improving the learning value of helping interactions, in the sense that they asked different questions when the relationship was safer. This would seem to fit with the existing research on trust, which has shown that in trusting relationships, people are more willing to give useful knowledge (Andrews and Delahay 2000; Penley and Hawkins 1985; Tsai and Ghoshal 1998; Zand 1972), and they are also more willing to listen to and absorb others' knowledge (Carley 1991; Mayer, Davis, and Schoorman 1995; Srinivas 2000). Thus, while safety might not be important in the decision to seek information from others, it does improve the learning from those interactions where it is present.

Alternatively, this result may indicate cultural differences between the two organizations studied. Although we did not attempt to measure cultural variables formally, it seemed clear ethnographically that in the organization in which we collected the quantitative data, there were strong cultural norms in place that encouraged asking questions and sanctioned negative reactions to being asked for advice. In contrast, in the organization in which we collected the qualitative data, there did not seem to be any particular norm about asking people questions. Supporting this interpretation is the fact that we obtained a low variance on the safety variable, with most people rating all others in the top two categories. This may not be true of other organizations so we do not wholly dismiss this variable.

Of course our study has limitations. First, the model we test is respondent centered. This is a strength in terms of grounding the constructs in respondent reality, but it is also a weakness with respect to generality and completeness. Since informants are not trying to build a systematic nomothetic model, it is possible that another model could be more parsimonious or account for a wider range of organizational phenomena.

Second, the model is incomplete in the sense that we do not take account of group or firmwide variables such as organizational climate. For example, we found that the safety variable had little variance and was not significantly related to advice seeking, but we believe that in large part this is a product of a firmwide cultural norm that makes information seeking acceptable behavior. From a theoretical point of view, this suggests a more complicated model in which some dyadic variables take on greater importance in the absence of cultural or other firm-level attributes.

Despite these limitations, we contend that the study offers some useful contributions. For one, this work offers a new perspective on the study of organizational learning by providing evidence of relational characteristics that are learned and related to information seeking. Given the extent to which relationships are critical for informational and learning purposes in organizations, we hope others will further consider the importance of "know who" in modeling various concerns. For example, this work holds potential to inform distributed cognition and transactive memory studies by identifying additional social variables that underlie information seeking in field-based settings. Hopefully, future research in this area will consider both the structure and content of relations among group members as well as how these relations affect the group's ability to leverage distributed expertise.

We also hope future research will consider these relations in terms of collective learning processes in organizations. For instance, assessment of these relational characteristics via social network techniques might help clarify sources of path dependence and absorptive capacity (e.g., Nelson and Winter 1982; Cohen and Levinthal 1990; March 1991). A firm's absorptive capacity has been defined in terms of the ability to "recognize the value of new information, assimilate it and apply it to new ends" (Cohen and Levinthal 1990, 129). The construct is often thought of in terms of the top-down cognitive biases resulting from existing knowledge or the lack of existing knowledge alone. Yet the core of the concept is more structural. As Cohen and Levinthal write, "To understand the sources of a firm's absorptive capacity, we focus on the structure of communication between the external environment and the organization, as well as among the subunits of the organization, and also on the character and distribution of expertise within the organization" (1990, 132). But what determines the structure of communication of information? Our study points to relational characteristics that enable information seeking and therefore index absorptive capacity.

Mapping knowledge, access, engagement, and (perhaps) safety relations makes it possible to predict whose knowledge will account for a disproportionate amount of an organization's absorptive capacity. For example, within the consulting firm where we completed the quantitative phase of this study, we conducted follow-up interviews with the three partners in charge of the group to get an understanding of how crucial our relations were in identifying people who were disproportionately important in terms of absorptive capacity. In terms of recognizing and realizing opportunities presented by the environment, we found that the five people most often mentioned with respect to knowledge, access, engagement, and safety relations were also considered critical to 78 percent of the project sales over the past year. In addition to being primarily responsible for recognizing and realizing environmental opportunities, an accounting of billable hours by project over the year prior to our study showed that this group of five also managed or provided subject matter expertise to all but three consulting engagements. As a result, they were key in operational decisions on these projects, informing what would be done and by whom—decisions having a significant impact on what the organization comes to know over time.

These relationships also hold importance for practitioners. Mapping information seeking as traditionally done in the social network tradition may be illuminating, but it does not necessarily provide a clear path to intervention. For example, we may find that one group has no information-seeking ties with another, but unless we know why, it is difficult to suggest interventions. Our model of the attributes of relationships underlying information seeking allows us to assess individually or in various combinations each kind of relation and its effect on knowledge exchange. Extremely different problems with quite different solutions might emerge from this view. If our concern lies with a group not "knowing what it knows," for example, we might suggest action learning, developmental staffing practices, or skill-profiling systems to help create knowledge of "who knows what." In contrast, if a group is having problems with, say, access, one might consider performance metrics that encourage people to be accessible to others or distributed technologies. Engagement might be developed via rich and synchronous media such as instant messaging and video-enhanced collaborative spaces. Safety might be developed via coaching or, more broadly, peer-feedback processes (for more detail on interventions, see Cross et al. 2001). By examining the four relations we have proposed, we can begin to diagnose a network and design an informed intervention with greater accuracy than if we just looked at the information network.

Acknowledgments

A prior version of this chapter was presented at the 2000 Vancouver Sunbelt Social Network Conference. We are grateful for the thoughtful comments received on this work from Paul Adler, Michael Johnson-Cramer, Candace Jones, Ron Rice, Lee Sproull, Kathleen Valley, and Mike Zack.

References

Adler, P. S., and S. Kwon. 2002. Social capital: Prospects for a new concept. *Academy of Management Review* 27, no. 1:17–40.

Allen, T. 1977. *Managing the flow of technology*. Cambridge: MIT Press.

Andrews, K. M., and B. L. Delahay. 2000. Influences on knowledge processes in organizational learning: The psychosocial filter. *Journal of Management Studies* 37:797–810.

Archea, J. 1977. The place of architectural factors in behavioral theories of privacy. *Journal of Social Issues* 33:116–137.

Argyris, C. 1982. *Reasoning, learning, and action*. San Francisco: Jossey-Bass.

Ashford, S. 1986. The role of feedback seeking in individual adaptation: A resource perspective. *Academy of Management Journal* 29:465–487.

Astley, G., and P. Sachdeva. 1984. Structural sources of intraorganizational power: A theoretical synthesis. *Academy of Management Review* 9:104–113.

Baker, F., and L. Hubert. 1981. The analysis of social interaction data. *Sociological Methods and Research* 9:339–361.

Bernard, H. R., P. Killworth, and L. Sailor. 1982. Informant accuracy in social network data V: An experimental attempt to predict actual communication from recall data. *Social Science Research* 11:30–66.

Borgatti, S., M. Everett, and L. Freeman. 1999. *UCINET 5 for Windows: Software for social network analysis*. Natick, Mass.: Analytic Technologies.

Brass, D. 1984. Being in the right place: A structural analysis of individual influence in an organization. *Administrative Science Quarterly* 29:518–539.

Brown, J. S., and P. Duguid. 1991. Organizational learning and communities-of-practice: Toward a unified view of working, learning, and innovation. *Organization Science* 2, no. 1:40–57.

Burkhardt, M., and D. Brass. 1990. Changing patterns or patterns of change: The effects of a change in technology on social network structure and power. *Administrative Science Quarterly* 35:104–127.

Burt, R. 1984. Network items and the general social survey. *Social Networks* 6:293–339.

Burt, R. 1992. *Structural holes*. Cambridge: Harvard University Press.

Carley, K. 1991. A theory of group stability. *American Sociological Review* 56:331–354.

Cohen, W., and D. Levinthal. 1990. Absorptive capacity: A new perspective on learning and innovation. *Administrative Science Quarterly* 35:128–152.

Cohen, M., J. March, and J. Olsen. 1972. A garbage can model of organizational choice. *Administrative Science Quarterly* 17:1–25.

Coleman, J. 1988. Social capital in the creation of human capital. *American Journal of Sociology* 94:S95–S120.

Cross, R., A. Parker, L. Prusak, and S. Borgatti. 2001. Knowing what we know: Supporting knowledge creation and sharing in social networks. *Organizational Dynamics* 3, no. 2:100–120.

Dougherty, D. 1992. Interpretive barriers to successful product innovation in large firms. *Organization Science* 3, no. 2:179–202.

Edmondson, A. 1996. Learning from mistakes is easier said than done: Group and organizational influences on the detection and correction of human error. *Journal of Applied Behavioral Science* 32, no. 1:5–28.

Edmondson, A. 1999. Psychological safety and learning behavior in work teams. Harvard Business School working paper.

Freeman, L. C., A. K. Romney, and S. Freeman. 1987. Cognitive structure and informant accuracy. *American Anthropologist* 89:310–325.

Good, G., D. Dell, and L. Mintz. 1989. Male role and gender role conflict: Relations to help seeking in men. *Journal of Counseling Psychology* 36, no. 3:295–312.

Granovetter, M. 1973. The strength of weak ties. *American Journal of Sociology* 78:1360–1380.

Granovetter, M. 1985. Economic action and social structure: The problem of embeddedness. *American Journal of Sociology* 91, no. 3:481–510.

Hanifan, L. J. 1920. *The community center*. Boston: Silver, Burdette, and Company.

Hansen, M. 1999. The search-transfer problem: The role of weak ties in sharing knowledge across organizational subunits. *Administrative Science Quarterly* 44:82–111.

Hollingshead, A. 1998. Retrieval processes in transactive memory systems. *Journal of Personality and Social Psychology* 74, no. 3:659–671.

Huber, G. 1991. Organizational learning: The contributing processes and literatures. *Organization Science* 2, no. 1:88–115.

Hutchins, E. 1991. Organizing work by adaptation. *Organization Science* 2, no. 1:14–29.

Hutchins, E. 1995. *Cognition in the wild*. Cambridge: MIT Press.

Ibarra, H. 1993. Personal networks of women and minorities in management: A conceptual framework. *Academy of Management Review* 18:56–87.

Krackhardt, D. 1988. Predicting with social networks: Nonparametric multiple regression analysis of dyadic data. *Social Networks* 10:359–382.

Krackhardt, D. 1992. The strength of strong ties: The importance of *philos* in organizations. In *Networks and organizations: Structures, form and action*, edited by N. Nohria and R. Eccles.

Kramer, R., and T. Tyler, eds. 1996. *Trust in organizations: Frontiers of theory and research*. Thousand Oaks, Calif.: Sage.

Lave, J., and E. Wenger. 1991. *Situated learning: Legitimate peripheral participation*. Cambridge: Cambridge University Press.

Lee, F. 1997. When the going gets tough, do the tough ask for help? Help seeking and power motivation in organizations. *Organizational Behavior and Human Decision Processes* 72, no. 3:336–363.

Leenders, R., and S. Gabbay. 1999. *Corporate social capital and liability*. Boston: Kluwer.

Lin, N. 1982. Social resources and instrumental action. In *Social structure and network analysis*, edited by P. Marsden and N. Lin. Beverly Hills, Calif.: Sage.

Lin, N. 1988. Social resources and social mobility: A structural theory of status attainment. In *Social mobility and social structure*, edited by R. Breiger. Cambridge: Cambridge University Press.

Lincoln, Y., and E. Guba. 1985. *Naturalistic inquiry*. Beverly Hills, Calif.: Sage.

March, J. G. 1991. Exploration and exploitation in organizational learning. *Organization Science* 2, no. 1:71–87.

March, J., and H. Simon. 1958. *Organizations*. New York: John Wiley.

Marsden, P. 1990. Network data and measurement. *Annual Review of Sociology* 16:435–463.

Mayer, R., J. Davis, and F. Schoorman. 1995. An integrative model of organizational trust. *Academy of Management Review* 20, no. 3:709–734.

McPherson, M., L. Smith-Lovin, and J. M. Cook. 2001. Birds of a feather: Homophily in social networks. *Annual Review of Sociology* 27:415–444.

Miles, M., and A. Huberman. 1994. *Qualitative data analysis*. 2d ed. Thousand Oaks, Calif.: Sage.

Monge, P., and N. Contractor. 2000. Emergence of communication networks. In *Handbook of organizational communication*, edited by F. Jablin and L. Putnam. 2d ed. Thousand Oaks, Calif.: Sage.

Monge, P., L. Rothman, E. Eisenberg, K. Miller, and K. Kirste. 1985. The dynamics of organizational proximity. *Management Science* 31:1129–1141.

Moreland, R., L. Argote, and R. Krishnan. 1996. Socially shared cognition at work: Transactive memory and group performance. In *What's social about social cognition*, edited by J. Nye and A. Brower. Thousand Oaks, Calif.: Sage.

Morrison, E. 1993. Newcomer information seeking: Exploring types, modes, sources, and outcomes. *Academy of Management Journal* 36, no. 3:557–589.

Nelson, R., and S. Winter. 1982. *An evolutionary theory of economic change*. Cambridge, Mass.: Belknap Press.

O'Reilly, C. 1982. Variations in decision-makers use of information sources: The impact of quality and accessibility of information. *Academy of Management Journal* 25:756–771.

Orr, J. E. 1996. *Talking about machines: An ethnography of a modern job*. Ithaca, N.Y.: Cornell University Press.

Penley, L. E., and B. Hawkins. 1985. Studying interpersonal communication in organizations: A leadership application. *Academy of Management Journal* 28:309–326.

Perrow, C. 1986. *Complex organizations: A critical essay*. New York: McGraw Hill.

Putnam, R. 1995. Bowling alone: America's declining social capital. *Journal of Democracy* 6, no. 1 (January): 65–78.

Rice, R., L. Collins-Jarvis, and S. Zydney-Walker. 1999. Individual and structural influences on information technology helping relationships. *Journal of Applied Communication Research* 27:285–309.

Rogers, E. 1995. *Diffusion of innovations*. 4th ed. New York: Free Press.

Rogers, E., and D. Kincaid. 1981. *Communication networks*. New York: Free Press.

Rulke, D., and J. Galaskiewicz. 2000. Knowledge distribution, group structure, and performance. *Management Science* 46, no. 5:612–625.

Scott, J. 1990. *Social network analysis*. Thousand Oaks, Calif.: Sage.

Shah, P. 1998. Who are employee's social referents? Using a network perspective to determine referent others. *Academy of Management Journal* 41, no. 3:249–268.

Simonin, B. 1999. Ambiguity and the process of knowledge transfer in strategic alliances. *Strategic Management Journal* 20:595–623.

Srinivas, V. 2000. Individual investors and financial advice: A model of advice-seeking behavior in the financial planning context. Ph.D. diss., Rutgers University.

Stevenson, W. 1990. Formal structure and networks of interaction within organizations. *Social Science Research* 19:113–131.

Stevenson, W. B., and M. Gilly. 1993. Problem-solving networks in organizations: Intentional design and emergent structure. *Social Networks* 22:92–113.

Strauss, A., and J. Corbin. 1990. *Basics of qualitative research: Grounded theory procedures and techniques.* Newbury Park, Calif.: Sage.

Szulanski, G. 1996. Exploring internal stickiness: Impediments to the transfer of best practices within the firm. *Strategic Management Journal* 17 (winter special issue): 27–43.

Tsai, W., and S. Ghoshal. 1998. Social capital and value creation: The role of intrafirm networks. *Academy of Management Journal* 41:464–476.

Tyre, M. J., and E. von Hippel. 1997. The situated nature of adaptive learning in organizations. *Organization Science* 8, no. 1:71–83.

Uzzi, B. 1997. Social structure and competition in inter-firm networks: The paradox of embeddedness. *Administrative Science Quarterly* 42:35–67.

Wagner, W., J. Pfeffer, and C. O'Reilly. 1984. Organizational demography and turnover in top-management groups. *Administrative Science Quarterly* 37:549–579.

Wasserman, S., and K. Faust. 1994. *Social network analysis: Methods and applications.* Oxford: Cambridge University Press.

Wegner, D. 1987. Transactive memory: A contemporary analysis of group mind. In *Theories of group behavior*, edited by B. Mullen and G. Goethals. New York: Springer-Verlang.

Weick, K., and K. Roberts. 1993. Collective mind in organizations: Heedful interrelating on flight decks. *Administrative Science Quarterly* 38:357–381.

Yin, R. K. 1994. *Case study research: Design and methods.* Rev. ed. Newbury Park, Calif.: Sage.

Zahra, S., and G. George. 2002. Absorptive capacity: A review, reconceptualization, and extension. *Academy of Management Review* 27, no. 2:185–203.

Zand, D. E. 1972. Trust and managerial problem solving. *Administrative Science Quarterly* 17:229–239.

Zenger, T., and B. Lawrence. 1989. Organizational demography: The differential effects of age and tenure distributions on technical communication. *Academy of Management Journal* 32:353–376.

7

Exploring the Eagerness to Share Knowledge: The Role of Social Capital and ICT in Knowledge Sharing

Bart van den Hooff, Jan de Ridder, and Eline Aukema

In today's knowledge-intensive economy, an organization's available knowledge is becoming an increasingly important resource. In the "resource-based" view of a firm, knowledge is considered to be the most strategically critical resource (e.g., Conner and Prahalad 1996; Grant 1996; Nahapiet and Ghoshal 1998; Pettigrew and Whipp 1993). The effective management of this resource is, consequently, one of the most significant challenges facing today's organizations (Davenport and Prusak 1998; Drucker 1993; Hansen, Nohria, and Tierney 1999; Weggeman 2000).

The sharing of knowledge between individuals and departments in an organization is considered to be a crucial process here (O'Dell and Grayson 1998; Osterloh and Frey 2000). Only when individual and group knowledge are translated to organizational knowledge can an organization start to effectively manage this resource. This is illustrated by a well-known statement by Hewlett-Packard chair Lew Platt: "I wish we knew what we knew at HP"—mirrored by Texas Instruments CEO Jerry Junkins's meditation, "If TI only knew what TI knows." (O'Dell and Grayson 1998, 154). In a number of studies we conducted, we found that knowledge sharing is indeed a major problem in many organizations (Van den Hooff and Vijvers 2001; Van den Hooff, Vijvers, and de Ridder 2002). A large part of an organization's knowledge remains implicit, "stored in people's heads," and connected to persons. Knowledge sharing is found to some degree within teams or departments, but much less so between these units. Therefore, determining which factors promote or impede the sharing of knowledge within groups and organizations constitutes an important area of research.

In this chapter, we focus on two such factors: social capital, and information and communication technology (ICT). Research and theory that take a social capital perspective on knowledge sharing (for instance, Faraj and McLure Wasko 2002; Nahapiet and Ghoshal 1998) often stress the significance of a *collectivist* approach. The central argument here is that a collectivist norm in a group, which means that group members let collective goals prevail over their individual interests, positively influences group members' *willingness* to share their knowledge with other group members. We explore the importance of collectivism with regard to knowledge sharing, as well as the specific role of ICT in such processes. In this chapter, we first discuss a number of existing theories concerning these variables, followed by the results of a field experiment we conducted. Based on these results, we reconsider our theoretical assumptions, and come up with some alternative conceptions of the role of norms and ICT in knowledge sharing.

Norms in Social Capital and Knowledge Sharing

In this section, we discuss a number of theoretical insights that led us to the expectation that a collectivist group norm positively influences group members' willingness to share their knowledge. We first examine social capital theory, stressing the importance of a collectivist norm for the creation of public goods in a social system. Next, we explore theories concerning group norms of collectivism and individualism, stating that a collectivist group norm fosters cooperation and the willingness to contribute to group interests. Knowledge sharing represents such a group interest in that this process can lead to the creation of a public good for the group: jointly owned intellectual capital. Hence, a collectivist group norm, based on these theories, is expected to contribute positively to knowledge sharing in a group.

Social Capital and Norms
Social capital, as it was introduced by Coleman, "constitutes a particular kind of resource available to an actor" (1988, S98). Next to financial capital, physical capital (tools, machines, and other productive equipment), and human capital (an individual's skills and capabilities),

social capital is a fourth form of capital available to actors. It differs from human capital in that it does not reside inside an individual but in the structure of relations among individual actors. Social capital is therefore not owned by individual actors; it is jointly owned.

The concept of *social capital* refers to the value of these social structures to actors in the form of resources that they can use to achieve their interests. Bourdieu and Wacquant define social capital as "the sum of the resources, actual or virtual, that accrue to an individual or group by virtue of possessing a durable network of more or less institutionalized relationships of mutual acquaintance and recognition" (1992, 119).

Coleman (1988) distinguishes three different ways in which social relations can constitute useful resources for individuals:

1. *Obligations, expectations, and trustworthiness:* Mutual obligations, expectations, and trust among actors constitute the first form of social capital. As Coleman describes it, "If A does something for B and trusts B to reciprocate in the future, this establishes an expectation in A and an obligation on the part of B" (S102). Coleman compares such obligations to "credit slips" held by A, constituting capital possessed by A. This form of social capital is determined by, on the one hand, the trustworthiness of the social structure (i.e., how realistic is A's expectation that the favor will be reciprocated?), and on the other hand, the actual extent of obligations held (i.e., how many credit slips does A have?).

2. *Information channels:* Each social relation carries a certain potential for information (or knowledge)—other actors are potential sources of information that one does not possess oneself. Although Coleman says that from this perspective, the relations are not valuable for the credit slips they provide (in the form of obligations), there is of course a similarity: providing information or knowledge to others can also be seen as establishing an expectation that others will reciprocate.

3. *Norms and effective sanctions:* Effective norms within social structures constitute, as Coleman puts it, "a powerful, though sometimes fragile, form of social capital" (S104). Concerning such norms, Coleman especially stresses the significance of a collectivist norm as a form of social capital: "the norm that one should forgo self-interest and act in the interests of the collectivity" (S104).

Group Norms: Collectivism versus Individualism

This last point warrants some further elaboration. The dimensions of collectivism and individualism have been frequently studied (Geertz 1974). Triandis (1989), for instance, distinguishes between the individualist and collectivist dimension of the "self." Wagner and Moch (1986) define *individualism* as a condition in which personal interests prevail over group needs. Triandis (1989) even claims that individualists will give priority to their own personal goals when this conflicts with important group goals. *Collectivism*, in contrast, describes a situation where there is a general orientation on group goals and an inclination toward cooperation (Wagner 1995), where collective goals prevail over individual ones (Triandis 1989). According to Coleman, only when a collectivist norm is salient within a social structure can the creation of a public good be expected to be successful.

Knowledge sharing can be conceived of as contributing to the creation of such a public good in the form of shared intellectual capital (Nahapiet and Ghoshal 1998). The degree to which any actor contributes to the collective intellectual capital depends on the gains that this actor expects to derive from it—that is, on the extent to which this actor expects others to reciprocate the favor. The norm of reciprocity, in other words, is considered to be crucial here—as Putnam phrases it, the norm that "I'll do this for you now, knowing that somewhere down the road you'll do something for me" (1993, 182–183). Adler and Kwon have a comparable view: reciprocity "transforms individuals from self-seeking and egocentric agents with little sense of obligation into members of a community with shared interests, a common identity and a commitment to the common good" (2002, 25). Here, reciprocity is seen as an obligation, which (when collectivist norms prevail) creates further reciprocal arrangements.

This description of social capital offers some insights on the relationship between collectivism and knowledge sharing: providing knowledge to others can be conceived of as creating an obligation on the part of the others to reciprocate the favor—the credit slip mentioned earlier. And as Coleman puts it, a collectivist norm in a social structure can be important in overcoming the public goods problem existing in collectives—that is, the question of how to stimulate individual actors to contribute their individual resources (in this case, their knowledge) to the collective good. When collectivism prevails, and group members are primarily oriented toward the group's goals and not their individual ones, reciprocity is almost automatically expected; as all group members have the common goal as their primary orientation, there is a general expectancy that anybody's individual contribution will be equaled by other group members.

Conclusion: Collectivism and Knowledge Sharing
Hence, the conclusion would be that a collectivist norm in a group would positively influence group members' willingness to share their knowledge. Various studies concerning the collectivism-individualism dimension seem to support this conclusion: collectivism promotes cooperation (Wagner 1995), leads to more effective knowledge processes (Gladstein 1984), encourages more communication in the group (Moorman and Blakely 1995), and results in better group performance (Eby and Dobbins 1997). On the whole, the theories and results discussed in this section lead to the expectation that people will be more inclined toward knowledge sharing (i.e., more willing to share what they know with other group members) in a collectivist condition than in a group where individualism prevails, and will also show more knowledge-sharing behavior (i.e., more actively share knowledge) under such conditions. So when there is a norm of collectivism in a group, individuals in that group will have a positive attitude toward knowledge sharing in the sense that they are more willing to share their knowledge. Consequently, in a group with a collectivist norm, more knowledge sharing will take place than in a group with an individualist norm.

ICT, Efficiency, and Collectivism

In the previous section, we looked at the relationship between social capital and knowledge sharing. Since ICT plays an increasingly critical role in knowledge-sharing processes, it is essential to analyze the effects that ICT can have on such processes. In this section, we discuss a number of existing theories concerning the effects of ICT on efficiency and social processes. Based on these theories, we expect that the use of ICT will positively contribute to knowledge sharing in groups.

Effects of ICT: Two Levels
In order to map the different effects of ICT on work, communication, and knowledge sharing, the distinction made by Sproull and Kiesler (1991) between the first-level and second-level effects of ICT is useful. First-level effects relate to the consequences of ICT use for individual tasks: gains in productiveness and efficiency are typically first-level effects. As Sproull and Kiesler themselves state: "First level effects of communication technology are the anticipated technical ones—the planned efficiency gains or productivity gains that justify an investment in new technology" (1991, 4). Second-level effects, on the other hand, relate to the social structure of the organization: changes in the communication structures in and between organizations as a result of ICT use. The fact that people assume different roles in computer-mediated communication along with the emergence of new and more external communication patterns are examples of second-level effects.

Contribution of ICT to Knowledge Sharing: Efficiency and Collectivism
Based on this distinction, two kinds of ICT effects are relevant to knowledge sharing: efficiency effects, and social effects.

Efficiency effects are related to Sproull and Kiesler's first-level effects: the contribution of ICT to making communication more efficient by removing certain barriers (of time, space, and convenience). Examples of efficiency effects include the facts that

• people no longer have to get together physically in order to be able to share their knowledge;

• collective knowledge is automatically stored in an easily retrievable format; and

• participants have more control over the point in time at which they communicate with others.

Efficiency of communication processes can be expected to positively influence the efficiency of the knowledge-sharing process, as research by (among others) Zhao, Kumar, and Stohr (2000) indicates.

Social effects are related to Sproull and Kiesler's second-level effects, and concern the effects of ICT on the social climate within a group, which in turn, can be expected to influence knowledge sharing. In the previous section, a collectivist norm in a group was identified as a potentially beneficial condition for knowledge sharing—but the modes of communication used can be expected to influence such a norm. Factors such as identification with a group can positively contribute to the emergence of such a norm. Theory and research suggest that ICT does influence the degree to which people identify with their communication partners as well as the extent to which collectivism becomes salient in groups.

With regard to the social effects of communication, the lack of "social cues" (such as tone of voice, facial expressions, gestures, and the like) in communication via ICT is often expected to negatively influence the social richness of this communication (Short, Williams, and Christie 1976; Daft and Lengel 1984, 1986; Trevino, Daft, and Lengel 1990). Thus, the lack of social cues could be expected to lead to less identification with those with whom communication takes place than in a face-to-face setting, and consequently to less attention to common goals over individual ones.

Empirical results, however, contradict this (Walther 1992; Walther and Burgoon 1992; Postmes, Spears, and Lea 1998). In a search to explain such results, Walther (1996) argues that computer-mediated communication can lead to *hyperpersonal* interactions—indeed, communication with a richer level of social relationships than found in face-to-face circumstances. His conclusion is that specific characteristics of ICT (such as reduced social cues and asynchronous communication) can even lead to socially richer communication, stronger identification with the group, and thus more collective behavior.

A related perspective is the Social Identification of De-individuation Effects (SIDE) model (Reicher, Spears, and Postmes 1995; Spears and Lea 1992). As Postmes, Spears, and Lea (1998) argue, communication through ICT has a number of typical characteristics, such as anonymity of communication partners, group immersion, and the limited transfer of social cues. These characteristics can lead to strong and consistent social effects as well as normative influences. SIDE proposes that social cues can facilitate the individuation of communication partners—in other words, forming impressions of them as idiosyncratic individuals. Under computer-mediated conditions, where social cues are relatively scarce, group characteristics are likely to be attributed to individuals—that is, their social identity is likely to become more salient than their individual one. Thus, provided that the relevant social group and its attributes are known, the lack of social cues in ICT can "accentuate the unity of the group and cause persons to be perceived as group members rather than as idiosyncratic individuals" (Tanis and Postmes in press, 8). SIDE theory, then, contends that group interests become group members' primary focus when ICT is their main means of communication—in other words, the more ICT is used for communication in the group, the more a collectivist norm will prevail. In conclusion, these theories lead to the expectation that the social effects of ICT will be in the direction of more collectivism.

Conclusion: Two Routes from ICT to Knowledge Sharing

Based on these existing theories concerning ICT effects, we expect ICT to have a positive contribution to knowledge-sharing processes in two respects: ICT offers opportunities to enhance the *efficiency* of such processes; and it leads to a more *collectivist* norm within a group, which in line with the arguments made in the previous section, can positively influence knowledge sharing. Given this, two "routes" of ICT effects on knowledge sharing can be distinguished: the efficiency route, and the collectivism route. Our central assertion is that both routes lead to a positive contribution of ICT to knowledge sharing.

Knowledge Sharing in Practice: A Field Experiment

Based on a number of accepted theories concerning social capital, group norms, ICT, and knowledge sharing, we have come to the following theoretical expectations:

• A collectivist group norm will positively contribute to individuals' willingness to share their knowledge.

- Consequently, more knowledge sharing will take place in groups with a collectivist norm than in groups with an individualist one.
- The use of ICT will positively contribute to a collectivist norm in a group.
- The use of ICT will also positively contribute to the efficiency of knowledge sharing.
- Hence, the use of ICT is positively related to both the willingness to share knowledge and actual knowledge sharing.

These expectations were the basis for a field experiment in a Dutch government department. In this experiment, members of the organization were assigned to groups that were given a knowledge-sharing task within an ICT environment—a "virtual office." This environment enabled project groups or teams to create their own ICT environment, in which they could store and share documents, have discussions, manage a shared calendar, and so forth. Our design was a field experiment, combining the advantages of experimental research (control over variables and participants) with those of a case study (studying a phenomenon in a "real-life setting"—in this case, studying knowledge sharing in an actual organizational setting). Fifty-four persons, divided among seventeen groups, participated in the experiment. Each group had its own virtual office, which it was asked to use for a number of tasks.

People were not assigned to groups entirely at random: they were first asked a number of questions (based on Kashima et al. 1995) in order to determine how they scored (individually) on the collectivism-individualism dimension. The scale used was homogeneous (Cronbach's alpha = .61), and a range of one to ten. Based on the range and division of scores, the groups were divided into "collectivists" (with a score higher than six on this scale) and "individualists" (with a score equal to or less than six). This led to an exact fifty-fifty division among the participants who completed the experiment: twenty-seven people were labeled collectivists, and twenty-seven were labeled individualists.

Subsequently, persons were assigned to groups with comparable participants—that is, collectivists with collectivists, and individualists with individualists. In order to strengthen this orientation and make it a group norm, the participants were informed (in an e-mail, in which they were also assigned to a group) of whether their group was primarily individualist or collectivist. So the group norm was assumed to be the sum total of the individual orientations, strengthened by the manipulation in the form of the e-mail message in which participants were told what the norm in their group was.

Finally, the knowledge-sharing task consisted of a number of activities that for each group, should result in a jointly written vision document on the value of virtual offices. In order to get to this vision, a number of subtasks were distinguished:

- Each participant was given a number of documents (primarily journal articles) concerning virtual offices and related subjects that other group members did not have, and was asked to select one document to share with the others. The group members all had to archive their document in the virtual office for the others to read.
- Then, each participant was asked to formulate their own vision on virtual offices based on the available documents and their experiences during the experiment.
- Finally, groups were asked to produce a document in which a joint vision on the value of virtual offices was formulated.

Based on the activities that the participants engaged in during the experiment, their actual knowledge-sharing behavior was scored on a five-point scale. Scores were given by the researchers based on observed behavior in the virtual offices, to which we had full access.

The results of this experiment were ambiguous at best. We found that individualist groups scored higher on knowledge-sharing behavior, but not significantly so. There was a considerable difference in the mean scores of collectivist and individualist groups (the latter score was more than 10 percent higher, or 0.56 points on a five-point scale), but we cannot conclude that this is a significant difference. It is, however, a clear trend: in this experiment, individualist groups seemed to be more active in their knowledge-sharing behavior than collectivist groups. These results contradicted our expectations. This, of course, warrants further explanation, which we will seek in the next section as we present a more detailed view on the relationship between ICT use, social capital, and knowledge sharing.

The Eagerness to Share: A Different View of the Role of Social Capital

The results of the field experiment discussed above required us to reconsider our theoretical assumptions. In this section, we will first look at an alternative explanation for our results with regard to the relationship between group norms and individuals' attitudes toward knowledge sharing (and the resulting behavior). Next, we will revisit theories concerning ICT for an alternative explanation.

Willingness and Eagerness to Share Knowledge

First of all, the discussion thus far has focused on an individual's willingness to share their knowledge with other group members. We would define this *willingness* as the extent to which an individual is prepared to grant other group members access to their individual intellectual capital. Group members who are willing to share their knowledge with others are ready to participate in creating and enhancing the group's shared intellectual capital in the sense that they contribute their individual intellectual capital to this public good. The results of our experiment led us to believe that another attitude can be distinguished as well—one that might play an equally important part in knowledge-sharing processes: an individual's *eagerness* to share knowledge, which we define as the extent to which an individual has a strong internal drive to communicate their individual intellectual capital to other group members.

These definitions imply that although both willingness and eagerness are attitudes that are positive toward knowledge sharing, there are some key differences. Willingness implies a positive attitude to other members of a group, a readiness to reply to colleagues kindly. So the willingness to share is related to a somewhat passive way of knowledge sharing. Actors are willing to provide access to their personal knowledge, but they do not have an internal drive to do so. They will not take the initiative to actively tell others what they know if others have not done so before. Eagerness, on the other hand, implies a positive attitude to actively show knowledge. An actor who is eager to share will spout his or her knowledge, invited or uninvited. We use eagerness to indicate a proactive way of sharing knowledge. People are eager to let others know what they know because they themselves consider it valuable and expect their individual performance to be appreciated by others.

This distinction can be related to theoretical insights concerning the different motivations people might have to engage in knowledge sharing (Hall 2001; Hinds and Pfeffer 2003; Osterloh and Frey 2000). Such theory indicates that there is more to that phenomenon than the somewhat economically rational behavior implied in much of social capital theory. This economically rational behavior entails the fact that people will only contribute to a collective good in a collectivist climate where they can count on their counterparts to reciprocate the favor. In practice, this perspective is contradicted by many collective processes—for instance, in the world of open source software (Von Hippel 2001). Another example of such a process is provided by Butler, Sproull, Kiesler, and Kraut (2002), who describe the process of on-line community building. In both open source software and on-line communities, the process is initiated by actors who contribute much of their knowledge, effort, and time without getting any visible benefit in return. Von Hippel (2001) claims that the "free revealing" of knowledge and expertise does not make any sense according to conventional economic wisdom. But, he argues, the actual costs of revealing knowledge are relatively low, and the benefits can take a variety of forms. According to Von Hippel, one important benefit is "elevated reputations"—the fact that distributing his or her expertise has a positive influence on an actor's reputation. This is similar to Butler, Sproull, Kiesler, and Kraut's explanation for why people should choose to contribute to community building without any directly visible benefit: "being seen as skilled, knowledgeable or respected may have psychic payoffs" (2002, 10).

Based on various literature (e.g., Boer, Van Baalen, and Kumar 2002; Hall 2001; Hinds and Pfeffer 2003; Lerner and Tirole 2000; Von Hippel and Von Krogh in press), knowledge sharing can be compared to such processes. In knowledge-sharing processes, a number of actors can be found whose reasoning may be somewhat less sound in conventional economic terms—who are eager to share their knowledge, regardless of any directly tangible benefits they can expect from it. For these actors, the benefits of an elevated reputation, peer recognition, or "being seen as skilled, knowledgeable or respected" are sufficient to freely distribute their knowledge. For those eager to share, reciprocity does not entail receiving others' knowledge in return—these "soft benefits" are what they expect in return (Boer, Van Baalen, and Kumar 2002; Butler et al. 2002; Hall 2001; Hinds and Pfeffer 2003; Lerner and Tirole 2000).

Group Norms and Individual Attitudes

Willingness and eagerness are two distinct attitudes an individual can have toward knowledge sharing. An individual can be willing as well as eager—or neither willing nor eager—to share knowledge. An individual's actual attitude toward knowledge sharing is likely to be influenced by a number of individual characteristics, the actual knowledge-sharing task at hand, and so on. In our discussion, however, we focus on the role of group norms in relation to an individual's willingness and/or eagerness to share knowledge. In line with the theories presented before, we claim that willingness will be fostered by a collectivist norm in a group. Yet we would argue that eagerness is stimulated when there is an individualist norm in a group.

Where individual interests are prominent (i.e., an individualist group norm is salient), an individual's attitude toward knowledge sharing can still be positive. Although individuals within such a group do contribute to the public good of collective intellectual capital, their rationale for doing so is not that they "forgo self-interest and act in the interests of the collectivity (Coleman 1988, S104)." Their prime motivation is the feeling that they have something to contribute as an individual that they themselves consider to be valuable and for which they will be appreciated. As mentioned earlier, the norm of reciprocity is salient within a group with an individualist norm—not in the sense that the knowledge "donor" expects others to return the favor but in the sense that he or she wants to be appreciated and recognized for it.

Eagerness before Willingness

So both a collectivist and an individualist group norm can in some way positively influence knowledge sharing—through positively influencing individuals' willingness and eagerness to share, respectively. The nature of willingness and eagerness, and the motivations lying beneath both attitudes, lead us to expect that different attitudes can play different roles in various stages of the knowledge-sharing process. On the one hand, collectivism creates an environment where cooperation and sharing are nurtured. On the other hand, it can hamper the individual initiative that is necessary for the process of collective intellectual capital creation to take off. In an extremely collectivist situation, actors could be willing to share knowledge, but they might not be eager to do so. Consequently, they might all wait for the other actors to take the initiative. In an environment with a predominantly individualist norm, however, individual initiative prevails over the collective interest, and such initiative will be found much more. The reverse side of this is that under extremely individualist conditions, group members will all actively spout their knowledge, but this will not lead to the creation of shared intellectual capital—as everyone shouts, and no one listens.

Ultimately, then, both attitudes (willingness and eagerness) are important in order for a knowledge-sharing process to continue. The process we studied here started from zero, with people who had not met face-to-face before the process started. Therefore, we would maintain that in the take-off phase of knowledge sharing, individualism and eagerness to share are significant influences. Once the process is under way, though, the collective interest should become more salient in order for knowledge sharing to lead to a certain collective performance—and not just the exchange of individual views and opinions, which could be the case in an exclusively individualist situation. So the eagerness to share could be necessary for the process to take off, and a willingness to share could be a precondition for the process to continue and lead to enduring collective intellectual capital. Modifying the group norm could influence such attitudes. Gradually establishing a more collectivist norm in an initially individualist group, for instance, might help in making those primarily eager to share at first, more willing to do so as well. Conversely, establishing a more individualist norm might induce eagerness in those mainly willing to at first share.

ICT and the Eagerness and Willingness to Share Knowledge

We have identified eagerness and willingness as two distinct attitudes toward knowledge sharing. These concepts give us the opportunity to be more precise about the role of ICT in knowledge sharing. Although the contribution of ICT to knowledge sharing was not explicitly measured in our experiment (all groups shared knowledge under an ICT-only condition), the results do raise questions about our theoretical assumptions concerning this contribution. Our argument would be that the contributions of ICT to knowledge sharing are of differing importance to those who are eager or willing to share knowledge, respectively. When we consider the role of ICT, we return to our contention that ICT makes a dual contribution to knowledge sharing: through the enhanced efficiency of communication, and through the creation of a more collectivist norm within a group.

Eagerness and Willingness, and Two ICT Routes to Knowledge Sharing
The efficiency route (ICT's contribution to the efficiency of communication) is primarily of interest to those whose initial attitude toward knowledge sharing is one of eagerness. ICT gives them the possibility of enhancing the efficiency with which they can communicate with others, and hence, share their knowledge with others. In line with Chen, Chen, and Meindl (1998), it can be argued that those eager to share are more geared toward sending information, prefer verbal communication, and have more interest in presenting their own views than creating a common view (Ting-Toomey 1988). As Chen, Chen, and Meindl (1998) assert, ICT fosters such styles of communication. For people whose initial attitude is mainly one of willingness to share knowledge, the efficiency of communication will be of minor importance in comparison with creating mutual understanding and a common view.

The collectivism route (ICT's contribution to a collectivist group norm) would seem to be more central to those initially primarily willing to share—although it might also lead to those primarily eager to become more willing as well. Nevertheless, the relevance of the collectivism route can be debated. As Postmes and Spears (2000) contend, individuals do indeed identify stronger with the group under conditions of anonymity and de-individuation promoted by communication via ICT. There is, however, a critical distinction here: this effect occurs only when a group *has* a social identity, a common characteristic with which individual members identify themselves. If such a common identity does not exist,

Postmes and Spears (2000), argue, group members tend to focus on interpersonal relationships instead of this common identity.

The groups in our field experiment did not have a common identity. They were not characterized in any group-specific way; each group was a relatively random collection of individuals whose only similarity was their score on the collectivism-individualism scale. For a social identity to become salient, to become a ground for a more collectivist norm in a group, there has to be a clearly identifiable social identity to begin with—and that was not the case in this experiment. Also, the groups conducted only a single task, within a period of one month. As Postmes, Spears, and Lea (2000) maintain, along with Walther (1996), a social identity or collectivist norm takes time to emerge. Our groups had insufficient interaction time to develop either a group identity or a higher degree of collectivism.

Our conclusion would be that for a truly collectivist climate to emerge, a group based on "common bonds" (i.e., the interpersonal relations among group members—actual relationships) is much more fertile than one based on a "common identity" (i.e., a common label or characteristic—an abstraction) (Postmes and Spears 2000). Therefore, the contribution of ICT to a more collectivist climate is debatable at best, and is more likely to occur in a setting where other, richer communication media are used as well—so that these common bonds are facilitated.

Conclusion: Eagerness and Efficiency

We can say that ICT fosters knowledge sharing by those mainly eager to share. ICT enhances the efficiency with which they can communicate their knowledge to other group members, regardless of time and geographic location. Moreover, ICT facilitates a communication style that fits their attitude toward knowledge sharing: primarily oriented toward sending, and not toward creating a common view. The contribution of ICT to a more collectivist climate is much less clear, and is expected to be dependent on the degree to which other, richer means of communication are used, too. If a group in which a collectivist norm is salient only uses ICT to share knowledge, this can easily be counterproductive. As no common bonds are developed, no "group feel" emerges, there is a mismatch between communication needs and medium characteristics.

ICT can be a useful tool in knowledge sharing, but it should always be part of a much wider strategy (in which sufficient attention is paid to face-to-face interactions and other means of communication) in order to create a climate in which knowledge sharing really takes place (Huysman and de Wit 2002). As Brown and Duguid (2001) argue, where practices are shared and a common body of implicit knowledge is thus created, ICT can be a useful medium for knowledge sharing—under the condition that richer modes of communication (such as face-to-face exchanges) are applied to create such common know-how, to create cohesiveness and communality in a group.

Implications for Theory, Research, and Practice

The concept of eagerness and the specific role of ICT in knowledge sharing both have implications for theory and research in the area of social capital. These are notions that warrant further theorizing and research.

Theoretical Implications

The eagerness to share sheds a different light on the role of social capital in knowledge sharing. Traditionally, social capital is seen as positively related to collectiveness, cooperation, and the creation of shared intellectual capital (Adler and Kwon 2002; Hargadon and Sutton 1997; Nahapiet and Ghoshal 1998). While this is certainly true, we think that the primarily collectively oriented social capital discussions concerning knowledge sharing tend to have a "collectiveness bias," and focus largely on the willingness to share knowledge. Our argument is that a more individually oriented perspective on knowledge sharing could be a valuable addition that might help obtain a richer view of how such processes take off and continue. This individually oriented perspective is the one that stresses the importance of the eagerness to share—of more individual motivations to engage in knowledge sharing.

In our view, social capital can be conceptualized as a characteristic of a group. As far as knowledge sharing in such a group is concerned, willingness and eagerness to share can be thought of as two related dimensions of social capital. So instead of only seeing collectivist norms and the willingness to share as part of social capital (as social capital theory often seems to do), we would assert that individualist norms and the eagerness to share are equally critical elements of social capital. Ultimately, this would mean that collectivism as such should not be considered to be the panacea for knowledge-sharing problems—high degrees of collectivism do not automatically lead to high degrees of knowledge sharing. Although Nahapiet and Ghoshal, for instance, "recognize that social capital may also have significant negative consequences" (1998, 147), their central thesis remains that the relational dimension of social capital (i.e., norms of collectivism and reciprocity) is positively related to the motivation to combine and exchange intellectual capital. We would, of course, certainly agree that this is the case, but would also like to state that such norms can be counterproductive in getting such processes started. As the eagerness to share (i.e., an individualist motivation) is crucial to initiating the process of combination and exchange that leads to the creation of intellectual capital, it is in our view of equal importance to the willingness to share.

Therefore, a richer view of social capital is called for. Social capital is created not by certain norms presiding over others (i.e., collectivism) but by the combination of and interaction between different norms: both collectivism and individualism. These, then, are norms not of the group as a whole but of subgroups. The interaction between such subgroups is facilitated by the fact that they have different normative perspectives.

Further Research
In line with the observations in the previous paragraph, more research is necessary that studies the sets of norms within a social system as well as the interaction among subsystems with different sets of norms. A more elaborate study of how (and by whom) knowledge-sharing processes are initiated, and how (and by whom) they are made continuous, would be an interesting contribution to research in this area. Longitudinal research, studying group norms and individual attitudes as antecedents of knowledge sharing, is called for here.

The specific role of ICT also deserves further attention. At best, ICT's role in knowledge sharing is ambivalent. Our distinction between the eagerness and willingness to share points toward interesting avenues of research. As our results seem to indicate, ICT's contribution to knowledge sharing primarily lies in facilitating communication efficiency in groups in which people are already eager to share. The collectivism route discussed earlier seems less relevant—for a more collectivist climate to emerge, richer media are more useful than only ICT—although ICT can certainly play this part when positioned within a broader range of modes of communication. New research could focus on ICT's role as a facilitator of continuous sharing within and among already existing groups. Such research could also consider ICT as a medium through which knowledge sharing can be forcefully initiated because of the way it supports the communication styles and efficiency goals of those eager to share—provided, again, that such individuals are members of the group. The specific characteristics of ICT along with their consequences for social interaction, group identification, and collective behavior should be incorporated into such research. Therefore, the comparison between different modes of knowledge sharing (both in combination and alone) is an intriguing avenue of research as well.

Practical Implications

Finally, the ideas presented in this chapter have some practical implications as well. When creating teams, organizations may benefit from a balanced composition in which both eager and willing individuals are included—each group having its specific role in initiating and continuing knowledge sharing. Another practical implication has to do with the ICT tools that are being used to support knowledge sharing. Although some of these tools may in name and nature seem to be oriented toward the collective (such as a virtual office or a community of practice), our view is that without parallel face-to-face interactions, such tools will still mainly support the efficiency of communication, and not contribute to collectivism. Organizations and teams should not put their expectations of ICT tools too high—such tools are not the way to constitute true cooperation, consensus, and collectivism, especially when there is no shared practice (or common bonds) at the outset. True collaboration warrants a mix of communication modes, in which face-to-face meetings play an important role. ICT can be useful in lifting some barriers (of time and place) in such collaborations, but it can hardly be expected to create these collaborations.

Conclusion

The somewhat ambiguous results from our field experiment in a large, Dutch government organization inspired us to rethink our initial ideas on the relationship between social capital, ICT, and knowledge sharing. Based on the fact that we found a trend in which groups with a primarily individualist norm shared more knowledge than groups with a mainly collectivist one, we have argued for the importance of distinguishing the eagerness to share knowledge from the willingness to share knowledge. Where the latter indicates a more passive readiness to share one's knowledge with other group members, the concept of eagerness concerns more proactive behavior, a strong internal drive to communicate what one knows to others. We have maintained that both eagerness and willingness are dimensions of social capital, and that a richer view of social capital is called for—not taking the group norm as its central focus as far as norms are concerned but the variety of norms within a group, and the way these have differing values in differing stages of the knowledge-sharing process. Eagerness, we allege, is valuable in the take-off phase of knowledge sharing, whereas willingness is crucial for knowledge sharing to continue.

The role of ICT has been related to these concepts in terms of two routes via which ICT was hypothesized to contribute to knowledge sharing: efficiency and collectivism. The first route relates to the contribution of ICT in making knowledge sharing independent of time and place, as well as easier and faster. The second route concerns the contribution of ICT to the identification of group members with the group as a whole (and thus, to a collectivist norm in the group). We have contended that the efficiency route is the one where the primary contribution of ICT to knowledge sharing is realized, and that ICT's role in the collectivism route is limited—and dependent on the broader range of communication instruments of which it is a part.

References

Adler, P. S., and S. Kwon. 2002. Social capital: Prospects for a new concept. *Academy of Management Review* 27, no. 1:17–40.

Boer, N. I., P. van Baalen, and K. Kumar. 2002. The implications of different models of social relationships for knowledge sharing. In *Proceedings of the third European conference on organizational knowledge, learning, and capabilities*. Athens, Greece: Athens Laboratory of Business Administration.

Bourdieu, P., and L. Wacquant. 1992. *An invitation to reflexive sociology*. Chicago: University of Chicago Press.

Brown, J. S., and P. Duguid. 2001. Knowledge and organization: A social-practice perspective. *Organization Science* 12, no. 2:198–213.

Butler, B., L. Sproull, S. Kiesler, and R. Kraut. 2002. Community efforts in online groups: Who does the work and why? In *Leadership at a distance*, edited by S. Weisband and L. Atwater. Retrieved April 2002, <http://opensource.mit.edu/papers/butler.pdf>.

Chen, C., X. Chen, and J. R. Meindl. 1998. How can cooperation be fostered? The cultural effects of individualism-collectivism. *Academy of Management Review* 23, no. 2:285–304.

Coleman, J. S. 1988. Social capital and the creation of human capital. *American Journal of Sociology* 94:S94–S120.

Conner, K. R., and C. K. Prahalad. 1996. A resource-based theory of the firm: Knowledge versus opportunism. *Organization Science* 7, no. 5:477–501.

Daft, R. L., and R. H. Lengel. 1984. Information richness: A new approach to managerial behavior and organizational design. In vol. 6 of *Research in organizational behavior*, edited by L. L. Cummings and B. M. Staw. Homewood, Ill.: JAI Press.

Daft, R. L., and R. H. Lengel. 1986. Organizational information requirements, media richness, and structural design. *Management Science* 32:554–569.

Davenport, T. H., and L. Prusak. 1998. *Working knowledge: How organizations manage what they know*. Boston: Harvard Business School Press.

Drucker, P. 1993. *The post-capitalist society*. Oxford: Butterworth-Heineman.

Eby, L. T., and G. H. Dobbins. 1997. Collectivistic orientation in teams: An individual and group-level analysis. *Journal of Organizational Behavior* 18:275–295.

Faraj, S., and M. McLure Wasko. 2002. The web of knowledge: an investigation of knowledge exchange in networks of practice. Retrieved April 2002, <http://opensource.mit.edu/papers/Farajwasko.pdf>.

Geertz, C. 1974. From the native's point of view: On the nature of anthropological understanding. In *Culture theory: Essays on mind, self, and emotion*, edited by R. A. Shweder and R. A. LeVine. Cambridge: Cambridge University Press.

Gladstein, D. L. 1984. Groups in context: A model of task group effectiveness. *Administrative Science Quarterly* 29:499–517.

Grant, R. M. 1996. Toward a knowledge-based view of the firm. *Strategic Management Journal* 17:109–122.

Hall, H. 2001. Input-friendliness: Motivating knowledge sharing across intranets. *Journal of Information Science* 27, no. 3:139–146.

Hansen, M., N. Nohria, and T. Tierney. 1999. What's your strategy for managing knowledge? *Harvard Business Review* (March–April): 106–116.

Hargadon, A., and R. I. Sutton. 1997. Technology brokering and innovation in a product development firm. *Administrative Science Quarterly* 42:716–749.

Hinds, P., and J. Pfeffer. 2003. Why organizations don't "know what they know": Cognitive and motivational factors affecting the transfer of expertise. In *Beyond knowledge management: Sharing expertise*, edited by M. Ackerman, V. Pipek, and V. Wulf. Cambridge: MIT Press.

Huysman, M. H., and D. de Wit. 2002. *Knowledge sharing in practice*. Dordrecht: Kluwer Academics.

Kashima, Y., U. Kim, M. J. Gelfand, S. Yamaguchi, S. Choi, and M. Yuki. 1995. Culture, gender, and self: A perspective from individualism-collectivism research. *Journal of Personality and Social Psychology* 5:925–937.

Lerner, J., and J. Tirole. 2000. The simple economics of open source. National Bureau of Economic Research working paper 7600. Retrieved April 2002, <http://opensource.mit.edu/papers/Josh Lerner and Jean Triole—The Simple Economics of Open Source.pdf>.

Moorman, R. H., and G. L. Blakely. 1995. Individualism-collectivism as an individual difference predictor of organizational citizenship behavior. *Journal of Organizational Behavior* 16:127–142.

Nahapiet, J., and S. Ghoshal. 1998. Social capital, intellectual capital, and the organizational advantage. *Academy of Management Review* 40, no. 2:242–266.

O'Dell, C., and C. J. Grayson. 1998. If only we knew what we know: Identification and transfer of internal best practices. *California Management Review* 40, no. 3:154–174.

Osterloh, M., and B. S. Frey. 2000. Motivation, knowledge transfer, and organizational forms. *Organization Science* 11, no. 5:538–550.

Pettigrew, A., and R. Whipp. 1993. *Managing change for competitive success*. Cambridge, Mass.: Blackwell.

Postmes, T., and R. Spears. 2000. Redefining the cognitive definition of the group: Deindividuation effects in common bond vs. common identity groups. In

SIDE issues center stage: recent developments in studies of de-individuation in groups, edited by T. Postmes, R. Spears, M. Lea, and S. Reicher. Amsterdam: Royal Netherlands Academy of Arts and Sciences.

Postmes, T., R. Spears, and M. Lea. 1998. Breaching or building social boundaries? SIDE-effects of computer-mediated communication. *Communication Research* 25:689–715.

Postmes, T., R. Spears, and M. Lea. 2000. The formation of group norms in computer-mediated communication. *Human Communication Research* 26, no. 3:341–371.

Putnam, R. D. 1993. The prosperous community: Social capital and public life. *American Prospect* 13:35–42.

Reicher, S., R. Spears, and T. Postmes. 1995. A Social Identity Model of Deindividuation Phenomena. In vol. 6 of *European Review of Social Psychology*, edited by W. Stroebe and M. Hewstonde. Chichester, U.K.: Wiley.

Short, J., E. Williams, and B. Christie. 1976. *The social psychology of telecommunications*. London: John Wiley.

Spears, R., and M. Lea. 1992. Social influence and the influence of the social in computer-mediated communication. In *Contexts of computer-mediated communication*, edited by M. Lea. Hemel Hempstead, UK: Harvester Wheatsheaf.

Sproull, L., and S. Kiesler. 1991. *Connections: New ways of working in the networked organization*. Cambridge: MIT Press.

Tanis, M., and T. Postmes. In press. Social cues and impression formation in CMC. *Journal of Communication*.

Ting-Toomey, S. 1988. Social categorization, social identity, and social comparison. In *Differentiation between social groups*, edited by H. Tajfel. London: Academic Press.

Trevino, L. K., R. L. Daft, and R. H. Lengel. 1990. Understanding manager's media choices: A symbolic interactionist perspective. In *Organizations and communication technology*, edited by J. Fulk and C. W. Steinfield. Newbury Park, Calif.: Sage.

Triandis, H. C. 1989. The self and social behavior in differing cultural contexts. *Psychological Review* 96:506–520.

Van den Hooff, B., and J. Vijvers. 2001. Tussen initiatie en integratie: de ontwikkeling en toepassing van een kennismanagementscan (Between initiation and integration: The development and application of a knowledge management scan). *Tijdschrift voor Communicatiewetenschap (Journal of Communication Science)* 29:288–308.

Van den Hooff, B., J. Vijvers, and J. de Ridder. 2002. Knowing what to manage: The development and application of a knowledge management scan. In *Proceedings of the third European conference on organizational knowledge, learning, and capabilities*. Athens, Greece: ALBA.

Von Hippel, E. 2001. Open source shows the way: Innovation by and for users—no manufacturer required! Retrieved April 2002, <http://opensource.mit.edu/papers/evhippel-osuserinnovation.pdf>.

Von Hippel, E., and G. Von Krogh. In press. Open source software and the private-collective innovation model: Issues for organization science. In *Organization science*. Retrieved February 2003, <http://web.mit.edu/evhippel/www/PrivateCollectiveWP.pdf>.

Wagner, J. 1995. Studies of individualism-collectivism: Effects on cooperation in groups. *Academy of Management Journal* 1:152–172.

Wagner, J., and M. K. Moch. 1986. Individualism-collectivism: Concept and measure. *Group and Organization Studies* 11:280–303.

Walther, J. B. 1992. Interpersonal effects in computer-mediated communication: A relational perspective. *Communication Research* 19:52–90.

Walther, J. B. 1996. Computer-mediated communication: Impersonal, interpersonal, and hyperpersonal interaction. *Communication Research* 23, no. 1:3–43.

Walther, J. B., and J. K. Burgoon. 1992. Relational communication in computer-mediated interaction. *Human Communication Research* 21:460–487.

Weggeman, M. 2000. *Kennismanagement: de praktijk (Knowledge management: Practice)*. Schiedam, Netherlands: Scriptum Management.

Zhao, J. L., A. Kumar, and E. A. Stohr. 2000. Workflow-centric information distribution through e-mail. *Journal of Management Information Systems* 17, no. 3:45–72.

8
Design Requirements for Knowledge-Sharing Tools: A Need for Social Capital Analysis

Marleen Huysman

Experiences with tools for knowledge-sharing purposes have demonstrated the negative consequences of using a managerial and technology-oriented perspective (Huysman and de Wit 2002). Most tools usually in the form of intranet applications fail the test of institutionalization. The assumption underlying this chapter is that ignoring the informal, non-canonical nature of the social networks—including people's motivation, ability, and opportunity to share knowledge—is one of the key causes of resistance to using knowledge-sharing tools. In order to improve IT-based knowledge sharing, electronic networks need to be embedded in the social networks of which they are a part. This has implications for our knowledge on the design requirements of such socially embedded network tools.

In this chapter, the conditions that influence the match between social networks such as communities and electronic networks such as groupware tools are analyzed. Ideas on social capital are used as a framework to guide the knowledge-sharing requirement analysis. This chapter starts with a discussion of the fallacies of the first generation of knowledge management. We will argue that one of the main shortcomings of the first generation is the ignorance and mismatch between informal knowledge-sharing groups, on the one hand, and IT tools to support this knowledge sharing, on the other. The chapter then explores if and how the concept of social capital can be used to improve this match by introducing knowledge management tools that are more socially embedded within the social system of which they form a part.

Knowledge Management Fallacies

Knowledge management has become a key concern to organizations, not only because of the growing importance of the value of knowledge work and the increase in IT possibilities but also because of the increasing complexity of work and the speed at which change takes place. Changing organizational boundaries and identities, as well as the growth of virtual organizations, e-lancers, teleworkers, and geographically dispersed teams, increase the difficulty of monitoring and controlling knowledge. As a result, organizations, especially larger ones, try hard to manage knowledge. This urge to get a grip on knowledge and leverage it to create competitive advantage is understandable, and constitutes a critical topic for discussion among practitioners and theorists alike. The currently accepted or most generally talked about and practiced way of coping with knowledge, however, meets with resistance at various levels.

A research project that studied the experiences of several larger companies with knowledge management came to the conclusion that most initiatives are confronted with three fallacies or traps (Huysman and de Wit 2002). These fallacies relate to the tendency of organizations to concentrate too much on:

- the role of IT in facilitating knowledge sharing—the *IT trap*;
- imposing managerial needs on knowledge sharing—the *management trap*; and
- individual learning as the purpose of knowledge sharing—the *individual learning trap*.

The IT Trap

Many knowledge management projects have their origins in the IT world. With the rise of advanced technology, opportunities to facilitate knowledge sharing within organizations are on the rise. Lately, such an IT-oriented approach has received more criticism. Science and consultancy both embrace the view that if the emphasis is placed on technology, knowledge sharing will be threatened. At the same time, there is an overall consensus that technology can fill an important role here. Certainly in practice, most knowledge management initiatives are (still) technology driven (Huysman and de Wit 2002).

The trap lurks particularly in the assumption that IT can positively support and improve knowledge sharing while ignoring the social conditions that trigger or hinder the sharing of knowledge among people. The tendency to perceive IT as independent from the social environment of which it is a part has caused disappointing acceptance rates. It is not the technology itself but the way people use it that determines whether it will be utilized. Or in the words of Zack and McKenny, "The strategic advantage associated with these technologies will not derive from having the technical skills to evaluate and implement communication technologies (or even be first mover), but rather will come from having the appropriate social context, norms, politics, reward systems and leadership to take advantage of electronic communication technology" (2000, 212).

This tendency to de-contextualize IT also feeds into the assumption that all knowledge can be transformed into data and stored in systems. Echoes from the legacy of the information management era resound here. When knowledge is saved in a system, it becomes transformed into explicit, codified knowledge (Zack 1999).[1] Knowledge usually has a large tacit dimension and appears to be difficult to share with others, let alone store in a system. In the case of implicit knowledge, the human being is both the knowledge carrier and the vehicle through which the knowledge is passed on. Next to the producer, the potential consumer of knowledge might also create pockets of resistance to using IT tools for knowledge sharing. Knowledge only has meaning if it can be related to people (Brown and Duguid 2000). People want to know from whom they learn as this provides important "metaknowledge." This is one of the reasons why recorded knowledge is often not reused (Huysman and de Wit 2002). Aids to knowledge sharing such as intranets and knowledge bases that are geared toward codifying knowledge are not effective enough. When sharing experiences, people prefer to look for support from personal networks rather than electronic ones. Metaknowledge cannot be recorded in technical networks, and it requires the support of social, personal networks. In that case, tacit knowledge does not need to be transformed into explicit knowledge in order to share it with others. What seems more promising is the support of social networks and knowledge connections to enable transfer (Huysman and de Wit 2002; Leonard and Sensiper 1998, Davenport and Prusak 1998).

The Management Trap
Another fallacy of many knowledge management initiatives is caused by the dominance of a management bias (Huysman and de Wit 2002). Most initiatives are born out of a managerial need to control and monitor knowledge that is, in some places, perceived by management as too sticky or, in other places, too leaky. The increase of workers' expertise, mobility, professionalism, and so forth stimulates managers to think of ways to extract as well as collect people's knowledge and make it accessible to others. The need to improve the efficiency of knowledge sharing is further stimulated by the growing awareness of the financial-economic importance of knowledge. Many publications illustrate that a firm's intellectual capital is usually worth much more than its intrinsic value (Bontis 1999; Edvinsson and Malone 1997; Roos et al. 1998; Stewart 1997; Sveiby 1997). Top management and shareholders are becoming increasingly aware that core competences and competitive ability are embedded in the knowledge of an organization's members. The urge to manage knowledge is also heightened by the perceived opportunities that technologies can offer to support knowledge sharing.

This management attention to knowledge sharing is, on the one hand, a positive sign as it expresses managerial support for and confidence in initiatives. The downside, however, is that most organizations tend to ignore the workers' added value of sharing knowledge. A basic question is often overlooked—namely, "Why would people share knowledge in the first place?" Clearly, this tendency is risky as management fully depends on the active involvement of workers to share their knowledge. For successful knowledge management, an initiative must be beneficial to both the organization and the knowledge worker.

The Individual Learning Trap
A third risk that the study encountered is the tendency to perceive knowledge sharing as an activity to support learning at the individual level. Most initiatives are focused on providing the individual with knowledge by collecting it from other individuals in an organization. In order to

enhance this support of knowledge collection and provision, repository systems such as knowledge-based systems are introduced. Besides the problems mentioned above, tools to capture and disseminate knowledge are mainly helpful in supporting the exchange of individual rather than social or community knowledge (Zack 1999).

Next to these tools, organizations introduce intranets and other groupware applications with the intention of supporting the flow of knowledge among individuals. These systems, however, mainly contribute to individual knowledge as opposed to organizational knowledge or learning (Newell, Swan, and Scarbrough 2001; Huysman and de Wit 2002). To support collective learning such as community learning, knowledge management initiatives need to acknowledge that most knowledge has a socially situated nature and cannot be uncoupled from the social community of which it is a part.

Despite the fact that most knowledge management initiatives are confronted with problems of implementation, supporting knowledge sharing with IT remains necessary and will only become more important with the increase in people working over various distances. Nevertheless, IT for the support of knowledge sharing needs to be based on different assumptions than in the past.

If we take the legacies of the first generation of knowledge management seriously, the design of knowledge-sharing tools must take the following into account:

• Avoiding the IT trap implies that "information systems aimed at knowledge management need to maintain the integrity of the social communities in which knowledge is embedded" (Boland and Tenkasi 1995, 359). This requires the design and introduction of socially embedded electronic networks.

• Avoiding the management trap means steering clear of systems that suggest imposing knowledge sharing rather than supporting people's need to share knowledge. This implies an in-depth analysis of the social nature of the knowledge-sharing community in order to understand how and why community members share knowledge.

• Avoiding the individual learning trap implies that tools need to support social relationships and connections among people rather than introducing knowledge repository systems that offer individuals the results of knowledge sharing (data extracted from knowledge).

In the rest of this chapter, we will explore what this implies for the design requirements of electronic networks. To do this, we first review the literature on tools for knowledge sharing, on the sociotechnique, and on the requirements for IT design. The chapter will then discuss the potentials of using a requirement analysis based on social capital framework.

Literature

When we refer to *electronic networks, e-networks for knowledge-sharing purposes*, or simply *knowledge-sharing tools*, we think of electronic applications that organizations use to support knowledge sharing and that connect organizational members through network technology. The most frequently used e-networks are intranets, Web-based groupware, and Lotus Notes.

Various user services facilitated by electronic networks have been proposed,[2] such as communication, collaboration, navigation, and access to information (Boettcher 1998); information provision, supporting communication and creating community awareness (Stenmark 2002); or communication, content, and collaboration (Choo, Detlor, and Turnbull 2000).[2]

Choo, Detlor, and Turnbull propose an intranet as a shared information work space: a content space to facilitate information access and retrieval; a communication space to negotiate collective interpretations and shared meanings; and a collaboration space to support cooperative work action. The latter function is similar to groupware systems. With communication facilities, organizational members can share knowledge, exchange thoughts, and brainstorm new ideas. The media used can vary from audio, to text, to video, and can be synchronous (e.g., video conferencing) and asynchronous (e.g., e-mail). With information or knowledge content reference is made to knowledge bases, or repository systems containing information about knowledge possessed by one member or a group of members. Many companies try to use e-networks to increase the sharing of experiences (e.g., reports) or best practices. Reference systems such as yellow pages referring to people having certain expertise also fall under the heading of content applications. A third functionality of most e-networks is collaboration. Some e-networks take the form of groupware systems and support the work processes of group members—for example, through work-flow applications.

With the development of network technology along with the requirement for second-generation knowledge management as discussed in the previous sections, two formerly distinct fields of interest, social networks and electronic networks, increasingly need to converge. Some authors have argued as well as empirically illustrated that electronic networks such as intranets cannot survive without a corresponding and coexisting social network (e.g., Huysman and de Wit 2002; Blanchard and Horan 1998; Newell, Swan, and Scarbrough 2000).[3]

Converging the social with the technical is the domain of sociotechnique. Within the field of IT research, sociotechnical studies usually focus on the continuous interactions among IT, people, the organizational context, as well as the negotiation taking place during the design, implementation, and use of IT. The theory is broad and covers various research approaches, such as actor network theory (e.g., Walsham 1997), social contructivism (Bijker, Hughes, and Pinch 1987; MacKenzie and Wajcman 1999), hermeneutics (e.g., Boland 1991), structuration theory (e.g., Orlikowski and Robey 1991), activity theory (e.g., Kuutti, 1991), and adaptive structuration theory (De Sanctis and Poole 1994).

Research has focused, for example, on the interrelation between the social and technical components of e-networks: the group versus the software. Because e-network applications usually have a high degree of interpretive flexibility (Bijker, Hughes, and Pinch 1987), researchers agree that the design, implementation, and use are subject to a continuous sociotechnical negotiation in order to appropriate the technology for personal needs. Ciborra (1996) refers to "drifting," the tendency of network technology to follow its own path over time as a result of unplanned and context-based usage. These ideas of drifting and appropriation already stress the importance of taking the functional requirements of the group seriously in designing network systems. Ciborra pleads for openness, care taking, and hospitality when introducing groupware.

Several authors advocate a more anthropological or ethnographic perspective on the design and use of such network tools (Brown and Duguid

2000). In fact, perhaps the most common methodology used in computer-supported cooperative work derives from ethnography, building on the notion that work environments are idiosyncratic group cultures (e.g., Jordan 1996). The aim of ethnographic studies is to carry out detailed observations of work processes in their natural settings. Yet there seems to be a gap between the empirical findings and the corresponding design constraints and requirements of the application. One reason might be that people who conduct ethnographic research usually do not develop applications (Ehrlich and Cash 1994). Translating the inherently descriptive nature of this research into technical requirements is complicated. Several techniques have been offered to circumvent these complicated issues, such as the use of videos and photos, checklists, and multidisciplinary design teams (see Ehrlich 2000). Although highly valuable, effective means of translating ethnographic studies into design requirements are not enough to match knowledge-sharing tools with social networks. It seems that present sociotechnical models used to study the interplay between IT and the social system are not yet appropriately adapted to the special requirements of electronic knowledge-sharing tools.

The need to include social requirement analysis for electronic tools to support knowledge sharing has been recently brought to bear on the concept of *info-culture analysis*, as first introduced by Bressand and Distler (1995). Some researchers have argued that the disappointing results of knowledge-sharing tools such as intranets is due to the fact that designers traditionally analyze the infrastructure and info-structure, but neglect the underlying info-culture (Choo, Detlor, and Turnbull 2000; Newell, Swan, and Scarbrough 2000; Ciborra 1996). The infrastructure relates to the hardware/software that enables the physical/communicational contact among network members. The info-structure relates to the formal rules governing the exchange among actors in the network. The info-culture relates to the stock of background knowledge actors take for granted, and it is embedded in the social relationships surrounding work-group processes.

Designing networks by analyzing the infrastructure only would result in a technology-driven implementation of these networks. The limitation of this technology-driven approach was accepted years ago. Infrastructure analysis has been succeeded by an approach that also looks at the info-structure. In terms of knowledge sharing and knowledge-based systems or networks, info-structure analysis implies examining, for example, formal business processes, hierarchies, coordination rules, and knowledge-sharing strategies (Choo, Detlor, and Turnbull 2000). Info-structure analysis relates to canonical knowledge-sharing processes. Without being more specific, various authors stress the need for an additional info-cultural analysis when designing knowledge networks like intranets (Choo, Detlor, and Turnbull 2000; Ciborra 1996; Newel, Swan, and Scarbrough 2000; Kumar, van Dissel, and Bielli 2000).

Including an analysis of the info-culture of a social group corresponds to what Kumar, van Dissel, and Bielli (2000) refer to as the third rationality of IT. Their research on the merchants of Prato inspired them to argue that traditional IT development approaches need to be augmented with additional strategies that as a precursor to development, examine the existing patterns of culture, relationships, and trust (or distrust) in the development situation, and take them into account for devising a development and implementation strategy. This third rationality introduces trust, social capital, and collaborative relationships as key concepts: "It is our belief that this third rationality will not only provide yet another way of examining the role of IT in organizations (different from rationalism and pluralism), it will provide guidance for developing a new generation of collaborative technologies for an increasingly interdependent, multi-cultural, globalized world" (Kumar, van Dissel, and Bielli 2000, 222).

The literature on IT and/or knowledge management has not yet analyzed in more detail what it means to include these various layers of development in the knowledge requirement analysis of knowledge management tools. Below, we will explore the potentials of using the ideas taken from the field of social capital to improve knowledge requirement analysis.

The Concept of Social Capital

Several authors have pointed to the importance of acknowledging social capital when investing in knowledge management (e.g., Lesser 2000) as well as the knowledge benefits derived from high levels of social capital

(Cohen and Prusak; Nahapiet and Ghoshal 1998). We believe that emphasizing social capital as the key ingredient in knowledge sharing not only relaxes the individual learning bias but also the managerial and technological ones. By scrutinizing communities' degree of social capital and improving the level of social capital, tools for knowledge sharing will likely be more in line with people's opportunity, motivation, and ability to share knowledge. People will be more inclined to use electronic networks if they are motivated and able to share knowledge with others, and if they have the opportunity to do so (Adler and Kwon 2002). These three aspects of knowledge sharing are considered essential for social capital (Adler and Kwon 2002; Nahapiet and Goshal 1998).

As usual with emerging new concepts, every contribution to this growing literature on social capital seems to use its own definition. In general, *social capital* refers to those stocks of social trust, norms (mutual reciprocity), and networks that people can draw on to share knowledge. The notion of social capital should be seen as an additional aspect of the already-well-known economic conditions or elements that make up organizational capital: physical capital, financial capital, and human capital. Whereas human capital refers to individual ability (Becker 1964), social capital refers to collective abilities derived from social networks. The "traditional" types of capital only partially determine the process of economic growth, and overlook the way in which economic actors interact and organize themselves to potentially generate growth and development. It is becoming increasingly common to view social capital as the missing link.

A focus on social capital in terms of knowledge sharing shifts the attention from individuals sharing knowledge to communities as knowledge-sharing entities. Communities and social networks are seen as the prime sources of a sense of membership as well as commitment, of mutuality as well as and trust, and of the places in organizations where people feel most at home and responsible for one another (Wenger 1998). Investing in social capital thus implies a more significant role for communities of practice (e.g., Snyder and Wenger 2000). In communities, people not only invest in their own learning but also in the learning of others. Communities are the main places where knowledge develops. The driving forces within communities and the key conditions that help communities stay active are mutual trust, a sense of mutuality, and recognition by one's peers (Lesser 2000); in other words, a high degree of social capital.

Nahapiet and Ghoshal (1998) introduce three dimensions of social capital: structural (network ties, network configurations and organization), cognitive (shared codes and language, shared narratives), and relational (trust, norms, obligations, identification). Studying the degree of social capital requires an analysis of the existing social networks and corresponding ties (a structural analysis); the existing shared language, frames of meaning, and stories (a cognitive analysis); and the existing level of trust and reciprocity (a relational analysis). For example, if a social system scores low, as it were, on the structural dimension, this would mean that its members lack the opportunity to share knowledge through network ties. If the group scores low on the cognitive dimension, this means that the group lacks a shared cognitive frame, and consequently is not able to understand and develop new ideas with each other. This becomes even more problematic when this group scores low on the relational dimension; there is no motivation to share knowledge because of a lack of trust and norms of reciprocity.

In their review article, Adler and Kwon (2002) also introduce a three-dimensional framework, in which they use the classification of opportunity, ability, and motivation. Given the similarity with Nahapiet and Ghoshal's classification, it is striking to note that Adler and Kwon do not refer to this often-cited article. In table 8.1, we have tried to connect the two classifications with each other.

Both the opportunity and structural dimensions refer to an analysis of "who" shares knowledge and "how" they do that. Both dimensions concern the existing opportunity, or the lack of one, to connect with each other, which can be analyzed on the infrastructure level. We combined the two dimensions into the *structural opportunity* dimension. Both the cognitive and ability dimensions correspond in combination to an analysis about "what" is shared. This *cognitive ability* dimension concerns the ability to cognitively connect with each other in order to understand what the other is referring to when communicating. Analyzing the infostructure will provide information about this cognitive ability. The relational and motivational dimensions both refer to the questions of "why" and "when" people share knowledge. Both dimensions concern the

Table 8.1
Conditions for knowledge sharing and knowledge requirement analysis

Knowledge-sharing research questions	Who shares knowledge, and how is knowledge shared?	What knowledge is shared?	Why and when is knowledge shared?
Dimensions (Nahapiet and Ghoshal 1998)	Structural dimension	Cognitive dimension	Relational dimension
Social capital sources (Adler and Kwon 2002)	Opportunity	Ability	Motivation
Content	Network ties, configuration, organization	Shared codes, language, stories	Trust, norms, obligation, identification, respect, generalized reciprocity
Layers of requirement analysis	Infrastructure	Info-structure	Info-culture
Conditions for knowledge sharing	Structural opportunity to share knowledge	Cognitive ability to share knowledge	Relation-based motivation to share knowledge

motivation to share knowledge based on the socially attributed characteristics of the relationship, such as trust, mutual respect, and generalized reciprocity (Putnam 2000). Analysis of the info-culture of the system will provide more insight into this *relation-based motivation* to share knowledge.

Social Capital Analysis as Part of Design

We assume that an analysis of the conditions for knowledge sharing based on a social capital analysis as depicted in table 8.1 provides more in-depth insight into the functionalities of tools for knowledge sharing and its embeddedness within the social network. We argue that the analysis of social capital by means of the conceptual framework (table 8.1) need to be included when designing tools that are intended to be used for knowledge-sharing purposes.

The structural dimension of social capital focuses mainly on the density of networks and bridging structural holes (Wasserman and Faust 1994; Burt 1992). The density of a network refers to the extent to which network actors are interconnected. Studying the density of a social network would reveal with whom people share knowledge. Next to who communicates with who, attention is also paid to the question of how they do that. Connecting to people in order to share knowledge brings an instrumental perspective to the fore. Different network tools exist that support people's opportunity to connect with each other. Next to these tools, various applications exist to analyze and map the structural dimension of knowledge sharing (see, e.g., Fesenmaier and Contractor 2001). Such an analysis of the structural opportunity of a social network to share knowledge will not only reveal the existing flows of knowledge (who is connected to who) but also the methods and tools used by the group to connect to each other. This "who" and "how" analysis forms an important part in surfacing the design requirement of knowledge-sharing tools. Yet analyzing the structural opportunity dimension, for example, by existing network tools only informs us about the structural embeddedness of the system.

The higher a social group's cognitive dimension, the more the members are able to share (tacit) knowledge. A group's cognitive ability depends on its ability to understand each group member. This cognitive ability can be analyzed by focusing on shared stories, language, communication regimes (Orlikowski and Yates 1994), and so forth. The social capital's cognitive dimension may enable knowledge sharing in the sense that stories, shared language, customs, and traditions can bridge the tacit-explicit division as well as divisions such as, say, old-timers–newcomers (Hinds and Pfeffer 2002).

The cognitive dimension of knowledge sharing has received less attention relative to the other two dimensions. Cognitive barriers to sharing knowledge highly influence the use of electronic networks. A cognitive barrier is, for example, the difficulty of bridging the distance between expert and novice, or the difficulty of expressing the tacit dimension of knowledge (Hinds and Pfeffer 2002). The assumption is that the higher the cognitive barrier, the more people will rely on social or personal networks instead of electronic ones (Leonard and Sensiper 1998). For instance, the use of expertise requires validation, and validation requires contacting the person. Likewise, the tacit dimension of knowledge requires access to a social network to transfer the tacit part of the knowledge (Brown and Duguid 2000).

In the case of designing and implementing a content space within e-networks, an analysis of knowledge requirements is needed. This implies an in-depth analysis of the knowledge needs and requirements of community members. Electronic networks are usually designed based on the espoused information needs of formal groups—teams within the canonical hierarchy of the organization. This would, for example, point to the need to access knowledge about the organization, customers, products, each other's experiences, and best practices. Frequently used methodologies are questionnaires and interviews. The methodology for gaining more in-depth insight also takes into account the theories-in-use and situated cultural nature of knowledge. We could think of the methodologies used within cultural studies such as ethnography, narrative, and pattern recognition and matching to support the reflectivity of a community of practice (Lanzara 1983).

If a network has a high relation-based motivation, this implies that members are intrinsically motivated to share knowledge with each other because of their willingness to contribute to the relationship. Shared norms, a sense of mutual trust and respect, and reciprocity stimulate people to share knowledge with each other (assuming they have the cognitive ability and structural opportunity to do so). Analyzing the relation-based motivation to share knowledge addresses the questions of why and when people share knowledge. As mentioned at the beginning of this chapter, a lot of knowledge management initiatives fail the test of institutionalization because they ignore the question, What is in it for the knowledge worker? Therefore, in analyzing tools to support knowledge sharing, one should take seriously the potential motivational barriers to share knowledge. Motivational barriers are not only problems due to a lack of personal benefits but they can also result from, for example, status differences, lack of trust, lack of perceived reciprocity, and lack of respect (e.g., Hinds and Pfeffer 2002; Huysman and de Wit 2002).

Trust is an important aspect of social capital. It is generally accepted that mutual trust positively influences the possibility of knowledge transfer (e.g., Dodgson 1993). Trust is needed to safeguard against opportunism and the obstruction of knowledge sharing (Szulanski 1996). Trust is also needed because a large dimension of the knowledge that is to be shared is of a tacit nature. People high on the relational dimension share a sense of mutuality, meaning that people not only want to learn but also want to help others learn. In turn, these people are more likely to contribute their knowledge to the electronic knowledge network or repository (Cohen and Prusak 2001). In contrast to the structural aspects of networks that address the density of ties, the relational elements refer to the "strength of ties" (Granovetter 1985). An analysis of the strength of ties offers insight into the strategies people employ to share knowledge (Hanson 1999). Strong ties are important for the exchange of tacit knowledge while weak ties are crucial for the sharing of explicit knowledge. Strong ties imply a high degree of trust, which makes the entire process flow more smoothly.

Because of the delicate nature of the topic, interviews and questionnaires would probably only reveal the espoused theories. Therefore, ethnomethodological studies of the knowledge-sharing culture of the social group are best suited to reveal the motivations of people to contribute to the relationship.

The literature on the appropriation of IT tells us that the design of a system—especially group-based systems—doesn't say much about its actual use. The structural properties, relation-based motivations, and cognitive abilities of the network will continuously change, thereby also changing the use of technology. Although the assumption is that a close connection with the social network will ensure a continuous perceived usefulness by the members of the social network, some organizational structural redesign is needed to support the continuous natural flow of knowledge between the two systems. One valuable strategy to do so is to institutionalize the activities of the knowledge brokers in the organization. Case-based research (e.g., Hargadon 1998; Wenger 1998) has illustrated that some organizations introduce knowledge brokers whose main role is to bridge various social communities. Various knowledge-broker roles can be identified. Wenger (1998) identifies "boundary spanners," who take care of one specific boundary of a knowledge community; "roamers," who go from place to place, creating an informal web of connections; and "outposts," who bring back news from the front and explore new territories. These types of knowledge brokers are believed to connect various communities. Strikingly less attention is given to knowledge brokers whose main role is to connect the social and technical, and sustain this connection. Research on knowledge sharing in practice has revealed different types of such "network brokers," such as "reviewers" of the content of the knowledge base, "boosters" whose main role is to get people connected and contributing, a "commuter" who runs between the front office where the users of the e-network are located and the back office where content producers operate, and "experts" who are used to gain knowledge beyond what is in the repository (Huysman and de Wit 2002). It is interesting to note that most of these roles were introduced by the broker him- or herself out of a need to fill a gap between the two networks. Most organizations were not aware of these noncanonical roles.

Clearly, more research is needed to explore and study the effect of organizational design requirements, such as the various roles of network brokers in sustaining a connection between the social and technical systems.

Concluding Remarks

This chapter can be seen as a call for the inclusion of sociotechnical principles in the design of knowledge-sharing tools. Whereas most literature on sociotechnique refers to the need to address the social when designing technology, this chapter goes one step further and argues that—at least in the case of knowledge-sharing tools—technology needs to be embedded in the social structure. The need for socially embedded e-networks has not yet been collectively accepted. Some researchers try to reveal the interdependency of social and electronic networks by, for example, concentrating on the sociology of networking (e.g., Wellman 2001) using a practice lens when studying the use of networks

(Orlikowski 2000), introducing a value-added model for design (Choo, Detlor, and Turnbull 2000), analyzing the information ecology of the organization before introducing intranets (Davenport 1997), or stressing the process of appropriation and drifting (Ciborra 1996; De Sanctis and Poole 1994). These researchers and others argue that the acceptance and use of e-networks highly depend on the extent to which the social aspects are taken into account. We agree with the sociotechnical approach, but think it needs to go further. Due to its dependency on a social network, the requirement analysis of knowledge-sharing tools needs to take into account the degree of social capital in a community. In other words, when designing IT tolls, it is important to consider if and how these tools support the structural opportunities, relation-based motivation, and cognitive ability to share knowledge. As noted earlier, people will be more inclined to use electronic networks if they are motivated and able to share knowledge with others, and if they have the opportunity to do so (Adler and Kwon 2002). Therefore, the higher the degree of social capital among people who share knowledge, the more an electronic network is used for that purpose.

Clearly, more empirical research is needed on the potentials of introducing social capital analysis into the design requirement for knowledge-sharing tools. Such empirical research needs to test the assumption that social capital analysis results in improving the social embeddedness of electronic networks.

Notes

1. We use Michael Polanyi's (1966) assumption that all knowledge has a tacit dimension, and that it exists on a spectrum. "At one extreme it is almost completely tacit, that is, semiconscious and unconscious knowledge held in peoples' head and bodies. At the other end of the spectrum, knowledge is almost completely explicit, or codified, structured, and accessible to people other than the individuals originating it. Most knowledge, of course, exists in between the extremes" (Leonard and Sensiper 1998).

2. Although these authors refer to intranets, electronic networks in general can be perceived of as providing these user services.

3. It might well be that this is a rather general observation, applicable to a large variety of electronic networks within various fields and disciplines, and at various levels, such as business (e.g., e-business and electronic data interchange),

consumers (e.g., e-commerce and virtual communities), and students (e.g., e-learning and communities of interest). The present chapter focuses on the relationship between electronic knowledge networks, such as intranets, and on personal networks, such as communities of practice within and across organizations, for the purpose of organizational learning.

References

Adler, P. S., and S. Kwon. 2002. Social capital: Prospects for a new concept. *Academy of Management Review* 27, no. 1:17–40.

Becker, G. 1964. *Human capital: A theoretical and empirical analysis with special reference to education.* New York: Columbia University Press.

Bijker, W. E., T. P. Hughes, and T. J. Pinch, eds. 1987. *The social construction of technological systems.* Cambridge: MIT Press.

Blanchard, A., and T. Horan. 1998. Virtual communities and social capital. *Social Science Review* 16, no. 3:293–307.

Boettcher, S. 1998. *The Netscape intranet solution, deploying a full-service intranet.* New York: Wiley Computer Publications.

Boland, R. J. 1985. Phenomenology: A preferred approach to research on information systems. In *Research methods in information systems*, edited by E. Mumford, R. Hirschheim, G. Fitzgerald, and A. T. Wood-Harper. Amsterdam: Elsevier Science.

Boland, R. J. 1991. Information system use as a hermeneutic process. In *Information systems research: contemporary approaches and emergent traditions*, edited by H.-E. Nissen, H. K. Klein, and R. Hirschheim. New York: North-Holland.

Boland, R. J., and R. V. Tenkasi. 1995. Perspective making and perspective taking in communities of knowing. *Organization Science* 6, no. 4:350–372.

Bontis, N. 1999. Managing organizational knowledge by diagnosing intellectual capital: Framing and advancing the state of the field. *International Journal of Technology Management* 18, 433–462.

Bressand, A., and C. Distler. 1995. *La planete relationelle.* Paris: Flammarion.

Brown, J. S., and P. Duguid. 2000. *The social life of information.* Cambridge: Harvard Business School Press.

Burt, R. S. 1992. *Structural holes: The social structure of competition.* Cambridge: Harvard University Press.

Choo, C. W., B. Detlor, and D. Turnbull. 2000. *Webwork, information seeking, and knowledge work on the World Wide Web.* Dordrecht: Kluwer Academic Publishers.

Ciborra, C. U., ed. 1996. *Groupware and teamwork.* New York: John Wiley and Sons.

Cohen, D., and L. Prusak. 2001. *In good company: How social capital makes organizations work*. Boston: Harvard Business School Press.

Davenport, T. H. 1997. *Information ecology*. New York: Oxford University Press.

Davenport, T. H., and L. Prusak. 1998. *Working knowledge: How organizations manage what they know*. Cambridge: Harvard Business School Press.

De Sanctis, G., and M. S. Poole. 1994. Capturing the complexity in advanced technology use: Adaptive structuration theory. *Organization Science* 5, no. 2:121–145.

Dodgson, M. 1993. Learning, trust, and technological collaboration. *Human Relations* 46:77–95.

Edvinsson, L., and M. S. Malone. 1997. *Intellectual capital: Realizing your company? true value by finding its hidden roots*. New York: Harper Business.

Ehrlich, K. 2000. Designing groupware applications. In *Knowledge, groupware, and the Internet*, edited by D. E. Smit. Boston: Butterworth-Heinemann.

Ehrlich, K., and D. Cash. 1994. Turning information into knowledge: Information finding as a collaborative activity. In *Proceedings of Digital Libraries 1994 Conference*. College Station, Texas.

Fesenmaier, J., and N. Contractor. 2001. Inquiring knowledge networks on the Web (IKNOW): The evolution of knowledge networks for rural development. *Journal of the Community Development Society* 32:160–175.

Granovetter, M. 1985. Economic action and social structure: The problem of embeddedness. *American Journal of Sociology* 91:481–510.

Hanson, M. T. 1999. "The Search-Transfer Problem: the role of weak ties in sharing knowledge across organization sub-units." *Administrative Science Quarterly*, pp. 82–111.

Hargadon, A. B. 1998. Firms as knowledge brokers: Lessons in pursuing continuous innovation. *California Management Review* 40, no. 3:209–227.

Hinds, P. J., and J. Pfeffer. 2002. Why organizations don't "know what they know": Cognitive and motivational factors affecting the transfer of expertise. In *Beyond knowledge management: Sharing expertise*, edited by M. Ackerman, V. Pipek, and V. Wulf. Cambridge: MIT Press.

Huysman, M. H., and D. de Wit. 2002. *Knowledge sharing in practice*. Dordrecht: Kluwer Academic Publishers.

Jordan, B. 1996. Ethnographic workplace studies and CSCW. *The design of computer supported cooperative work and groupware systems*, edited by D. Shapiro, M. Tauber, and R. Traunmuller. Human factors in information technology series 12. Amsterdam: Elsevier Science.

Kumar, K., H. G. van Dissel, and P. Bielli. 2000. The merchant of prato—revisited: Toward a third rationality of information systems. *MIS Quarterly* 22, no. 2:199–226.

Kuutti, K. 1991. Activity theory and its applications to information systems research and development. In *The information systems research arena of the '90s: Challenges, perceptions, and alternative approaches*, edited by H.-E. Nissen. Amsterdam: North-Holland.

Lanzara, G. F. 1983. The design process: Frames, metaphors, and games. In *Systems design for, with, and by the users*, edited by U. Briefs, C. Ciborra, and L. Schneider. New York: North-Holland Publishers.

Leonard, D., and S. Sensiper. 1998. The role of tacit knowledge in group innovation. *California Management Review* 40, no. 3:112–132.

Lesser, E. L., ed. 2000. *Knowledge and social capital: Foundations and applications*. Boston: Butterworth-Heinemann.

MacKenzie, D. A., and J. Wajcman, eds. 1999. *The social shaping of technology*. 2d ed. Buckingham: Open University Press.

Nahapiet, J., and S. Ghoshal. 1998. Social capital, intellectual capital, and the organizational advantage. *Academy of Management Review* 23, no. 2:242–266.

Newell, S., J. Swan, and H. Scarbrough. 2001. From global knowledge management to internal electronic fences: Contradictory outcomes of intranet development. *British Journal of Management*, 12, no. 2:97–111.

Orlikowski, W. J. 2000. Using technology and constituting structures: A practice lens for studying technology in organizations. *Organization Science* 11, no. 4:404–428.

Orlikowski, W. J., and D. Robey. 1991. Information technology and the structuring of organizations. *Information Systems Research* 2:143–169.

Orlikowski, W. J., and J. Yates. 1994. Genre repertoire: The structuring of communicative practices in organizations. *Administrative Science Quarterly* 39:541–574.

Polanyi, M. 1966. *The tacit dimension*. New York: Doubleday.

Putnam, R. 2000. *Bowling alone: The collapse and revival of American community*. New York: Simon and Schuster.

Roos, J., G. Roos, N. Dragonetti, and L. Edvinsson. 1998. *Intellectual capital: Navigating in the new business landscape*. New York: New York University Press.

Snyder, W., and E. Wenger. 2000. Communities of practice: The organizational frontier. *Harvard Business Review* Jan–Febr: 139–145.

Stenmark, D. 2002. Information vs. knowledge: The role of intranets in knowledge management. In *Proceedings of Hawaii International Conference on System Sciences–35*, 7–10 January, Hawaii.

Stewart, T. A. 1997. *Intellectual capital: The new wealth of organizations*. New York: Doubleday.

Sveiby, K. E. 1997. *The new organizational wealth: Managing and measuring knowledge based assets*. London: Berrett-Koehler.

Szulanski, G. 1996. Exploring internal stickiness: Impediments to the transfer of best practice within the firm. *Strategic Management Journal* 17:27–43.

Walsham, G. 1997. Actor network theory and IS research: Current status and future prospects. In *Information systems and qualitative research*, edited by A. S. Lee, J. Liebenau, and J. I. DeGross. London: Chapman and Hall.

Wasserman, S., and K. Faust. 1994. *Social network analysis: Methods and applications*. Cambridge: Cambridge University Press.

Wellman, B. 2000. Computer networks as social networks. *Science* 293, no. 14:2031–2034.

Wenger, E. 1998. *Communities of practice*. New York: Cambridge University Press.

Zack, M. H. 1999. Managing codified knowledge. *Sloan Management Review* 40, no. 4 (summer): 45–58.

Zack, M. H., and J. L. McKenny. 1995. "Social Context and Interaction in Ongoing Computer-Supported Management Groups", in Fulk, J. and G. De Sanctis "focused Issue on Electronic Communication and Changing Organizational Forms, Organizational Science, 6, no. 4:394–422.

Zack, M. H., and J. L. McKenny. 2000. Social context and interaction in ongoing computer-supported management groups. In *Knowledge, groupware, and the internet*, edited by D. E. Smith. Boston: Butterworth-Heinemann.

9
Explaining the Underutilization of Business-to-Business E-Commerce in Geographically Defined Business Clusters: The Role of Social Capital

Charles Steinfield

This chapter explores the intersection between two important trends related to the ambivalent relationship between IT and social capital. First, the growing use of the Internet to support business-to-business (B2B) trade has unquestionably focused attention on the opportunities for improved efficiencies in procurement processes (Segev, Gebauer, and Färber 1999; Kaplan and Sawhney 2000; Laudon and Traver 2001). With the stress on transaction efficiencies, the dominant theoretical paradigm guiding analyses of B2B electronic markets has been transaction-cost economics, highlighting the ability of such markets to reduce various search and monitoring costs for participating firms (Bakos 1997, 1998; Segev, Gebauer, and Färber 1999; Steinfield, Chan, and Kraut 2000; Garicano and Kaplan 2001). At the height of the dot-com euphoria, hundreds of third-party B2B marketplaces across many different industries were developed, establishing the B2B e-hub as one of the most prominent new business models in what has been called the "digital economy" (Timmer 1998). Nevertheless, despite widespread optimistic projections by industry consultants, academic analysts, and government policymakers (Katsaros, Shore, Leathern, and Clark 2000; Department of Commerce 2000), most third-party-provided B2B marketplaces have not met with much success, and many have failed entirely (Laudon and Traver 2001; Tedeschi 2001).

The second trend, generally ignored in the e-commerce literature, is the renewed fascination with the significance of business clusters in cities, regions, and nations as a critical facet of economic growth and vitality (Porter 1990, 1998, 2000; Breschi and Malerba 2001). Unlike most e-commerce research, which usually begins from the assumption that electronic networks make distance and physical location irrelevant (Cairncross 1997), those who study business clusters emphasize the crucial importance of proximity in encouraging knowledge sharing, reducing transaction costs, and stimulating innovation (Breschi and Malerba 2001). Indeed, explanations of successful local business clusters often focus on traditional social capital explanations, such as the importance of trust and social relationships, as a catalyst for knowledge sharing and innovation across firms that may not even be trading partners (Maskell 2001).

In this chapter, we bring these two disparate topics together, pointing out the fundamental disconnect that characterizes the current thinking. If geographically defined business clusters are of increasing significance, then electronic marketplaces that fail to take location and social relationships into account will be of little use in these contexts. In fact, research on electronic marketplaces has historically emphasized the need to free buyers and suppliers from the constraints of geography by enabling access to distant buyers and suppliers. This is hypothesized to occur because electronic networks reduce transaction costs that formerly served as a barrier to trade (Malone, Yates, and Benjamin 1987). Choice is broadened, the role for market governance is extended, and prices are lowered due to electronic markets. What, then, is the role of electronic commerce in such geographically defined business clusters? We suggest that there is ample evidence for the importance of location in electronic commerce (Steinfield and Klein 1999), and new work on electronic commerce communities explicitly examines the social elements of network marketplaces (Hummel and Lechner 2002). Since most B2B electronic markets ignore location and social capital as critical components of economic exchange, however, they will be underutilized in local business clusters. Rather, much as Kumar, van Dissel, and Bielli (1998) observed in their study of the Prato textile industry, members of local business clusters are likely to find little added value in transaction-oriented interorganizational systems, which may offer little improvement to the already-low transaction costs they face using interpersonal means of coordination.

The remainder of this chapter is organized as follows. The first section provides an overview of the literature on B2B marketplaces, noting key emphases and trends. The second section introduces perspectives on geographic business clusters, including economic and social capital views. The third section looks at the evidence to date on IT use and electronic markets associated with geographic business clusters. The fourth section explores emerging developments in the area of more socially aware e-commerce. Finally, this chapter concludes with some attention to the many research issues raised by the preceding discussions.

Business-to-Business Marketplaces on the Internet

E-commerce researchers generally expect the value of B2B electronic transactions to vastly exceed business-to-consumer (B2C) retail trade due to the enormous volume of goods and services traded among firms (Kaplan and Sawhney 2000; Subramami and Walden 2000; Garicano and Kaplan 2001; Laudon and Traver 2001). Laudon and Traver (2001), citing figures from a Jupiter Media Metrix report, estimated U.S. B2B trade at $12 trillion in 2001—a surprising figure in that it exceeds the estimated gross domestic product of the United States that year. The potential for even a small fraction of this trade to be conducted over the Internet has attracted hundreds of new B2B market entrants. As early as 2000, the U.S. Department of Commerce (2000) reported that more than 750 B2B e-markets were operating worldwide in a range of different industries.

Estimates of the scope of B2B trade are made difficult by definitional problems. Generally, B2B trade refers to all electronic trade between firms. Some researchers include electronic data interchange (EDI)-based transactions in their calculations, however, while others do not. Additionally, Subramami and Walden (2000) point out interesting definitional paradoxes. For example, when an employee at one company orders a book from Amazon.com, Subramimi and Walden do not consider it to be B2B e-commerce, despite the fact that the product is being transferred from one business to another. Rather, they suggest that B2B trade is a process involving the joint action of multiple firms.

B2B e-commerce typically is divided into two main categories based on the characteristics of the infrastructure and interfirm relationships of the participants: private industrial networks and Internet-based marketplaces (Laudon and Traver 2001). The control of the network and the extent to which it is biased or neutral are common distinctions (Steinfield, Kraut, and Plummer 1995). Early interorganizational ISs were often set up by large suppliers so that business customers could use terminals for their orders. Modern, Internet-based versions of these systems are simply B2B electronic catalogs or storefronts (Laudon and Traver 2001). These services are biased toward a single seller, and competitive strategy theorists note that such systems serve to increase buyer-switching costs and create "lock-in" (Bakos and Treacy 1986; Shapiro and Varian 1999). Early B2B electronic exchange was also often managed by a single buyer in an electronic data interchange (EDI) network or intranet organized in a hub-and-spoke structure. In these systems, powerful buyers required all of their suppliers to utilize the IS as a means of improving the buyers' procurement efficiencies. The threat of the loss of the buyers' business coerced suppliers to undertake the investment necessary to engage in electronic transactions.

The Internet is rapidly becoming the dominant platform for electronic B2B exchange. Third-party market makers have established a range of supposedly neutral (unbiased) markets offering many trade facilitation services to industry buyers and sellers. Moreover, single buyer and seller systems are evolving into private industry consortiums-based systems (Laudon and Traver 2001).

B2B Internet marketplaces have been classified according to two key dimensions of business purchasing: how businesses buy, and what businesses buy (Kaplan and Sawhney 2000). The "how" dimension distinguishes between spot purchasing to fill an immediate need, and systematic purchasing for planned, long-term needs. The former is often done using ephemeral, market-based transactions, without long-term contracts. The latter is frequently done after significant negotiation, and is generally used for purchasing in large volumes from trusted trading partners. The "what" dimension normally distinguishes between vertical (also called direct or manufacturing) inputs that relate to the core products of a firm and horizontal (often called indirect or MRO for maintenance, operating, and repair) inputs, such as office supplies, that are acquired by all firms. Laudon and Traver (2001) sketch out the following four types of Internet-based B2B marketplaces:

- E-distributors (supporting spot purchasing for horizontal inputs) such as Grainger.com or Staples.com offer electronic catalogs representing thousands of suppliers. Laudon and Traver (2001) call them the Amazon.com for industry since they operate much like retailers. The main benefit for buyers is simply the reduced search cost, although additional services like credit and account management are offered to help further reduce transaction costs.
- E-procurement services such as Ariba.com also offer MRO supplies, but focus on systematic rather than spot purchasing. Such B2B intermediaries provide a range of procurement services, including the licensing of procurement software that supports a variety of value-added services. They do not own the supplies but offer the catalogs of thousands of suppliers from whom they also obtain fees and commissions. They theoretically bring value by aggregating both buyers and sellers, decreasing search costs for both parties, and therefore are subject to significant positive network externalities (the more buyers they attract, the more suppliers will join, and vice versa).
- Exchanges such as ChemConnect are intermediaries that focus on bringing together buyers and sellers within a particular industry, and concentrate on the spot purchasing of manufacturing inputs. They charge commissions, but offer a range of purchasing services to buyers and sellers, supporting price negotiations, auctions, and other forms of bidding in addition to normal fixed-price selling. Buyers benefit from greater choice and lower prices, while sellers gain access to large numbers of buyers. These vertical markets are often used to unload surplus materials (for example, via auction). They are also subject to network externalities.
- Industry consortiums are best represented by Covisint, the electronic procurement system developed by the leading automobile manufacturers. These exchanges are typically jointly owned by large buying firms seeking to rely on electronic networks to support long-term relationships with their suppliers. Entrance is by invitation only, and the buying clout of the founders influences suppliers to make the investments needed to participate.

The high failure rate of third-party B2B e-hubs, coupled with the growth of industry consortiums, reflects an important dynamic. Businesses have established relations with their suppliers, and the trust engendered by reliable performance and commitment over the long run may be more valuable to firms than any short-term price advantages offered by the supposedly neutral marketplaces. Indeed, a new trend in the B2B electronic trade arena is the rise of "collaborative e-commerce" where networks are used for far more than simple transaction support. Joint product design, more tightly integrated inventory databases, and other forms of coordination between producers and suppliers occur over private intranets. In a sense, these developments are merely the latest manifestation of what Malone, Yates, and Benjamin (1987) refer to as electronic hierarchies, where firms rely on networks to facilitate outsourcing, but only to a small number of firms with which they are tightly integrated. Substantial empirical evidence exists suggesting that these interorganizational forms are more common and long lasting than the market exchanges (Steinfield, Kraut, and Plummer 1995; Kraut et al. 1998).

The rise of industry consortiums, and the growth of collaborative e-commerce indicate that social capital concepts do indeed have an important place in B2B electronic markets. The experience of B2B electronic marketplaces reflects a movement away from an emphasis on arm's-length transactions, to one where the networks are used to support existing relationships. Even in network marketplaces predicated on adding value through buyer and supplier aggregation, there is an emphasis on the role of social aspects and community. For example, several B2B e-hubs now offer support for community formation and communication in much more explicit ways than in the past. CommerceOne, for instance, advertises a "meeting support" service whereby members can engage in rich, multimedia interaction with each other. Still, many of the relational aspects of B2B e-hubs are necessitated precisely because of their virtual nature and the lack of social capital in the first place. These include:

- Emphasis on the value of communities from a network externalities perspective: That is, each new member adds new value for all members as a potential trading partner and source of market liquidity. This potentially creates a positive feedback loop, but fails to consider the potential damage that the introduction of new competition has to established partners. Also, given the downward pressures that new competitors impose on prices, it is not surprising that many suppliers shied away from joining supposedly neutral B2B exchanges.

• Provision of reputation systems: Because electronic exchanges bring together unfamiliar trading partners, feedback and ratings systems can be used as a proxy for prior experience to help guide buyers and sellers to trustworthy partners.

• Member qualification: In order to help reduce the potential for opportunistic behavior, B2B marketplaces may engage in various activities to qualify members, such as ensuring that suppliers have adequate capacity for meeting orders, verifying buyer credit ratings, admitting entry by nomination or invitation, and so forth.

Social Capital Perspectives on Geographically Defined Business Clusters

Rarely are the roles of location and social capital discussed in the literature on B2B electronic markets. Yet economists and geographers have documented the significant role that location plays in the formation and maintenance of business trading communities, primarily within the context of discussions about business clusters (Porter 1990, 1998, 2000).

Porter (1998, 10) defines a cluster as a "critical mass of companies in a particular field in a particular location." He further notes that they can include "a group of companies, suppliers of specialized inputs, components, machinery, and services, and firms in related industries." They can also include "firms in downstream industries, producers of complementary products, specialized infrastructure providers, and other institutions that provide specialized training, and technical support" as well as industry groups such as trade associations. This description parallels the structure of many of the electronic business trading communities established over the past several years, except that Porter's cluster members are physically colocated in a particular region.

Several of the primary economic benefits ascribed to business clusters are similar to the main benefits of participation in a B2B electronic market: improved access to specialized inputs, lower transaction costs, and access to complementary goods and services. Additionally, Porter argues that the more important benefit is an improvement in the rate of innovation in geographically defined clusters.

Rather than relying on electronic networks and automation to achieve these transactional and informational advantages, clusters capitalize on proximity. A concentration of skilled workers, for example, increases access to needed labor inputs. Proximity helps in many less formal ways, however. As has been shown repeatedly in analyses of such clusters as Silicon Valley, knowledge sharing can occur through spontaneous or chance encounters among professionals living in the same community, enhancing the overall innovation capacity (Rogers and Larsen 1984; Maskell 2001; Saxenian and Hsu 2001). Porter (1998) further refers to the advantages of common language, culture, and social institutions in reducing transaction costs, and notes that local institutions are likely to be more responsive to the specialized needs of a cluster (e.g., for creating public infrastructure). He even points to peer pressure and the presence of rivals as causes for the enhanced competitiveness of firms that are embedded in a local cluster.

A growing group of theorists now explicitly recognize the significance of social capital as a resource that enhances competitive advantage in organizational settings (for reviews, see Adler and Kwon 2002; Nahapiet and Ghoshal 1998). Although definitions of *social capital* differ somewhat, especially given the application of social capital across disparate contexts and disciplines, the primary focus is generally on the resources arising from personal relationships that individuals may draw on in various aspects of their social life (see chapter 8). Hence, it functions much like other forms of capital in that it can be accumulated, and the social capital from one relationship in one context may be beneficial in other contexts. A businessperson, for example, may be referred to a new supplier through a common acquaintance he or she met at an athletic club or church.

Recent theoretical work stresses the multidimensional nature of social capital (Adler and Kwon 2002; Nahapiet and Ghoshal 1998; Tsai and Ghoshal 1998). Nahapiet and Ghoshal (1998) develop a framework to explain how social capital can provide advantages to individuals and firms with three basic dimensions: structural, relational, and cognitive. The structural dimension refers to the pattern of social ties for a given individual. People are embedded in a network of ties, which can function as conduits to needed information and resources. The relational dimension underscores the importance of trust and the sense of obligation that arises from close, personal contacts. Empirical work has demonstrated that people often turn to trusted personal contacts, especially for high-risk transactions, in order to reduce their vulnerability to opportunism and other transaction costs (DiMaggio and Louch 1998). The cognitive dimension focuses on the notion that exchanges of information and other resources are facilitated by shared codes or knowledge that enable common goals as well as common understanding. It offers a distinct resource not tied directly to specific personal relationships, but develops as a public good in a particular social system based on interactions among its members. As discussed by Marleen Huysman (this volume), Nahapiet and Ghoshal's (1998) three dimensions are quite similar to Adler and Kwon's (2002) framework, in which social capital is viewed as operating by providing opportunities, motivation, and ability. Opportunities arise from participation in a network (structure), motivation arises from the qualities embedded in relationships (relational), and ability requires common understanding (cognitive).

Social capital theory supplies a powerful lens through which many of the advantages of geographically proximate business clusters can be understood and extended. Access to skilled labor may be enhanced, for example, when complemented by referrals from social contacts that help connect people searching for work with firms seeking employees. Spontaneous interactions, which Porter contends facilitate innovation, occur because social embeddedness and proximity afford opportunities for such encounters. Common language, culture, and social institutions represent the basis for a shared understanding and goals that comprise the cognitive/ability dimension emphasized by social capital theorists.

Although Porter does not dwell on the relational aspects of social capital, much of the cluster research provides ample evidence of the potential economic benefits arising from this dimension. The sense of obligation, goodwill, and reciprocity that emerges from strong relationships can have significant economic benefits. Social embeddedness researchers posit that at least at the extremes, there are basically two kinds of relationships through which economic transactions occur: arm's-length ties characterized by short-term and constantly shifting ties among loose collections of firms or individuals; and embedded ties characterized by stable and long-term relationships (Powell 1990; Uzzi and Gillespie 1997; Uzzi 1999). Transaction-cost theory considers the former to be marketlike and efficient, allowing self-interested actors to avoid opportunism through their ability to easily switch to a new buyer or seller (Williamson 1975). In a transaction-cost view, economic exchanges that are dependent on social networks can result in inefficiencies as social obligations prevent actors from pursuing transactions with higher-quality or lower-cost partners. In contrast, social embeddedness researchers have found distinct advantages to the reliance by organizational actors on a limited number of trusted relations for their most critical economic exchanges. These include reduced search costs to find appropriate trading partners, lower monitoring costs as trust arising from social obligation and the importance of maintaining a reputation within a social structure work against undue opportunistic behavior, time savings through personal advice and referrals, higher-quality information transfer among actors, greater emphasis on joint problem solving, and an increased likelihood that new transactions will remain within a relationship rather than be directed toward new partners (Granovetter 1985; Powell 1990; Uzzi and Gillespie 1997). Uzzi and Gillespie (1997) found empirical support for these benefits in their study of social embeddedness in the garment business in New York.

To the extent that such strong ties develop over a long period of time and are sustained by interactions in other social contexts such as community associations or social gatherings, then clearly proximity should be correlated with their incidence. Hence, social capital theory offers fertile ground for understanding many of the dynamics that create vitality in local business clusters.

Evidence for IT Use in Business Clusters

Research on IT use in business clusters is limited and inconclusive. Johnston and Lawrence's (1988) seminal work on value-adding partnerships concentrated extensively on the Prato-area textile industry.

Their analysis examined how the large textile mills formed in the 1930s had disaggregrated into small, specialized firms that focused on one part of the overall value chain in textile production (e.g., washing, coloring, cutting, etc.). They showed how networks of firms worked in concert to meet the market demands for the good of the network and pointed out how an interorganizational IS was being used to facilitate coordination (Johnston and Lawrence 1988). A decade later, though Kumar, van Dissel, and Bielli (1998) revisited the merchants of Prato, and found that the IS had been all but abandoned. The system offered no real added value in terms of transaction-cost reductions over the personal forms of coordination that had evolved over centuries of textile production in the region. Kumar, van Dissel, and Bielli (1998) suggest that trust and personal relationships—the social capital of the region—were effective substitutes for the interorganizational system, rendering it unnecessary.

Other research on interorganizational systems embedded within particular business communities also underscores the fallacy of ignoring the socially embedded nature of interfirm transactions. A case study of media buyers and sellers in France illustrates the sometimes oppositional nature of ISs built from a transaction-cost rationality and existing practices based on personal relationships (Caby, Jaeger, and Steinfield 1998). The market for television advertising had become more complex due to the liberalization of the market and the resulting increase in private channels. An electronic marketplace was created by the media industry, allowing media buyers to find available time slots and reserve them. Theoretically, this would reduce selling costs and improve transaction efficiencies. It was built on France's Minitel system, and so required minimal investment by the buyers. Yet it soon failed, largely because it prevented many of the relationship-based selling strategies that media representatives preferred. The buyers could not offer the best times and prices to their preferred customers, for example. Moreover, customers behaved strategically, often reserving time slots only to prevent competitors from obtaining them. Before long, the media representatives were bypassing their own system, only to return to their prior methods of selling media time.

Another study by Kraut and colleagues (Kraut et al. 1998) investigated personal and electronic forms of transaction coordination between producers and suppliers in several industries. Their research extends the

Kumar, van Dissel, and Bielli findings in important ways. In contrast to the Prato case, electronic networks were more likely to be used precisely when there were existing relationships among producers and suppliers, and greater use was associated with more tightly coupled producer-supplier relations. Kraut and Colleagues (1998) explain this by pointing out that to be able to conduct electronic transactions, investments are required by the participants. Suppliers are unlikely to make such investments unless they can expect a certain amount of business. Personal relationships still mattered, however, and indeed they were positively associated with electronic transactions. There was also an interesting interaction between the two: the more firms attempted to substitute electronic transactions for personal forms of coordination, the more errors and quality problems they experienced with transactions. If they complemented electronic transactions with personal coordination, such problems were mitigated.

A recent study by Schultze and Orlikowski (2002) provides further evidence of the damage that can occur when an Internet-based B2B market approach replaces direct personal relations among buyers and sellers. Their study of a health insurance intermediary firm identified the mechanisms by which the provision of Internet-based services can turn relationships that formerly were partnerlike into weaker, brokerlike ones. Agents in the firm complained that the reliance on information provided over the Web enabled clients to bypass them and reduced the sense of obligation that had formerly led clients to voluntarily funnel claims through them. This loosening of a sense of obligation meant a real loss of income, as their commissions were dependent on having served as the intermediary for such claims.

That electronic transactions might follow from physical proximity is further suggested by Castells's (2001) fascinating account of the geography of the Internet, where he points out the spatial concentration associated not only with producers of Internet content and infrastructure but among firms that use the Internet. The following quote illustrates Castells's thinking: "These advanced service centers are territorially concentrated, built on interpersonal networks of decision-making processes, organized around a territorial web of suppliers and customers, and increasingly communicated by the Internet among themselves" (2001,

228). Indeed, a study of Internet traffic by Kolko (2000) convincingly demonstrated that the majority of Internet protocol (IP) traffic flows within, rather than across, locations. B2B transactions are embedded in an enabling social and cultural context, yet in striving for transaction efficiencies, most efforts to create electronic networks to support transactions go to great lengths to ignore and even bypass this context.

Social and Locational Aspects of B2B Electronic Markets

Business strategists have recognized the value of on-line communities as a source of competitive advantage, viewing virtual communities as a form of a business model (Armstrong and Hagel 1996; Timmer 1998). Hummel and Lechner (2002) argue that despite the emphasis on transaction support, members of a B2B electronic market form a socioeconomic virtual community where a social atmosphere is created through two types of informational contributions: news or files that participants create and share with each other, and social information such as ratings, reviews, or recommendations. Following Hammann (2000), they analyze the social profile of B2B on-line communities according to the extent to which features are present that support four dimensions of community (Hummel and Lechner 2002). These four dimensions include a clearly defined group of actors, the nature of interactions among members, the bonding among members, and having a common place. Table 9.1 below illustrates the features defined by Hummel and Lechner that can be used to create a virtual community.

They rate a number of different virtual communities on these dimensions, scoring B2B virtual communities relatively low on the "common place" dimension due to the lack of analysis of the participants (for example, unlike B2C services, techniques such as collaborative filtering are less used), limited volunteerism, and the lack of rituals. Nevertheless, they do see B2B communities as relatively strong in three areas: the efforts to enhance interaction and knowledge transfer among members, having ties to off-line activities, and the efforts to create a trustworthy transaction environment.

Hummel and Lechner's analysis demonstrates that B2B electronic markets may indeed offer more than simple transaction support. They

Table 9.1
Features of the four dimensions of community

Clearly defined group	Interaction among members	Bonding among members	Common place
• Clear limitations • References to real communities • Entry rules • Primary authorization • Rules of treatment • Punishment for misconduct	• Chats/forums • Possibility for own postings • Screening of contributions • Active organization • Events • Regard to recent events	• Privacy protection • Individualizing • Subcommunity • User friendliness • Identification of organizer • Identification of members	• Archive • Analysis of participants • Voluntary work • Rituals • Role of members

Source: Adapted from Hummel and Lechner (2002)

are still relatively limited in their support for creating a sense of common place, however. Researchers are investigating many new strategies for incorporating social information into Web-based commerce that may help to enhance the social component of electronic commerce, including B2B e-commerce. Collectively, these techniques have been called "social navigation" (Dieberger et al. 2000). Dieberger and colleagues (2000) assert that people have a well-developed sense for relying on social information as an aid in day-to-day navigation. For example, we use crowds variously to tell us what places to avoid (when we are in a hurry) or what places to try out (as in a bustling restaurant). Social navigation tools are increasingly being built into e-commerce services, including collaborative filters, reputations, and ratings systems. The information traces left by others is used to guide buyers' decisions.

New work by human-computer interaction researchers attempts to build even more explicit social navigation tools into on-line communities and e-commerce by providing tools to enhance participants' awareness of the activity others (Erickson and Kellogg 2000; Jung and Lee 2000). Jung and Lee (2000), for instance, describe an electronic marketplace design that incorporates a range of tools to make the actions of other people visible to participants. Although they focus on B2C e-commerce, the concepts may also have relevance for B2B communities. Users know how many others are in the marketplace, whether particular people ("buddies") are present, and where they are in the market. Erickson and Kellogg (2000) incorporate a concern for privacy through the concept of social translucence in the design of social navigation systems. Translucent systems offer visibility and awareness (e.g., a user can see socially significant information and thus be aware of the presence or activity of others). Such systems also offer accountability because both parties are aware of the other's knowledge. There is a distinction, however, between translucent and transparent systems, much as in real life. An onlooker may see that two people are conversing, but cannot listen in on the conversation without explicit permission.

The goal of these researchers is to design techniques to improve the sense of "place" in on-line spaces (Harrison and Dourish 1996). In this perspective, places differ from spaces in that there are social meanings attached to them. There are socially and culturally rooted norms for behavior and action in places that, on the surface, may share similar spatial features. Harrison and Dourish (1996) suggest that the greater use of social information can help turn on-line spaces into on-line places.

Collectively, these new trends in the design of socially aware systems offer insights into how interorganizational ISs might be applied to support regional business clusters. Nevertheless, despite the new emphasis on social information, the very basic notion of location still seems underdeveloped. Recent research on the role of physical presence in e-commerce indicates that there can be synergies between on-line and off-line activities (Steinfield, Bouwman, and Adelaar 2002; Steinfield, Adelaar, and Lai 2002). Members of a physical community have opportunities for off-line interaction and exchange that can be augmented in a shared, collaborative e-commerce system, without assuming that the system has to serve as a substitute for in-person or other forms of interactions.

Conclusions and a Research Agenda

The essential arguments of this chapter can now be restated as follows. There is widespread agreement on the importance of local business clusters for the economic vitality of cities, regions, and nations. There is also agreement that the success of business clusters depends on the exploitation of social capital—proximity affords interaction opportunities, common language and culture enhance shared understanding, relationships facilitate knowledge sharing and thus innovation, and trust arising from relationships lubricates commerce and reduces transaction costs. Yet most efforts to improve B2B commerce focus on the construction of transaction support systems that are relatively opaque to—or even worse, attempt to substitute for—social information and assume that location is irrelevant. Hence, B2B electronic systems are underutilized by local business clusters. The current evolution from arm's-length B2B marketplaces to collaborative e-commerce implies new opportunities to better support local business clusters with on-line systems. Such systems, however, must be infused with more social and location awareness, including on-line tools for social navigation as well as recognition that there can be synergies between on-line and off-line activities.

From this set of assertions, a number of new research questions are briefly noted below.

• Homogeneous versus heterogeneous business clusters: Some clusters may be comprised of relatively similar firms, or at least firms that represent the vertical stages in a particular industry value chain. On the other hand, researchers point to the inclusion of a rather heterogeneous mix of capabilities in a location—such as universities, government agencies, and infrastructure providers—that helps to sustain the cluster (Porter 1998). Do these aspects of cluster composition influence the design and viability of collaboration and commerce systems that might be employed?

• New versus established clusters: Much of the discussion assumes the prior existence of a business cluster. Yet there is also much interest in forming new business clusters, and the dynamics of new clusters that have not yet built up a reservoir of social capital may be quite different (Bresnahan, Gambardella, and Saxenian 2001). It may be that different features are required for systems to support these two distinct needs. New clusters require more features to help form relationships.

• Explicit versus implicit social support: Social navigation systems gather social information unobtrusively and often have a degree of translucence designed to protect privacy (Erickson and Kellogg 2000). Hence, a system may tell a user how many others have also purchased some item, but not reveal who those others are, nor offer support for direct interaction with those other users. Yet if we assume a local business cluster already has trusted relationships, perhaps more direct and explicit social support would be more useful.

• Role of tacit versus explicit knowledge: There has been much discussion about the role of tacit knowledge in local business clusters (Brown and Duguid 1998; Breschi and Malerba 2001). It may explain the importance of proximity, in that direct personal interaction is needed to transfer tacit knowledge. On the other hand, it may also be that local business clusters strategically avoid codification to maintain their competitive advantage (Breschi and Malerba 2001). Can an on-line collaborative e-commerce system support the transfer of tacit knowledge? Or can these systems only facilitate exchange when knowledge is codified, which then might have the effect of dissipating the competitive advantage of local business?

• Open versus closed systems: Given the importance of trusted relationships, what should the membership policies be in collaborative e-commerce systems in local business clusters. Should these be open or closed to new members, and if open, can new members join without endorsement or approval by existing members? What role should customers, including nonlocal ones, play? As Porter (2000) notes, most business clusters exist to export goods to nonlocal markets. Shouldn't the e-commerce system in use support that?

• Metrics for performance: How can we measure the effectiveness of a B2B collaborative commerce? Steinfield et al. (2002) argue that too often, e-commerce systems are judged purely on the basis of sales, even though they may offer many other contributions not captured by such data.

• Where are the synergies with between on-line and off-line activity? Steinfield, Bouwman, and Adelaar (2002) argue that in both B2B and B2C e-commerce, there are opportunities for synergy among on-line and off-line activities. What types of synergies exist, and how might they be enhanced through system design?

- Relationships between local versus nonlocal social capital: Most studies of local business clusters treat them as isolated entities (Breschi and Malerba 2001). Yet there is evidence of the significance of nonlocal links for spurring innovation, especially for emerging clusters. For example, Saxenian and Hsu (2001) note how Taiwanese scientists and engineers educated in Stanford, California, maintained relationships with former peers now working in Silicon Valley. These relationships were conduits for social and economic links that facilitated the creation of the Hsinchu business cluster outside Taipei.

Clearly, the intersection of social capital theory, business cluster economics, and IT offers researchers abundant new opportunities for study.

References

Adler, P. A., and S. Kwon. 2002. Social capital: Prospects for a new concept. *Academy of Management Review* 27, no. 1:17–40.

Armstrong, A., and J. Hagel III. 1996. The real value of on-line communities. *Harvard Business Review* 74:134–141.

Bakos, J. Y. 1997. Reducing buyer search costs: Implications for electronic marketplaces. *Management Science* 43, no. 12:1676–1692.

Bakos, J. Y. 1998. The emerging role of electronic marketplaces on the Internet. *Communications of the ACM* 41, no. 8:35–42.

Bakos, J. Y., and M. E. Treacy. 1986. Information technology and corporate strategy: A research perspective. *MIS Quarterly* 10, no. 2:107–119.

Breschi, S., and F. Malerba. 2001. The geography of innovation and economic clustering: Some introductory notes. *Industrial and Corporate Change* 10, no. 4:817–834.

Bresnahan, T., A. Gambardella, and A. Saxenian. 2001. "Old economy" inputs for "new economy" outcomes: Cluster formation in the new Silicon Valleys. *Industrial and Corporate Change* 10, no. 4:835–860.

Brown, J., and P. Duguid. 1998. Organizing knowledge. *California Management Review* 40, no. 3:90–111.

Caby, L., C. Jaeger, and C. Steinfield. 1998. Explaining the use of inter-firm data networks for electronic transactions: The case of the pharmaceutical and advertising industries in France. In *Telecommunications and socio-economic development*, edited by S. MacDonald and G. Madden. Amsterdam: Elsevier.

Cairncross, F. 1997. *The death of distance*. Boston: Harvard Business School Press.

Castells, M. 2001. *The Internet galaxy*. Oxford: Oxford University Press.

Dieberger, A., P. Dourish, K. Hook, P. Resnick, and A. Wexelblat. 2000. Social navigation: Techniques for building more usable systems. *Interactions* (November–December): 37–45.

DiMaggio, P., and H. Louch. 1998. Socially embedded consumer transactions: For what kinds of purchases do people most often use networks. *American Sociological Review* 63:619–637.

Erickson, T., and W. Kellogg. 2000. Social translucence: An approach to designing systems that support social processes. *ACM Transactions on Computer-Human Interaction* 7, no. 1:59–83.

Garicano, L., and S. N. Kaplan. 2001. Beyond the hype: Making B2B e-commerce profitable. *Capital Ideas* 2: no 4, <http://gsbwww.uchicago.edu/news/capidens/win01/b2b.html>.

Granovetter, M. 1985. Economic action and social structure: The problem of embeddedness. *American Journal of Sociology* 91, no. 3:481–510.

Hammann, R. 2000. Computernetze als verbindendes element von gemeinschaftsnetzen. In *Virtuelle Gruppen: Charakteristika und Problemdimensionen*, edited by U. Thidecke. Opladen: Westdeutscher Verlag.

Harrison, S., and P. Dourish. 1996. Re-place-ing space: The roles of place and space in collaborative systems. In *Proceedings of computer-supported cooperative work*. Cambridge, Mass.: Association for Computing Machinery.

Hummel, J., and U. Lechner. 2002. Social profiles of virtual communities. In *Proceedings of thirty-fifth Hawaii international conference on systems sciences*. Kona, Hawaii: Institute of Electrical and Electronics Engineers.

Johnston, R., and P. Lawrence. 1988. Beyond vertical integration: The rise of the value-added partnership. *Harvard Business Review* (July–August): 94–101.

Jung, Y., and A. Lee. 2000. Design of a social interaction environment for electronic marketplaces. In *Proceedings of DIS 2000—designing interactive systems: Processes, practices, methods, techniques*. New York: Association for Computing Machinery.

Kaplan, S. N., and M. Sawhney. 2000. E-hubs: The new B2B marketplaces. *Harvard Business Review* 78 (May–June): 97–103.

Katsaros, H., M. Shore, R. Leathern, and T. Clark. 2000. *U.S. business-to-business Internet trade projections*. New York: Jupiter Research.

Kolko, J. 2000. The death of cities? The death of distance? Evidence from the geography of commercial Internet usage. In *Internet upheaval: Raising questions, seeking answers in communications policy*, edited by I. Vogelsang and B. Compaine. Cambridge: MIT Press.

Kraut, R., C. Steinfield, A. Chan, B. Butler, and A. Hoag. 1998. Coordination and virtualization: The role of electronic networks and personal relationships. *Journal of Computer Mediated Communication* 3, no. 4, <http://www.ascusc.org/jcmc/vol3/issue4/kraut.html>.

Kumar, K., H. van Dissel, and P. Bielli. 1998. The merchant of Prato—revisited: Toward a third rationality of information systems. *MIS Quarterly* 22, no. 2:199–226.

Laudon, K., and C. Traver. 2001. *E-commerce: Business, technology, society.* Boston: Addison-Wesley.

Malone, T., J. Yates, and R. Benjamin. 1987. Electronic markets and electronic hierarchies: Effects of information technology on market structure and corporate strategies. *Communications of the ACM* 30, no. 6:484–497.

Maskell, P. 2001. Towards a knowledge-based theory of the geographical cluster. *Industrial and Corporate Change* 10, no. 4:921–944.

Nahapiet, J., and S. Ghoshal. 1998. Social capital, intellectual capital, and the organizational advantage. *Academy of Management Review* 23, no. 2:242–266.

Porter, M. 1990. *The competitive advantage of nations.* New York: Free Press.

Porter, M. 1998. The Adam Smith address: Location, clusters, and the "new" microeconomics of competition. *Business Economics* 33, no. 1 (January): 7–13.

Porter, M. 2000. Location, competition, and economic development: Local clusters in a global economy. *Economic Development Quarterly* 14, no. 1:15–34.

Powell, W. 1990. Neither market nor hierarchy: Networked forms of organization. In vol. 12 of *Research in organizational behavior*, edited by B. Staw and L. Cummings. Greenwich, Conn.: JAI Press.

Rogers, E., and J. Larsen. 1984. *Silicon Valley fever: Growth of high-technology culture.* New York: Basic Books.

Saxenian, A., and J. Y. Hsu. 2001. The Silicon Valley-Hsinchu connection: Technical communities and industrial upgrading. *Industrial and Corporate Change* 10, no. 4:893–920.

Schultze, U., and W. Orlikowski. 2002. When complementarity is not enough: The implications for unbundling services in the Internet era. Paper presented at the International Conference on Information Systems, December. Barcelona.

Segev, A., J. Gebauer, and F. Färber. 1999. Internet-based electronic markets. *Electronic Markets* 9, no. 3:138–146.

Shapiro, C., and H. R. Varian. 1999. *Information rules: A strategic guide to the network economy.* Boston: Harvard Business School Press.

Steinfield, C., T. Adelaar, and Y. Lai. 2002. Integrating brick and mortar locations with e-commerce: Understanding synergy opportunities. In *Proceedings of the 34th Annual Hawaii international conference on systems sciences* (CD/ROM). Computer Society Press.

Steinfield, C., H. Bouwman, and T. Adelaar. 2002. The dynamics of click and mortar e-commerce: Opportunities and management strategies. *International Journal of Electronic Commerce* 7, no. 1:93–119.

Steinfield, C., A. Chan, and R. Kraut. 2000. Computer-mediated markets: An introduction and preliminary test of market-structure impacts. *Journal* of *Computer Mediated Communication* 5, no. 3, <http://www.ascusc.org/jcmc/vol5/issue3/steinfield.html>.

Steinfield, C., and S. Klein. 1999. Local vs. global issues in electronic commerce. *Electronic Markets* 9, nos. 1–2:45–50.

Steinfield, C., R. Kraut, and A. Plummer. 1995. The effect of networks on buyer-seller relations. *Journal of Computer Mediated Communication* 1, no. 3, <http://www.ascusc.org/jcmc/vol1/issue3/steinfld.html>.

Subramami, M., and E. Walden. 2000. Economic returns to firms from business-to-business electronic commerce initiatives: An empirical investigation. In *Proceedings of twenty-first international conference on information systems* (CD/ROM). Brisbane, Australia: Association for Information Systems.

Tedeschi, R. 2001. E-commerce report: Companies in no hurry to buy over the Internet. *New York Times*, 5 March, <http://www.nytimes.com/2001/03/05/technology/05ECOMMERCE.html>.

Timmer, P. 1998. Business models for electronic markets. *Electronic Markets* 8, no. 2:3–8.

Tsai, W., and S. Ghoshal. 1998. Social capital and value creation: The role of intrafirm networks. *Academy of Management Journal* 41, no. 4:464–476.

U.S. Department of Commerce. 2000. *The emerging digital economy*. Washington, D.C.: U.S. Department of Commerce.

Uzzi, B. 1999. Embeddedness in the making of financial capital: How social relations and networking benefit firms seeking financing. *American Sociological Review* 64:481–505.

Uzzi, B., and J. Gillespie. 1997. Knowledge spillover in corporate financing networks: Embeddedness and the firm's debt performance. *Strategic Management Journal* 23, no. 7:595–618.

Williamson, O. 1975. *Markets and hierarchies*. Englewood Cliffs, N.J.: Prentice Hall.

10

The Impact of Social Capital on Project-Based Learning

Mike Bresnen, Linda Edelman, Sue Newell, Harry Scarbrough, and Jacky Swan

Strategic management perspectives have shifted in recent years toward a more knowledge-based view of the firm, in which developing the capabilities for creating and sharing knowledge is important for value creation, and thus, for increasing a firm's competitive advantage (Kogut and Zander 1992, 1996). Creating, disseminating, and using firm-specific knowledge therefore becomes a primary goal of an organization and also a major source of competitive advantage (Kogut and Zander 1996). Capturing and exploiting firm-specific knowledge by transforming it into economically useful products and services, however, is not an easy task since knowledge resides in the experience as well as expertise of individuals and groups within the organization (e.g., Brown and Duguid 2001; Hansen, Nohria, and Tierney 1999; Szulanski 1996). Despite the importance of information and communication technologies for knowledge codification and sharing, it has long been realized that the creation and diffusion of knowledge within organizations relies on the development of social networks, shared systems of meanings, and the cultivation of shared values and norms (Spender 1996). Unless the knowledge concerned is explicit and easily codified, then it becomes difficult to capture it in ways that make it amenable to the use of information and communication technologies (Fahey and Prusak 1998; Hansen, Nohria, and Tierney 1999). A more social or community-based approach to understanding knowledge creation, codification, and sharing within organizations thus becomes essential (e.g., Swan et al. 1999). Such an approach requires an understanding of the ways in which knowledge is socially constructed and practically situated as well as enacted, and involves examining the tacit assumptions and meanings that underpin processes of creating, sharing, and applying knowledge (Brown and Duguid 1991, 2001; Wenger 2000).

As part of this emphasis on the social aspects of knowledge creation and sharing, considerable attention has recently been directed toward the role played by social capital (e.g., Adler and Kwon 2002; Burt 1997; Coleman 1988; Leenders and Gabbay 1999; Nahapiet and Ghoshal 1998; Portes 1998; Putnam 1993, 1995). *Social capital* derives from the network of relationships that connect people together, and refers to the "goodwill that is engendered by the fabric of social relations and that can be mobilized to facilitate action" (Adler and Kwon 2002, 17). According to many commentators, it represents a potentially valuable asset to a firm or organization—an asset that can be appropriated and exploited to increase intellectual capital and innovative potential, thereby helping a firm gain competitive advantage (Coleman 1988; Nahapiet and Ghoshal 1998). A good deal of attention has therefore been directed at exploring the ways in which social capital can be developed and mobilized, and the circumstances under which its benefits are more effectively realized (e.g., Burt 1997). Despite the acknowledgment that capturing the benefits of something as intangible as social capital can be difficult (Gabbay and Leenders 1999), it is only comparatively recently that there has been any recognition that social capital also has its negative side (Adler and Kwon 2002; Gargiulo and Benassi 1999; Leenders and Gabbay 1999; Locke 1999; Portes and Landolt 1996). Adler and Kwon (2002), for example, are more circumspect than many in emphasizing some of the drawbacks of social capital as well as its potential benefits (see also Locke 1999; Portes and Landolt 1996). Reviews such as theirs offer a useful corrective to less critical accounts, which while they accept that there may be problems with the appropriation and use of social capital, nevertheless tend to stress the benefits of seeking to increase it (e.g., Nahapiet and Ghoshal 1998). Unfortunately, however, such accounts do not always provide the detailed empirical evidence needed to substantiate claims made about the contradictory or negative effects of social capital. Moreover, as Adler and Kwon (2002, 34–35) themselves highlight in their call for more research in this area, there is not much recognition or analysis of the contingent effects of differences in task environments or organizational contexts on the development and use of social capital.

One such environment or context where one might expect there to be significant benefits to be gained from the use of social capital to enhance innovation (as well as difficulties in realizing these benefits) is that of projects and project organization. Project environments are a potentially rich arena for the study of social capital effects because they are often closely associated with product or process innovation, and yet pose particular difficulties for the capture and transfer of knowledge and learning (DeFilippi and Arthur 1998; Prencipe and Tell 2001). Not only are there significant task discontinuities from one project to the next that make the direct application of any "lessons learned" difficult, there are also potentially disruptive effects on social groups and networks as project teams move on, disband, and reform to pursue different projects with different objectives. Consequently, social capital may be more difficult to develop, sustain, and exploit than in other types of organizational environments. Yet at the same time, it may be more vital for innovation in project environments to the extent that social ties become important for capturing and passing on project-based knowledge and learning from one project to the next (cf. Bresnen et al. 2003; Hansen, Nohria, and Tierney 1999; Hansen 2002).

Project-based learning therefore provides a crucial and relatively unexplored context for the study of social capital effects. As well as being important in its own right, research in this context can also help throw further light on the effects of contingencies associated with project work (e.g., specific objectives, finite duration, and multiprofessional teamwork) on social capital formation. In so doing, it can contribute more generally toward a fuller understanding of the effects of different contingencies on the development, use, and effects of social capital. This chapter, then, aims to add to a richer understanding of the effects of social capital by reporting the results from a research project designed to explore social and organizational factors influencing cross-project learning. The research on which this chapter is based consists of case studies of projects being undertaken within three organizations in three different industrial sectors in the United Kingdom: telecommunications, construction, and health care.[1] In looking at the positive and negative effects of social capital on cross-project learning within the three case study projects, the analysis draws on the conceptual framework developed by Nahapiet and Ghoshal (1998) that distinguishes between the structural, cognitive, and relational dimensions of social capital. A number of findings emerge from the research that clearly show the costs as well as benefits of social capital, including many contradictory effects. For example, the research demonstrated how social capital simultaneously helped provide access to knowledge that was otherwise difficult to obtain, and encouraged an inward-looking perspective as well as the use of local search behaviors for knowledge generation and sharing. It also revealed the potential exclusionary effects of social capital, and the impact that social capital can have in reinforcing professional and occupational solidarity in ways that increase resistance to organizational change. The implications of these findings for understanding the effects of social capital on knowledge sharing in general and project-based learning in particular are therefore discussed.

Social Capital: Sources and Consequences

Definitions of social capital share in common the central idea that the networks of relationships in which people are embedded act as a key resource, and thus, are a source of competitive advantage to a firm. According to Nahapiet and Ghoshal, for example, social capital can be defined as "the sum of the actual and potential resources embedded within, available through, and derived from the network of relationships possessed by an individual or social unit. Social capital thus comprises both the network and the assets that may be mobilized through that network" (1998, 243).

Although social capital is said to share many of the features of other forms of capital, including financial (Nahapiet and Ghoshal 1998), it is also clear that it has a number of distinctive qualities (Adler and Kwon 2002). First, it is not a form of capital that is possessed individually. It is instead embedded within the structure of relationships that connect people together—a form of "collective good," as opposed to "private property" (Coleman 1988; Portes 1998; Putnam 1995). Second, and related to this, its sources and effects are inextricably linked to the development and maintenance of social ties, and hence, to associated patterns of inclusion and exclusion. One implication of this is that social capital requires "maintenance" of the social relationships on which it depends (Burt 1992; Adler and Kwon 2002, 22). Another implication is that the exclusionary effects of close social ties cannot be ignored (Adler and Kwon 2002, 22). Third, because of its intangible and ephemeral qualities, social capital is difficult, if not impossible, to quantify and measure (Gabbay and Leenders 1999). Social capital is emergent, arising out of day-to-day interactions, and so cannot easily be appropriated as an organizational resource. By the same token, at the outset of any innovative activity it may be difficult, if not impossible, to anticipate the social capital benefits accruing to a particular team or project.

Given these distinctive features—indicative of the highly complex nature of social interaction—it becomes useful to try to unpack the constituent elements or key dimensions of social capital. Nahapiet and Ghoshal (1998), in their comprehensive review of the conceptual literature on social capital, provide a useful framework for understanding social capital's main elements. They argue that social capital consists of three distinct, but closely interrelated dimensions, which they describe as structural, cognitive, and relational. The structural dimension of social capital refers to "the overall pattern of connections between actors—that is, who you reach and how you reach them" (Nahapiet and Ghoshal 1998, 244). These connections or network links are important because they provide channels to information or knowledge that is unevenly distributed within and across organizations (Burt 1992; Granovetter 1992). Features of networks that are of particular significance include the configuration of a network along with the patterns of connectivity and interdependence among network actors. Crucially, it is not the size of the network that is important but its diversity (Burt 1992). Not only do nonredundant contacts increase the likelihood of gaining access to new information; they are also likely to take much less time and effort to maintain than redundant contacts. Other key features of a network include the degree to which there are players occupying central or boundary-spanning roles. Being aware of the location of such players can be helpful in understanding communication patterns as well as related behaviors such as power positioning and knowledge flows (Brass and Burkhardt 1992).

The cognitive dimension of social capital refers to "those resources providing shared representations, interpretations, and systems of meaning among parties" (Nahapiet and Ghoshal 1998, 244). Aspects of note here are the shared meanings that are created and reinforced through common language and codes as well as shared narratives within a given social group. Shared meanings develop through an ongoing and self-reinforcing process of participation within a group. In other words, through an ongoing dialogue and the collective processes of sense making among a group of people (cf. Weick 1995), systems of meaning are developed, applied, and combined. For example, "communities of practice," in which individuals collaborate and share ideas through mechanisms such as narration and joint work (Brown and Duguid 1991, 2001), depend heavily on cognitive social capital.

The relational dimension of social capital refers to "the kind of personal relationships people have developed with each other through a history of interactions" (Nahapiet and Ghoshal 1998, 244). A key aspect of importance here is the normative basis of an exchange relationship (Coleman 1990). The extent to which there are relations of trust, the effects of norms and sanctions, the impact of obligations and expectations, and patterns of identification among group members are all relevant (Nahapiet and Ghoshal 1998, 244). Social capital theorists are particularly interested in norms that underpin cooperation and control within groups, such as trust and reciprocity. Norms of trust can lead to enhanced cooperation, which in turn leads to increased trust (cf. Ring and Van de Ven 1994). Similarly, norms of reciprocity can enhance the likelihood of further reciprocal exchange arrangements, if obligations are seen as being satisfactorily fulfilled.

Taken together, it is argued that these three dimensions of social capital, when coupled with the opportunity for knowledge combination and exchange, result in the creation of new intellectual capital (Nahapiet and Ghoshal 1998). Social capital therefore becomes a valuable asset for any organization to acquire. Empirical work on the effects of social capital is now extensive, and many empirical studies that use the social capital concept have recently been published, exploring its effects in a wide range of arenas and organizational practices. Included among these are human resource development practices within a firm such as recruitment processes, career development, and executive compensation (e.g., Belliveau, O'Reilly, and Wade 1996; Fernandez, Castilla, and Moore 2000; James 2000; Podolny and Baron 1997). Leana and Van Buren (1999), for instance, examine how employment practices can encourage or discourage the emergence of potentially beneficial social capital within organizations. Research attention has also been directed toward understanding the part played by social capital in various types of managerial activities such as resource allocation (Bouty 2000), innovation (Hansen 1999; Tsai and Ghoshal 1998), and managing change (Gargiulo and Benassi 2000). Burt (1997), for example, contends that there are significant information and control benefits when senior managers use social capital. Other researchers have studied the impact of social capital in a variety of organizational and interorganizational contexts—for example, in network formation (Walker, Kogut, and Shan 1997), project-based organization (DeFilippi and Arthur 1998), and in particular, the development of exchange relationships and strategic alliances among organizations (e.g., Chung, Singh, and Lee 2000; Koka and Prescott 2002; Kraatz 1998; Tsai 2000; Uzzi 1997). Research has even started to explore cross-national differences in the use of social capital (e.g., Burt, Hogarth, and Michaud 2000).

Exploring the Darker Side of Social Capital

Notwithstanding the difficulties involved in attempting to mobilize social capital or capture its benefits in any given situation, the predominant view in this literature is that the accumulation of social capital is fundamentally of benefit to an organization. The emphasis in much current research is on identifying and examining the benefits that can be achieved by participation in social networks, and exploring the factors that can facilitate and inhibit the emergence and development of social capital in different situations. At the same time, however, there are an increasing number of commentators who have begun to take a more skeptical view of social capital (Adler and Kwon 2002; Gabbay and Leenders 1999; Gargiulo and Bernassi 1999; Locke 1999; Portes 1998; Portes and

Landolt 1996; Portes and Sensenbrenner 1993; Prusak and Cohen 2001; Uzzi 1997).

In a recent review of the existing literature, Adler and Kwon (2002) develop a framework for studying the effects of social capital that pays attention not only to the benefits of social capital but also its risks and the circumstances under which they occur. Drawing on Sandefur and Laumann's (1998) distinction between the information, influence, and solidarity effects of social capital, Adler and Kwon (2002) outline the potential risks and costs of social capital from the perspective of both individuals and organizations. With regard to information effects, one of the main problems for individuals and organizations is that social capital can be a costly and inefficient means of obtaining information as it involves considerable effort in establishing as well as maintaining relationships. This is particularly the case with explicit types of knowledge, which are more efficiently transferred via weak tie relationships, as opposed to the strong ties that may be needed for the transfer of complex, tacit types of knowledge (Hansen 1999). Research has also suggested that there is the possibility that excessive information brokering, although it may benefit individuals, may actually hamper innovation within the wider organization (Gabbay and Zuckerman 1998).

With regard to influence effects, social capital can create conditions of dependency for an individual if there is asymmetry in a relationship among contacts, and also have an inhibiting effect on processes of individual discovery and knowledge creation (Locke 1999; Portes 1998). At the organizational level, the use of social capital for influence attempts can also have significant political consequences and costs (Adler and Kwon 2002, 30–31). Studies of network relationships among organizations, for example, often highlight the downside of overdependence on a few key network members (e.g., Uzzi 1997). Similarly, the consequences of power imbalances are a crucial feature of interfirm collaborative relationships (e.g., Elg and Johannson 1997).

With regard to solidarity effects, these can "backfire" as individuals become too embedded in and overdependent on particular social networks (e.g., Locke 1999). In other words, just as organizational learning may lead to "competence traps" (Levitt and March 1988), social capital may lead to "relationship traps," discouraging the use of other, more wide-ranging forms of search behavior. An extreme example of this can be seen in the occurrence of "groupthink," where group overcohesiveness can have disastrous effects on decision making (Janis 1982; Moorhead, Ference, and Neck 1991). For the organization as a whole, there is also the prospect that overidentification with particular groups leads to fragmentation within the wider organization or community (cf. Fine 1999). Developing a community is helpful for members as it can provide them with a sense of a common identity and cohesiveness. At the same time, however, it also marks out those who are outside of that community (Portes 1998; Prusak and Cohen 2001). Consequently, while social capital may have an integrative effect, it can also have an exclusionary effect that leads not only to internal fragmentation within the wider organization (cf. Meyerson and Martin 1987) but also to the rejection of new sources of knowledge that fall outside the existing social network.

Taken together, these problems suggest a number of potentially major dysfunctional consequences for an organization that stem from a reliance on social capital for knowledge sharing and innovation. Adler and Kwon (2002) themselves highlight the "not invented here" syndrome (cf. Katz and Allen 1982) as the result of strong internal ties combined with weak external ones. In their exploration of research and development, Katz and Allen (1982) found that strong social norms, coupled with shared experiences, led to a propensity for internal information searches and a lack of acceptance of new ideas that were externally generated. Over time, this inward-looking approach led to an overall decline in the level of innovation. The negative consequences of strong internal ties along with shared norms and outlooks are also emphasized by others who similarly identify the potentially dangerous insulating effect of social capital in situations where the organization needs to respond to complex, changing environmental conditions (Gargiulo and Benassi 1999; Prusak and Cohen 2001; Uzzi 1997).

The work of Adler and Kwon (2002) and others is important in proposing a number of potential problems with social capital in practice. As they themselves note, however, the amount of research undertaken to explore these issues is still comparatively limited, and there is still much to be gained by opening up the study of social capital in organizational settings to further, fine-grained empirical investigation. Moreover, it is clear that there is much to be learned by analyzing the contingencies that influence the impact of social capital in diverse organizational and environmental contexts (Adler and Kwon 2002, 34–35). Much of the available research in this area tends to ignore differences in the particular type of task environment and organizational context in which social capital develops. Yet there is good reason to suspect that such differences matter. For example, forms of organization that are temporary and transient in nature (e.g., those that are project based) may experience greater difficulties (at least initially) in the development of shared understandings and relational norms. Similar problems may occur where interprofessional teamwork is the norm (also common in project organization), in contrast to forms of organization that rely more on shared professional networks, norms, and values.

The research on which this chapter is based therefore looks at the effect of different forms of organization by taking as its focus the comparatively underresearched area of innovation and learning in project environments (DeFilippi and Arthur 1998; Prencipe and Tell 2001). Capturing project-based learning is critical for innovation, yet poses particular challenges for an organization. These challenges arise because of a number of characteristics of projects, including the variability and unique nature of many project tasks, the importance of cross-disciplinary work, and the many discontinuities in the flows of resources, information, and personnel that occur from one project to the next (Bresnen et al. 2003). Under such circumstances, it not only becomes difficult to transfer learning from one (dissimilar) project to the next, it also makes it difficult to develop and sustain the social networks on which social capital depends (DeFilippi and Arthur 1998). Consequently, it becomes essential to understand to what degree and in what ways social capital affects innovation and learning in such settings, and how these effects are moderated or otherwise influenced by the precise circumstances of project-based work. To what extent, for example, does a focus on project goals encourage the team cohesiveness that is important in developing relational social capital? Or to what extent does it instead encourage the team to be more inward-looking in its search for solutions, and resistant to external sources of information and knowledge? Studying the effects of social capital in project-based settings, then, can shed some light on the types of task contingencies that may influence the benefits realized and drawbacks encountered in practice in the use of social capital more generally (cf. Adler and Kwon 2002).

Research Aims and Methods

As mentioned earlier, the research undertaken to explore social capital effects on project-based learning consisted of an exploratory, in-depth study of three case study projects, drawn from three quite distinct sectors in the United Kingdom: telecommunications, construction, and health care (for the purposes of this study, called Teleco, Constructco, and Healthorg). These sectors were chosen not only because they are significant sectors of activity in their own right but also because they reflect a range of market and technological environmental conditions—presenting different constraints and opportunities for the capture of knowledge and learning. For example, Teleco was operating in a rapidly changing environment, undergoing major internal transformation as it moved into the wireless-communication market. Constructco's environment was relatively stable, although it too faced pressures and constraints produced by the inherently complex nature of construction work (e.g., Bresnen 1990). Healthco faced uncertainty due to changes in the regulatory environment and government policy.

Since the research was concerned with investigating the effects of social factors on the ways in which organizations capture and transfer knowledge and learning from project to project, a specific project was chosen in each organization as the focus of investigation and unit of analysis. Projects were chosen at initial meetings with senior managers, administrators, or directors of the organizations. Each organization was asked to nominate a "typical project" that met a set of guidelines established by the research team. Recognizing the difficulties in comparing projects at different phases of their life cycle (Leonard-Barton 1990), the companies were asked to nominate a mature project that was well established, and that involved a number and variety of project team members.

Interviewing was adopted as the main method of investigation because of its ability to capture rich detail, and also because there is a strong indication in the literatures on knowledge and organizational learning that understanding the context in which knowledge transfer occurs is extremely important (Argote 1999; Szulanski 1996). In addition, use was made of archival data about the project or organization. Web sites were accessed where available, and written documentation in the form of records of previous meetings, financial reports, and/or press releases was also collected where appropriate. To ensure data consistency and validity, triangulation across different sources of primary and archival data was used. Case study findings were also fed back to company representatives at a workshop held near the end of the research.

Interviews were conducted with twenty-five individuals across the organizations (eight each at Teleco and Constructco, and nine at Healthorg) over a seven-month period. The average length of an interview was 1.25 hours, although interviews varied in length from 0.5 to over 2 hours. One interview at Healthorg was conducted in a group format, including four project participants. In all cases, every attempt was made to have two researchers present at each interview, although this was not always possible due to scheduling difficulties. All interviews followed a predesigned interview protocol, which included questions about the barriers and enablers of knowledge transfer within and among project teams.

The Case Studies

The following section contains brief descriptions of each of the three case study organizations, the specific projects that were investigated, and the contexts in which they occurred, as a prelude to a fuller cross-case analysis.

Teleco: The Technology Watch Group

Teleco is a major player in the highly dynamic and competitive telecommunications market, and has recently been involved in a radical restructuring of its research and development facilities. The company was traditionally a voice service provider with a large, internal, state-of-the-art research and development department. Recent rapid changes in communications technology, however, had prompted the company to shift its strategy from making its own products to buying new technologies. The technology watch project consisted of a small core group, whose mandate was to identify new technologies that were of strategic interest to Teleco and champion these technologies with managerial decision makers. To identify new and promising technologies, the team drew heavily on internal expertise, bringing experts from across the company into the analysis process. Once an analysis was complete, a case was presented to upper management, who would then make the final decision as to how well the technology aligned with the company's overall strategic direction and how it could be incorporated into the company's product range (e.g., via purchasing the technology or partnering with the other firm).

Consequently, the technology watch team played a crucial role in the strategic shift from "make" to "buy" and the accompanying internal reorganization. Still, the process had not been without its problems, which stemmed in part from the enormous size and complexity of the company. As well as the formally constituted technology watch project team that the research focused on, individual technology teams within Teleco also had their own technology watch groups. This created redundancy and a good deal of confusion—both externally, in the marketplace for new technologies, and internally, in the company itself. In addition, Teleco did not have a central database in which to store information about previously rejected firms and/or technologies, which meant that on more than one occasion, a particular technology was evaluated by multiple groups. Moreover, the feedback received from senior management about the work of the technology watch team was often inadequate. This created considerable frustration within the team as promising technologies were presented, only for them to be rejected without adequate justification. The team was often not made aware of the outcomes of the cases they presented as well. Finally, it was also suggested that there was a good deal of frustration inherent in the goals of the project team, in that team members identified promising technologies, but were then not supported when they attempted to push them forward and integrate them into the wider Teleco organization.

Constructco: The Regional Engineering Manager

Constructco is a large, national contractor with offices in four regions across the United Kingdom. It undertakes commercial building and civil engineering work, much of it on a design and build basis, where the company is involved in all aspects of the design and construction process. Like many other large contractors, Constructco employs little direct labor, farming out most of the actual work to other general building and specialist subcontractors. Each of its projects is bid for and managed from a regional office by a team of project managers, estimators, engineers, and other support staff, and then by a resident site team responsible for construction.

Recognizing the need for communication across the disparate construction projects within a region, the company introduced the role of regional engineering manager (REM). The responsibilities of the REM were to include value engineering (i.e., more cost effective engineering) of new bids and existing projects, and the mentoring, training, and developing of the construction site engineering staff. At the time of the research, there were ten REMs in the company, two in each region (one for civil engineering and one for commercial building projects), and two who worked out of the company's head office.

While the overall mandate of the REM was clear, the implementation of the role varied considerably between regions. For example, in the Midlands region, the REMs divided their time fairly evenly between value engineering and communication and mentoring responsibilities. In this region, it was not unusual for the REM to provide extensive "best-practice" information to the construction sites. In the Southeast region, in contrast, the REMs spent most of their time value engineering projects and rarely had the opportunity to mentor or communicate with site-based engineering staff. This inconsistency was due in part to differences in acceptance of the REM role: initially, it was considered by some regions to be an unnecessary expense and an additional overhead burden borne by the region. While much of this resistance had dissipated with the realization of cost savings, the more immediate and tangible benefits of the value-engineering part of the role meant that in some regions, the emphasis still tended to be on this aspect of the role.

As the REM did not have direct responsibility over any individual engineers, it meant that they had to rely on informal means to accomplish their tasks. This worked better in some regions than in others, and also depended on the interpersonal skills and personal relationships of individual REMs. It did create considerable frustration for many of the REMs, however. On the other hand, the lack of direct reporting requirements, coupled with quarterly company-wide REM meetings, served to create a strong sense of community among the REMs. This was evidenced in the frequency and regularity with which they communicated with each other (by phone and e-mail), which enabled the sharing of engineering information and knowledge across the company.

Healthorg: Reengineering Cataract Treatment at a National Health Service (NHS) Trust

This case study was conducted at one of the United Kingdom's NHS trust hospitals. The specific project investigated was the "reengineering" of the cataract diagnosis and treatment process. Cataract surgery, which is a twenty-minute procedure, represents 96 percent of the ophthalmology workload. Traditionally, cataract diagnosis and treatment had taken a number of visits to various specialists. Patient diagnosis would typically begin at the optometrist, who would identify the problem and refer the patient to his or her general practitioner (GP). After a visit to the local GP, the patient would be referred to the hospital, which would involve separate appointments with the consultant and hospital nurse for further diagnosis and a physical examination. Only when all of these visits were complete would the patient be put in the queue for obtaining a date for cataract surgery. In many trusts, lead time for cataract surgery was over twelve months. After surgery, another visit to the consultant was scheduled to check on the patient, who would then be referred back to the optometrist for new glasses. In total, it took patients at least six visits and often over a year to have a routine, twenty-minute, outpatient, surgical procedure.

Given the resource-intensive and time-consuming nature of this process, a more efficient cataract diagnostic and treatment process could have potentially significant benefits in maximizing theater usage and minimizing waiting times. A member of the hospital's "transformation team"—a group of eight staff whose remit was to reengineer hospital processes and who were responsible at any one time for a number of service delivery projects—was assigned to the process to facilitate the change. The transformation team members gathered a group of eye experts from both the hospital and community to discuss ways in which to cut surgery lead times as well as improve patient satisfaction. Members of the cataract team included the head nurse in the eye unit, a hospital administrator, GPs, local optometrists, and a surgical consultant, who was instrumental in championing the change and leading the change process. In total, approximately five project-team meetings were held over a six-month period.

A number of changes to the existing process were made. Nonessential visits to the GP, consultant, and nurse were eliminated. Optometrists were empowered and trained to diagnose patients; they also filled out a detailed information form and called the hospital to book a time for the patient's surgery. The preoperation physical examination was replaced with a self-diagnostic questionnaire that each patient filled out and returned to the hospital before surgery. Under the new procedure, nurses now telephoned patients before surgery to check health information and answer any questions. Postoperation consultant appointments were also replaced with follow-up telephone calls. The net result was a number of crucial performance gains. Surgery lead times were radically reduced from over twelve months to just six to eight weeks. Theater utilization rates improved with the addition of an administrator whose sole responsibility was scheduling theater use. Finally, and most important, follow-up phone conversations with cataract project patients indicated that patient satisfaction had improved dramatically.

Findings and Discussion

The following section presents our findings from each of the three case studies according to Nahapiet and Ghoshal's (1998) three dimensions of social capital. A summary of the benefits and drawbacks identified in each case can be found in tables 10.1 and 10.2 below.

Table 10.1
Benefits of social capital: Examples from the case studies

	Teleco	Constructco	Healthorg
Structural Access to information that is difficult to otherwise obtain	Extensive use of internal personal social networks for information gathering	Network of regional engineering managers provides access to engineering knowledge and expertise	Networks used to obtain information about cataract processes in other National Health Service trusts
Cognitive Dialogues of shared meaning	Remaining employees had long-term employment and had worked with numerous colleagues through several reorganization schemes	Shared engineering language facilitated communication	Group brought together who created their shared meanings and understandings of the need for change
Relational Facilitates access for exchange—that is, trust, reciprocity	Radical change in corporate culture reinforced a strong sense of cohesion among remaining employees; high levels of trust and reciprocity	Strong norms of reciprocity and trust established among REMs, and between REMs and site staff	Context created in which disparate professionals learned to trust one another

Table 10.2
Drawbacks of social capital: Examples from the case studies

	Teleco	Constructco	Healthorg
Structural Redundancy of contacts makes it difficult to obtain novel information	Significant employee exodus created gaps in organizational knowledge base and left networks incomplete	Redundancy of information due to shared backgrounds and experience of REMs	Process reengineering team comprised eye care specialists who had previously worked with the trust
Cognitive Exclude new sources of knowledge from outside the community	Need to overcome previous tendencies to act as "technologists" and embrace an internal marketing approach	Network of REMs used to the exclusion of other sources of knowledge within or outside the company	Need to overcome professional perceptions and demarcations in order to create new knowledge
Relational Norms of control and compliance leading to resistance to innovation	Remnants of former superior culture led to "not invented here" attitudes with respect to information-gathering efforts	Pressure to succumb to norms emphasizing need for immediate commercial returns	Need to overcome resistance by administrative staff to proposed changes in cataract process

The Structural Dimension of Social Capital and the Problem of Knowledge Redundancy

Clear examples of some of the advantages and benefits as well as drawbacks and risks of structural social capital were found in each of the case studies. At Constructco, the REMs used each other as a first point of contact for information when they had a problem to solve. Knowing who to contact to access specialist experience and expertise was obviously useful in helping the REMs to deal with immediate problems on-site or in the engineering design process. One REM described the mechanisms and benefits of the network as follows: "If you have information, you can instantly contact the other REMs [by e-mail] and say, 'Look, we have a problem, or we have used this and this isn't so good'.... If I have a good idea, I do that on a report sheet, then I will put it on the e-mail to each of the REMs in all regions. They have then got the same information I've got. It's up then to the REMs to say, 'That's important to my business, I'll circulate it.'" On the other hand, there was a good deal of redundancy involved: each REM tended to be privy to the same sorts of information, and therefore, found it more difficult to find solutions to difficult or unusual problems. The REMs would then be forced to expand their search for new knowledge outside the company, though this could lead to problems with inadequate contacts, inaccurate information, or suboptimal solutions. As the company's senior engineer explained, accessing specialist knowledge outside the company could prove difficult due to contractual constraints: "If we have a problem in paint, we don't have paint experts inside [the company] but there are paint experts out there. One of the problems that we recognized was that many external experts have got professional indemnity insurance to carry. They can't afford to make mistakes. So therefore when we are going to them for a solution, they are going to give us the same solution, which might not be very economical, it might not be the best for us."

The net effect was to encourage an emphasis on local search behaviors—an effect made more pronounced by the felt dependence of the REMs on one another for information and support across the company. Moreover, the preferred means of contact were very traditional—namely, direct contact, telephone, and e-mail. Although e-mails were used intensively, other (electronic) mechanisms for accessing a wider knowledge base (namely, the company's intranet and wider Internet access) were not so well used. This was due partly to practical constraints (e.g., limited access on some sites and difficulties in updating the database), but also to personal contact being clearly the preferred mode of communication. According to the company's senior engineer: "I find it's easier to dial one of our regional offices than it is to get on to the intranet"; and "In these days of electronic wizardry and technology, my opinion is that you can't beat a face-to-face, eyeball-to-eyeball meeting."

In the case of Teleco, the technology watch team used its extensive internal network to draw on the expertise of Teleco employees when deciding if a company was potentially interesting. Nevertheless, using this internal network was problematic in that it also limited other types of search activity that might potentially have been employed. In particular, it limited attempts to access sources external to Teleco—at a time when drawing on external networks for knowledge and advice was increasingly important in making new technology recommendations (cf. Prusak and Cohen 2001). One senior manager indicated his frustration with the system, stating that "I would hope that my [internal] customers would be not only XXX, which are my current customers, but YYY as well. The reason why my current customers are XXX is because the people who have moved into XXX came from AAA [the business unit in which this manager works]." Within Teleco, a sophisticated information retrieval system had been developed to help overcome local search behaviors, but according to one project manager, it was not widely used because people preferred "personal e-mail, the water cooler, and meetings." This overreliance on strong internal connections became a problem for the team's way of working when a significant employee exodus from the company occurred, leaving considerable gaps in internal knowledge and expertise as well as huge holes in individuals' networks of contacts (cf. Uzzi 1997). As the project manager said:

> [The team] don't have the time to [really investigate a particular technology] and they don't have the expertise, but they have to know whom they can draw on. ... It's a huge tacit kind of network of intelligence they depend on. People that they know there. Of course, it's a problem for them now if this team is getting reduced, heavily losing its technological expertise. This is a gaping hole in terms of [Teleco's] investments. Because what do you do in the e-trading space if you don't have the knowledge down in here?

The importance of the difference between strong and weak structural ties is perhaps best illustrated by the Healthorg case. Here, there was a similar reliance on extended networks, as efforts were made to obtain information about alternative cataract treatment processes elsewhere in the National Health Service (NHS). Yet the significant point about this case was that the successful development and deployment of a new system fundamentally depended on the social processes involved in arriving at a solution within the team itself (Newell et al. 2003). Extended contacts giving access to information elsewhere were in this case less valuable than the strong, tacit understandings within the multidisciplinary team that developed in the process of drawing up a new process for cataract diagnosis and treatment. As one of the medical staff explained, there was a general awareness that the cataract process could be improved, but there was still a need to find a way of bringing people together to effect a change: "I think the problem that people have found is that [medical] consultants are keen to change things, but feel that the clinical load is so great that they just get on and work to the best of their ability within the current system. It requires management facilitation to enable them to change. It is very difficult to just change on your own." In this case, that facilitation was provided, enabling different professional groups to come together and use their knowledge as well as expertise to introduce the change process. By the same token, however, these mechanisms and processes were less useful as a way of disseminating knowledge about how to develop a similar system elsewhere.

The Cognitive Dimension of Social Capital and the Problem of Social Closure

Again, there were examples from across the cases of benefits and drawbacks of cognitive social capital. Perhaps the strongest findings emerge again from the Healthorg case, where the reengineering process brought together a wide range of different specialists with quite diverse perspectives and interests (doctors, nurses, administrators, optometrists, and clerical staff). Crucially, the success of this initiative depended on establishing a common frame of reference for viewing the problem, and in particular, a willingness to relax professional and functional boundaries and jurisdictions (doctors accepting optometrists' ability to diagnose;

clerical staff accepting less control over consultants' appointment bookings). According to one of the opticians, for example, "because it is breaking down the barriers between GPs, optometrists, and the hospital, it opens up avenues of referral from, say, GPs to optometrists that weren't there before." Indeed, the team devoted considerable time and energy to achieving a common purpose (including holding project meetings outside of working hours). Breaking down existing barriers and *avoiding* the problems caused by an initial lack of a shared perspective were the main messages from this case. In other words, in the process of achieving the change, well-established social capital—in the form of professional belief systems—was a barrier to, rather than an enabler of, change.

In the other cases, the impact of cognitive social capital was perhaps less strong, though still significant. At Teleco, the effects were mainly positive: staff who remained with the company had long-standing connections with colleagues throughout the company and beyond with whom they shared similar professional perspectives. On the other hand, one of the issues confronted in this case was that staff were expected no longer to be simply "technologists" (writing codes and building systems) but more commercially aware, with the ability to "market" their ideas internally and find project sponsors. In other words, existing cognitive social capital was no longer in tune with the needs of the organization. As one manager put it, "If you think you have got a good service, you have got to sell it internally."

At Constructco, there were also mixed results. It was certainly the case that a shared engineering culture and language helped the REMs communicate effectively among themselves and function cohesively as a group within the organization. This is apparent in the following comments from the company's senior engineer:

Just as we have these engineering forums within the business, then the REMs themselves and myself, we get together every three months so that we can, face-to-face, discuss our collective concerns and problems, successes and failures, and such like.... I would like to get on to the site more, but when I do get on to the site, nearly every time I go on to the site, there may be an engineer I've spoken to before, or an engineer that I haven't spoken with before. I probably get more out of him sitting down for an hour than the site manager knows. He'll say, "I can't do this, or I am struggling a bit with this," but he won't say it to the line manager.

At the same time, shared orientations were undoubtedly offset by the fact that there were quite different expectations of the role in different regional offices. According to the senior engineer who established the role, "Each of our businesses to some degree runs itself differently.... [Each has] REMs, and they are all doing different things.... It's first of all train the engineers and then put them in the right place at the right time. [Then] it's sort out the technical problems, technical issues at tenders. [Then] it's to make sure that when we have won the job at the contact stage that they are best engineered... so depending on the needs of each business, they need to put resources into all those." Moreover, it could also be argued that their shared perspective inevitably meant that problems and solutions were still looked at mainly in pure engineering terms, despite the espoused emphasis on value engineering within the company. In this case, the effects of shared understandings and orientations were therefore equivocal.

The Relational Dimension of Social Capital and the Problem of Resistance to Change

In the literature on social capital, most interest has been directed at norms of reciprocity and trust as the underlying basis for processes of exchange. Even though these might be the most widely discussed norms with respect to social capital, others are potentially important too. In particular, norms of conformity, control, and compliance—for example, to professional codes or bureaucratic rules—can deter individuals from embracing or accepting innovative organizational activities if they involve any significant deviation from the established norms.

At Constructco, reciprocity norms were important in governing relationships among the REMs and construction site managers. The REMs relied on site managers to carry out or support any engineering initiatives that were introduced. Site managers were willing to do this, even though it meant extra work, because they knew that the REM was acting as an advocate for them at the head office. Similar norms of reciprocity and trust were apparent in the relationships among the REMs themselves as well as the REMs and site engineers. One site engineer, who expressed satisfaction with the way in which the role was working, remarked that "in the [past] two months, I have seen [the REM] three or four times. Also I have spoken to him ten or fifteen times regarding specifications on-site.... So I was after his technical ability as well as asking for his opinion on certain things that he [would] do on-site.... But I know he has been round to speak to all the engineers.... He does spend, I think, three or four days traveling to the sites, speaking to people, asking if there are things wrong or do [they] need anything." Conversely, the existence of within-company differences in how the REMs were used did demonstrate how the normative bases of social capital within the firm could impede initiatives such as this, which cut across established organizational systems and practices. Specifically, the variable balance of emphasis between short-term project engineering and long-term staff development needs did reveal how different subcultural norms (themselves important sources of relational social capital) could have a negative impact on new initiatives. These norms initially led to resistance or, at best, reluctant acceptance of the new role. Later, they could still lead to significant distortions in performance of the REM role across the company as a whole.

Norms of reciprocity were also critical in the information search process that the technology watch group undertook at Teleco. Moreover, such norms were heightened among the remaining employees and helped create a strong sense of cohesion. Yet the norms of reciprocity were often abused at Teleco due to the influence of the dominant hierarchy. It was reported, for example, that junior employees would pass on information about a promising technology to a more senior organizational member, but without receiving credit for their involvement. This created defensive strategies, one being the avoidance of social networks to disseminate information until the investigation of the technology was substantially completed. Other powerful norms also had negative consequences, including a strong "not invented here" attitude (cf. Katz and Allen 1982) associated with traces of the dominant culture that had existed prior to the reorganization of the company.

At Healthorg, the picture was quite different in that traditional normative assumptions about the role of medical professionals needed to be altered to some extent at least before the new process was developed and accepted. Through extensive contact and consensus building, perceptions about professional qualifications were slowly modified and preexisting normative barriers to innovation were broken down (Newell et al. 2003). This process was by no means entirely free of conflict, and significant problems were reported in gaining acceptance of the new system by particular groups of staff. As well as skepticism by some medical staff, the main obstacle encountered was a fear that the centralization of appointments would threaten clerical staff's jobs and/or status and autonomy (Portes 1998). In this instance, norms of solidarity and collective interest were more important underlying influences, and the social capital that was mobilized to resist change proved a major obstacle to implementing the new procedure. As one of the transformation team members recalled, "[Changing the secretaries' role] did meet a lot of resistance. It was very odd really because they saw it as a great threat, and one of the comments was that they hadn't become secretaries to do typing, and they saw the waiting list management as a big part of their role. They felt that we were undermining their role by taking this away, and if we took it away from them, they also felt that we were taking away their patient contact."

In order to overcome this resistance, meetings were held to try and get the secretaries to see that the new role would not be inferior. Nevertheless, as the same interviewee noted, "That was the other thing: a lot of the meetings you had with them, they didn't speak. There was never any two-way [communication]; it was really hard, there was never anything coming from them. They would sit there stony-faced, and it would all come out afterward. They would get together afterward and carry out a postmortem. We gave them plenty of opportunities to tell us how they felt about it, but they were not forthcoming on a face-to-face basis at all." To overcome this resistance, the transformation team arranged for some of the secretaries to go to another hospital where the secretarial role had already been rearranged. As the interviewee recalled, however: "Yes, the secretaries that went were really enthusiastic at the time and they seemed really into it, and then they came back and met with the other secretaries, and we lost it." In this case, then, there was a clear sense in which the social capital that existed within the secretarial group actually raised barriers to change (Portes 1998; Portes and Sensenbrenner 1993).

The Interaction Effects of the Structural, Cognitive, and Relational Dimensions

The findings from our study also suggested that there were important interaction effects between the structural, cognitive, and relational dimensions of social capital that had both positive and negative effects. Where groups had developed a shared set of understandings (cognitive dimension) and strong norms of trust and reciprocity (relational), then they were also likely to have developed strong and multiple social linkages (structural). Whereas this might have created a level of cohesion that facilitated knowledge movement within the group, it could also create barriers around the group, shielding it from potentially beneficial knowledge and information that was simply outside the reach (structural), understanding (cognitive), and/or acceptance (relational) of group members. Such barriers would make it difficult for community members to access information from outside the group, and could lead to distorted perceptions about the value and efficacy of accessing or searching for information and knowledge beyond the immediate community (cf. Uzzi 1997).

Similar barriers to innovation were identified in the cases explored here, although their effects were different depending on the particular combination of circumstances. Perhaps the main barrier facing the transformation team at Healthorg was the need to overcome a lack of trust among the health care professionals who were involved in the process, but who came from different professional disciplines. Once the consultants had accepted that optometrists were capable of undertaking a greater role in the diagnosis of cataract problems, however, then the new procedure for diagnosis and treatment could be implemented fairly easily. The point here is simply that social capital in the form of professionally based norms and practices would normally have inhibited the development and implementation of this innovative practice. Introducing the new procedure necessitated the reconfiguration of contacts to form a multiprofessional team (structural), the development of a shared understanding of the problem and possible solutions (cognitive), and the breaking down of barriers to trust and acceptance among the various health care specialists (relational). Conversely, where problems were experienced in implementation (i.e., resistance from the secretaries), this reflected the self-reinforcing combination of strong internal structural ties, shared orientations and concerns about the effects of the change, and a strong bond of solidarity within the group (cf. Gabbay and Leenders 1999; Gargiulo and Benassi 1999).

In the case of Constructco, it has already been noted how the creation of a cohesive network of REMs was beneficial in providing access to information and knowledge throughout the company. At the same time, though, members came from within a single discipline and tended to search across the same internal networks. These factors promoted a somewhat inward-looking perspective and created problems where there was a need to access new information or different expertise to address novel problems (Bresnen et al. 2003). In other words, the ability to network with other internal REMs and engineers (structural) on the basis of both a shared professional understanding (cognitive) and a two-way relationship premised on norms of reciprocity and trust (relational) created a cohesive internal network for the flow of information and advice. At the same time, however, there were suggestions that its very internal coherence may have meant that it did not necessarily engage fully with other streams of information, knowledge, and advice within the organization as a whole, or indeed within a wider professional and institutional field (cf. Uzzi 1997).

Interaction effects can also work to the detriment of an individual firm's efforts at value creation where innovation depends on a wider network of interfirm linkages. As organizational members leave the focal firm, they take their preexisting social capital ties with them. Individual network ties with current and former colleagues, which are based on norms of trust and reciprocity as well as a common affiliation with a broader community of professionals, are often stronger than individual ties to the focal organization (Liebeskind et al. 1996). This can lead to potentially valuable information leaving the focal firm through preexisting social networks. The Teleco case was the prime example of these types of problems. Against a backdrop of radical organizational change and the uncertainties that this created, employees were leaving the firm in large numbers to seek out more fulfilling and potentially more lucrative positions in other companies. The effects of this were quite complex. On the one hand, this employee exodus could be beneficial to Teleco, leading to the creation and use of wider social networks to access new ideas and information. On the other hand, it could be potentially damaging in that preexisting social capital ties could let vital organizational knowledge "leak" outside the firm's boundaries (cf. Brown and Duguid 2001). In this case, we see a complex relationship among the three dimensions of social capital, brought about by a change in the pattern of relationships, which also had complex cognitive and relational effects as a result of the tensions among professional, commercial, and organizational norms, values, and interests.

Conclusion

The case study findings above lend weight to the argument that social capital is of benefit to organizations and an important potential source of value creation within firms. Indeed, the cases supplied important examples of how innovative processes were facilitated by, or even depended on, extended networks of social contacts, the creation of dialogues of shared meaning, and the establishment of appropriate norms governing interaction. Yet they also shed some light on the problematic nature of social capital as well as the contradictory tendencies it can create in terms of knowledge creation and sharing within and among organizations in a project-based setting. Here, examples were given of the inhibiting effects of closed social or occupational networks, how shared systems of meaning could exclude alternative views or sources of knowledge from outside the community, and the power of group or occupational norms in creating social solidarity and resistance in the face of pressures to change. It was further noted how these less beneficial aspects of social capital could interact with one another and be mutually reinforcing, thereby creating powerful forces that are able to cancel out any beneficial value creation efforts. Not only does this suggest, therefore, that the effects of social capital may be negative as well as positive (cf. Adler and Kwon 2002), it also indicates that they may be ambivalent or equivocal—creating contradictory tendencies that make social capital at the same time both beneficial *and* costly.

The findings thus point to a number of themes that can be recast as propositions for further research, and as the basis for a more complete model of the benefits and drawbacks of social capital. First, that structural social capital provides access to asymmetrically distributed information, while at the same time discouraging a wider information search. This reflects not only the perceived costs of attempting to access knowledge through weak external ties and the tendency to rely instead on strong internal ties (cf. Granovetter 1973; Hansen 1999) but also the positive attractions of relying on internal ties in terms of reinforcing close network relationships. The result is an overdependence on particular local networks, and the danger of insulation from wider, potentially more valuable sources of knowledge and information (Prusak and Cohen 2001; Uzzi 1997). Second, that cognitive social capital creates close ties among community members, while simultaneously excluding new information and knowledge that originates from outside the social network and/or that does not align closely with existing shared understandings (cf. Portes 1998). The net effects include the possible creation of competence traps (Levitt and March 1988) and/or the reinforcement of a not invented here attitude (Katz and Allen 1982). Third, that relational social capital creates strong normative forces that in turn create more cohesive relationships, but that at the same time, deter individuals from deviating from expected behaviors, and potentially isolate groups from wider social norms. Not only do such effects encourage individual conformity (cf. Locke 1999), they may also exacerbate any between-group differences or conflict as a result of their exclusionary and/or solidarity effects (cf. Fine 1999). Fourth, that structural, cognitive, and relational dimensions of social capital interact in ways that can magnify the benefits that are realized or the drawbacks that are experienced.

These findings also indicate that there is much to be gained from continuing with a cross-sector approach to exploring social capital effects. The ability to draw comparisons and contrasts in the use of structural, cognitive, and relational social capital across three quite distinct case studies emphasizes the effect that the social context has on the ways in which social capital develops and is used within organizations. One particular theme, common across the cases, relates to the attempt to create intellectual capital both within and across professional boundaries. As a social capital perspective would imply, professional boundaries can facilitate the development of shared meanings and norms of cooperation and trust. Yet they can also simultaneously create powerful barriers to the sharing of knowledge within the organization where this involves sharing across functional or professional boundaries. Also important across the cases were strong internal cultural barriers associated with well-established professional/functional or even divisional structures that inhibited attempts to introduce pan-organizational structural change initiatives. Even where the actors in our cases made significant headway in overcoming such barriers, the new structures and processes that were implemented were often adversely affected by traces of prior structures or cultures—whose tenacious grip reflected the impact of deeply entrenched values and norms along with well-established social capital.

The finding that there are these double-edged effects of social capital, which stem from professional/functional and subunit boundaries within the organization, highlights too the value of examining social capital effects in project-based settings. Given the fragmented and discontinuous nature of project-based work (Prencipe and Tell 2001), an organization faces many difficulties in developing its core competencies, harnessing human capital, and transferring knowledge and expertise from one project to the next (DeFilippi and Arthur 1998). Moreover, project-based organization is one setting in which professional diversification and the problems of subunit differentiation referred to above are the norm. Consequently, there may be specific challenges and issues faced in the generation and harnessing of social capital benefits in situations where project-based learning is important. In such a situation, there is perhaps even more of a premium to be placed on social capital as it helps to improve the network of subunit relations and enhances knowledge relatedness across projects (Hansen 2002). On the other hand, the findings reported here suggest that there are circumstances under which social capital can also have negative consequences for innovation and learning. If the forces driving social capital formation and use in the case of projects center on project objectives and processes, then it is not difficult to see how social capital might be of benefit to the performance of individual projects. It is also not difficult, however, to see how social capital might *inhibit* cross-project learning and innovation to the extent that it reinforces an overdependence and overemphasis on the specific local networks surrounding particular project teams. Such a combined effect creates, of course, a potential paradox in which social capital has its most deleterious effects under circumstances where it is more important to develop it to encourage innovation (i.e., project-based environments). In other words, the development of social capital for innovation in particular types of environments and in particular combinations of circumstances may be more harmful than beneficial.

Understanding the effects of social capital in project-based settings also has something to say about the types of contingencies that are likely to enable and inhibit learning and innovation more generally. For example, it seems likely that the sources, dynamics, and effects of social capital formation will be contingent on a number of other factors that are commonly encountered in project environments, and that may also have similar contradictory and paradoxical effects. These include whether the environment is dynamic as opposed to static, whether the scale of operations is small or large, to what extent activity depends on interorganizational relationships, and the stage the firm is at in the product or organizational life cycle. The archetypal small, newly established firm operating in a high-tech environment (e.g., biotechnology) is one obvious instance of where these factors combine to make social capital via network collaboration more difficult to achieve, yet at the same time, more vital for the development of innovative products and processes (Oliver and Liebeskind 1998; Powell 1998). Research has also indicated that the transience of project work in more established and mature industries, such as construction, inhibits the establishment of networks of contacts and relational norms that can be critical in the development and maintenance of collaborative interorganizational relationships (Bresnen and Marshall 2002). More research is needed in order to identify the precise effects of such contingencies. Nevertheless, it is clear from the research reported here that the contingencies associated with project work may have a significant bearing on understanding the impact of social capital more widely.

Last, but by no means least, the research also reinforces the idea that the diffusion of knowledge and learning throughout an organization is crucially dependent on the social dimensions of knowledge creation, sharing, and capture (Brown and Duguid 2001; Hansen, Nohria, and Tierney 1999; Wenger 2000). It was striking that in the two cases in which IT was potentially of great importance in encouraging knowledge sharing (Teleco and Constructco), much more of a premium was placed on social interaction as a way of building networks as well as sharing knowledge and learning. If the social aspects of knowledge creation and sharing are in many respects more vital than new technology in enhancing innovations within organizations (Fahey and Prusak 1998), then it is clear from this research that the same holds true for social capital. Of course, information and communication technologies play a key part in enabling the social interaction on which social capital depends. Still, the research suggests that there are limitations in their use in enhancing the quality of social interaction necessary for the diffusion of complex forms of knowledge and the tacit understanding associated with social capital (especially the cognitive and relational aspects). Moreover, there are limitations in their ability to overcome some of the barriers associated with social capital insofar as these derive from shared assumptions and understandings along with integrative and/or discriminatory relational norms. As demonstrated in the Teleco and Constructco cases, it may even be the case that those shared assumptions and norms partly revolve around the appropriateness or otherwise of different modes of communication. In this case, the use of IT may be seen as antithetical to the social interaction needed for the development of social capital. If that is the case, then it becomes important to understand the circumstances under which information and communication technologies support or inhibit the development of social capital. Furthermore, it poses a significant question about whether such technologies exacerbate or ameliorate the negative consequences of social capital—helping to build up or break down barriers to the social interaction necessary for the emergence and use of social capital. These issues and questions are critical topics for further research and debate. As the general tenor of the findings reported in this chapter would suggest, though, the relationship between such technologies and social capital is by no means as symbiotic as is often hoped, and is perhaps just as likely, if not more likely, to be antagonistic.

Note

1. The research was supported by the Engineering and Physical Sciences Research Council, grant number GR/M73286. Cases in pharmaceuticals and social services were also examined in the research, but are not included here owing to space limitations.

References

Adler, P. S., and S.-W. Kwon. 2002. Social capital: Prospects for a new concept. *Academy of Management Review* 27, no. 1:17–40.

Argote, L. 1999. *Organizational learning: Creating, retaining, and transferring knowledge.* Boston: Kluwer Academic Publishers.

Belliveau, M., C. O'Reilly, and J. Wade. 1996. Social capital at the top: Effects of social similarity and status on CEO compensation. *Academy of Management Journal* 39, no. 6:1568–1584.

Bouty, I. 2000. Interpersonal and interaction influences on informal resource exchanges between R&D researchers across organizational boundaries. *Academy of Management Journal* 43, no. 1:50–66.

Brass, D. J., and M. E. Burkhardt. 1992. Centrality and power in organizations. In *Networks and organizations: Structure, form, and action*, edited by N. Nohria and R. G. Eccles. Boston: Harvard Business School Press.

Bresnen, M. 1990. *Organizing construction: Project organization and matrix management.* London: Routledge.

Bresnen, M., L. Edelman, S. Newell, H. Scarbrough, and J. Swan. 2003. Social practices and the management of knowledge in project environments. *International Journal of Project Management* 21, no. 3:157–166.

Bresnen, M., and N. Marshall. 2002. The engineering or evolution of cooperation? A tale of two partnering projects. *International Journal of Project Management* 20, no. 7:497–505.

Brown, J. S., and P. Duguid. 1991. Organizational learning and communities of practice: Towards a unified view of working, learning, and innovation. *Organization Science* 2, no. 1:40–57.

Brown, J. S., and P. Duguid. 2001. Knowledge and organization: A social practice perspective. *Organization Science* 12:198–213.

Burt, R. S. 1992. *Structural holes: The social structure of competition.* Cambridge: Harvard University Press.

Burt, R. S. 1997. The contingent value of social capital. *Administrative Science Quarterly* 42, no. 2:339–366.

Burt, R. S., R. M. Hogarth, and C. Michaud. 2000. The social capital of French and American managers. *Organization Science* 11, no. 2:123–147.

Chung, S., H. Singh, and K. Lee. 2000. Complementarity, status similarity, and social capital as drivers of alliance formation. *Strategic Management Journal* 21, no. 1:1–22.

Coleman, J. S. 1988. Social capital and the creation of human capital. *American Journal of Sociology* 94:S94–S120.

Coleman, J. S. 1990. *Foundations of social theory*. Cambridge: Harvard University Press.

DeFilippi, R., and M. Arthur. 1998. Paradox in project-based enterprises: The case of filmmaking. *California Management Review* 40, no. 2:125–140.

Elg, U., and U. Johannson. 1997. Decision-making in inter-firm networks: Antecedents, mechanisms, and forms. *Organisation Studies* 16, no. 2:183–214.

Fahey, L., and L. Prusak. 1998. The eleven deadliest sins of knowledge management. *California Management Review* 40, no. 3:265–276.

Fernandez, R. M., E. J. Castilla, and P. Moore. 2000. Social capital at work: Networks and employment at a phone centre. *American Journal of Sociology* 105:1288–1356.

Fine, B. 1999. The developmental state is dead—long live social capital? *Development and Change* 30, no. 1:1–19.

Gabbay, S. M., and R. Th. A. J. Leenders. 1999. CSC: The structure of advantage and disadvantage. In *Corporate social capital and liability*, edited by R. Th. A. J. Leenders and S. M. Gabbay. Boston: Kluwer Academic Publishers.

Gabbay, S. M., and E. W. Zuckerman. 1998. Social capital and opportunity in corporate R&D: The contingent effect of contact density on mobility expectations. *Social Science Research* 27:189–217.

Gargiulo, M., and M. Benassi. 1999. The dark side of social capital. In *Corporate social capital and liability*, edited by R. Th. A. J. Leenders and S. M. Gabbay. Boston: Kluwer Academic Publishers.

Gargiulo, M., and M. Benassi. 2000. Trapped in your own net? Network cohesion, structural holes, and the adaptation of social capital. *Organization Science* 11, no. 2:183–196.

Granovetter, M. 1973. The strength of weak ties. *American Journal of Sociology* 78:1360–1380.

Granovetter, M. 1992. Problem of explanation in economic sociology. In *Networks and organizations: Structure, form, and action*, edited by N. Nohria and R. G. Eccles. Boston: Harvard Business School Press.

Hansen, M. T. 1999. The search transfer problem: The role of weak ties in sharing knowledge across organizational sub-units. *Administrative Science Quarterly* 44:82–111.

Hansen, M. T. 2002. Knowledge networks: Explaining effective knowledge sharing in multiunit companies. *Organization Science* 13, no. 3:232–248.

Hansen, M. T., N. Nohria, and T. Tierney. 1999. What's your strategy for managing knowledge? *Harvard Business Review* 77:106–117.

James, E. H. 2000. Race-related differences in promotions and support: Underlying effects of human and social capital. *Organization Science* 11, no. 5:493–508.

Janis, I. L. 1982. *Victims of groupthink.* 2d ed. Boston: Houghton Mifflin.

Katz R., and T. J. Allen. 1982. Investigating the not invented here (NIH) syndrome: A look at the performance, tenure, and communication patterns of fifty R&D project groups. *R&D Management* 12, no. 1:7–19.

Kogut, B., and U. Zander. 1992. Knowledge of the firm, combinative capabilities, and the replication of technology. *Organization Science* 3, no. 3:383–397.

Kogut, B., and U. Zander. 1996. What firms do? Coordination, identity, and learning. *Organization Science* 3, no. 3:383–397.

Koka, B. R., and J. E. Prescott. 2002. Strategic alliances as social capital: A multidimensional view. *Strategic Management Journal* 23, no. 9:795–816.

Kraatz, M. S. 1998. Learning by association? Interorganizational networks and adaptation to environmental change. *Academy of Management Journal* 41:621–643.

Leana, C. R., and H. J. Van Buren. 1999. Organizational social capital and employment practices. *Academy of Management Review* 24, no. 3:538–555.

Leenders, R. Th. A. J., and S. M. Gabbay, eds. 1999. *Corporate social capital and liability.* Boston: Kluwer Academic Publishers.

Leonard-Barton, D. 1990. The intraorganizational environment: Point-to-point versus diffusion. In *Technology transfer: A communication perspective,* edited by F. Williams and D. V. Gibson. Thousand Oaks, Calif.: Sage Publications.

Levitt, B., and J. G. March. 1988. Organizational learning. *Annual Review of Sociology* 14:319–340.

Liebeskind, J. P., A. L. Oliver, L. Zucker, and M. Brewer. 1996. Social networks, learning, and flexibility: Sourcing scientific knowledge in new biotechnology firms. In *Managing in times of disorder: Hypercompetitive organizational responses,* edited by A. Y. Ilinitch, A. Y. Lewin, and R. D'Aveni. Thousand Oaks, Calif.: Sage Publications.

Locke, E. A. 1999. Some reservations about social capital. *Academy of Management Review* 24, no. 1:8–9.

Meyerson, D., and J. Martin. 1987. Cultural change: An integration of three different views. *Journal of Management Studies* 24, no. 6:623–647.

Moorhead, G., R. Ference, and C. P. Neck. 1991. Group decision fiascos continue: Space shuttle Challenger and a revised groupthink framework. *Human Relations* 44, no. 6:539–550.

Nahapiet, J., and S. Ghoshal. 1998. Social capital, intellectual capital, and the organizational advantage. *Academy of Management Review* 23, no. 2:242–266.

Newell, S., L. Edelman, H. Scarbrough, J. Swan, and M. Bresnen. 2003. "Best practice" development and transfer in the NHS: The importance of process as well as product knowledge. *Health Services Management Research* 16:1–12.

Oliver, A. L., and J. P. Liebeskind. 1998. Three levels of networking for sourcing intellectual capital in biotechnology: Implications for studying interorganizational networks. *International Studies of Management and Organization* 27, no. 4:76–103.

Podolny, J. M., and J. N. Baron. 1997. Resources and relationships: Social networks and mobility in the workplace. *American Sociological Review* 62:673–693.

Portes, A. 1998. Social capital: Its origins and applications in modern sociology. *Annual Review of Sociology* 23:1–24.

Portes, A., and P. Landolt. 1996. The downside of social capital. *American Prospect* 94, no. 26:18–21.

Portes, A., and J. Sensenbrenner. 1993. Embeddedness and immigration: Notes on the social determinants of economic action. *American Journal of Sociology* 98:1320–1350.

Powell, W. W. 1998. Learning from collaboration: Knowledge and networks in the biotechnology and pharmaceutical industries. *California Management Review* 40:228–240.

Prencipe, A., and F. Tell. 2001. Inter-project learning: Processes and outcomes of knowledge codification in project-based firms. *Research Policy* 30:1373–1394.

Prusak, L., and D. Cohen. 2001. How to invest in social capital. *Harvard Business Review* 79, no. 6:86–93.

Putnam, R. D. 1993. The prosperous community: Social capital and public life. *American Prospect* 13:35–42.

Putnam, R. D. 1995. Bowling alone: America's declining social capital. *Journal of Democracy* 6:65–78.

Ring, P. S., and A. H. Van de Ven. 1994. Developmental processes of cooperative interorganizational relationships. *Academy of Management Review* 19:90–118.

Sandefur, R. L., and E. O. Laumann. 1998. A paradigm for social capital. *Rationality and Society* 10:481–501.

Spender, J.-C. 1996. Organizational knowledge, learning, and memory: Three concepts in search of a theory. *Journal of Organizational Change* 9, no. 1:63–78.

Swan, J. A., S. Newell, H. Scarbrough, and D. Hislop. 1999. Knowledge management and innovation: Networks and networking. *Journal of Knowledge Management* 3:262–275.

Szulanski, G. 1996. Exploring internal stickiness: Impediments to the transfer of best practices within the firm. *Strategic Management Journal* 17:27–43.

Tsai, W. 2000. Social capital, strategic relatedness, and the formation of intra-organizational linkages. *Strategic Management Journal* 21, no. 9:925–940.

Tsai, W., and S. Ghoshal. 1998. Social capital and value creation: The role of interfirm networks. *Academy of Management Journal* 41, no. 4:464–476.

Uzzi, B. 1997. Social structure and competition in interfirm networks: The paradox of embeddedness. *Administrative Science Quarterly* 42:35–67.

Walker, G., B. Kogut, and W. Shan. 1997. Social capital, structural holes, and the formation of an industry network. *Organization Science* 8, no. 2:109–128.

Weick, K. E. 1995. *Sensemaking in organizations*. London: Sage Publications.

Wenger, E. 2000. Communities of practice and social learning systems. *Organization* 7, no. 2:225–246.

III
Applications of IT

Part III explores a variety of computer applications that have the potential to foster social capital. The main question here is how to support social capital by means of appropriately designed functionality. It should be noted that while the computer applications and design approaches examined in these four chapters are interesting as well as innovative, an evaluation of their (long-term) effects on the level of social capital among their communities of users is still missing.

In chapter 11, Mark S. Ackerman and Christine Halverson survey important parts of their own work. They suggest moving from the metaphor of knowledge management to a new metaphor, expertise sharing, which promotes a focus on the inherently collaborative and social nature of these activities. The authors show how the application of standard mechanisms for expertise sharing and knowledge management suffers from various problematic social issues. They state that there is still a considerable gap between what we do socially and what computer science as a field knows how to support technically. Overcoming this sociotechnical gap is one of the intellectual challenges facing design-oriented research on social capital. Ackerman and Halverson explain how they have tried to bridge this gap in their work. They classify their approaches to fostering expertise sharing and social capital into three types: tying together repositories with networks; self-feeding expertise locators; and lightweight social spaces. Ackerman's seminal work on the Answer Garden is presented as an example of tying together repositories and social networks. While the Answer Garden has not yet solved the problem of locating expertise in an organization, more empirical and technical work has also been conducted on the notion of building self-feeding expertise locators. Finally, the authors describe their work on lightweight social spaces. Findings concerning the design and use of the Zephyr messaging system along with chatlike communication tools such as Babble and Loops are detailed and discussed.

In chapter 12, Robbin Chapman looks at community support among after-school students. Like Ackerman in the Answer Garden approach, she ties together a shared repository with technical support for social networks. The chapter outlines a study at the Computer Clubhouse, a network of technology centers where youth participate in constructionist design activities. Within the Computer Clubhouse, social capital provides the framework that supports the process of learning through interaction. The challenge for communities like the Computer Clubhouse has been to determine how shared repositories can support the connections, networks, and reciprocal behavior so crucial to their sustained functioning. Chapman has taken into account the importance of social capital in the design of a suite of software tools, called Pearls of Wisdom (PoW), to support asynchronous knowledge sharing among community members. PoW offers features to motivate individual participation in the creation and use of knowledge artifacts. From a design point of view, it is interesting to note that though her basic approach is similar to that of the Answer Garden, the specific implementation is different. So community specifics—such as the assumed-knowledge distribution, incentives for knowledge sharing, availability of actors, or cultural homogeneity—have led to different implementations. When building applications to impact social capital, one has to take these specifics into account.

In the next chapter, Andreas Becks, Tim Reichling, and Volker Wulf work out a framework for expertise location and matchmaking. Such a system can be applied to make actors who are little known or even unknown to each other—but who share similar backgrounds, interests, or needs—aware of each other. Especially in virtual space, such a system can help build social capital in compensating for the lack of a physical context. The authors describe their framework, which allows one to apply different algorithms to match personal data describing the actors' behavior, background, qualifications, or interests. Design principles for matching algorithms, a general architecture for an expertise-matching, and the implementation of these functionalities are presented. The authors also show how their framework was applied to supplement a learning platform with an expertise-matching functionality. In this case, the establishment of a community of colearners can be supported. The authors also discuss future challenges in the field of expertise matching.

Finally, in chapter 14, Gerhard Fischer, Eric Scharff, and Yunwen Ye survey their work in the field of social creativity, discussing the role of social capital in this context. Social creativity is an important research field since complex design problems require more knowledge than any single person can possess, and the knowledge relevant to a problem is often distributed among different stakeholders. The authors show how social creativity can be supported by innovative computer applications. In fostering social creativity, however, appropriate technology has to be complemented by a concern for social capital. In this chapter, the authors first analyze existing success models (open source and knowledge sharing by means of an Internet portal) for social creativity. They then present their own work in creating social capital–sensitive applications (e.g., Code Broker, Envisionment and Discovery Collaboratory, and courses-as-seeds) that support collaborative design, problem solving, and knowledge co-construction. These systems show the importance of encouraging users to act as active contributors and illustrate some of the motivational challenges on which these systems rely. The authors conclude that without an appropriate emphasis on social capital, the impact of these new technologies will be negligible.

11
Sharing Expertise: The Next Step for Knowledge Management

Mark S. Ackerman and Christine Halverson

There are numerous ways to handle knowledge within organizations. Indeed, knowledge management has been a flourishing commercial area for almost ten years, and one can point to many precursors within organizations as well. Knowledge management—regardless of its title or position in history—has always been an important, though not necessarily frequent, aspect of organizational life. It would be difficult to imagine a modern corporation that did not occasionally reflect on and improve its methods of handling communications, data, and information—or try to learn from its experience. In this chapter, we move from the metaphor of *knowledge management* to a new metaphor, *expertise sharing*, which promotes a focus on the inherently collaborative and social nature of the problem.

In our view, knowledge management subsumes a number of differing strategies. What all of these strategies share—as do many information-access strategies—are interactions with or foundations in the social setting of an organization or institution. This is made explicit in descriptions of social capital. While social capital itself can be many things, we consider here its use within an organizational knowledge management definition (Cohen and Prusak 2001; Lesser, Fontaine, and Slusher 2000; Stewart 2001; Wenger 1998). (See the excellent review of the definitions and connotations of social capital in chapter 2 of this volume.)

Social capital, as described by Marleen Huysman in this volume (following Nahapiet and Ghoshal 1998), has three dimensions: structural, cognitive, and relational. As Eric Lesser and Laurence Prusak point out, "The structural dimension of social capital reflects the need for individuals to reach out to others within an organization to seek out resources that they may not have at their disposal" (2000, 126). It consists of network ties along with their configuration and organization. The cognitive dimension includes shared language and common narratives; hence, it includes the social-cognitive aspects of an organization. The relational dimension includes trust, norms, obligations, and a shared identification. As Lesser and Prusak note, "Social capital is developed and fostered when individuals believe that their actions will be appropriately reciprocated and that individuals will meet their expected obligations" (ibid., 127).

Our contention here is that the structural, shared cognitive, and relational dimensions are the social aspects that must be incorporated into next-generation knowledge management systems. The structural, shared cognitive, and relational dimensions of an organization allow knowledge and expertise to be shared. Expertise sharing, then, requires a deep consideration of how social capital must unfold throughout knowledge management. Only in considering these components of an organization can anything approaching knowledge management or organizational memory be achieved in practice.

Accordingly, we will first show how the four standard mechanisms for sharing expertise and managing knowledge suffer from various collaborative and social issues. Underlying these issues is one of the intellectual challenges facing computer-supported cooperative work (CSCW, or groupware) as a field. The gap between what we know we have to do socially and what computer science as a field knows how to do technically (what we call here the *social-technical gap*) has led the two of us to reflect on potential systems designs in order to ameliorate this social-technical gap. Accordingly, the last half of the chapter is a review of our research that has evolved into a more organizationally attentive direction. Our description of these research systems focuses on their incorporation of, or augmentation to, the structural and relational aspects of social capital. (We will largely leave the cognitive aspects aside, as we believe that much of knowledge management assumes these. In addition, we will use the terms *social-structural* and *social-relational* to differentiate them from the technology terms *structural* and *relational*.) We conclude with some potential future research directions.

Current Technical Possibilities for Knowledge Management

Broadly speaking, there are four technical directions that knowledge management or expertise-sharing systems take at present. They align along a dimension that ranges from "objectified" knowledge decontextualized and separated from individuals and placed into repositories, to "embedded" and "community" knowledge found in groups of individuals. These should not be viewed as a progression since each has its own place. One of the purposes of this chapter is to understand the trade-offs involved in handling expertise or knowledge in each manner.

The first technical possibility is a *repository*. Typically, this consists of a data store of "knowledge" fragments. These are similar to, or sometimes the same as, corporate databases. An *expertise locator* is a recommendation engine or "yellow pages" directory that helps people find other people with the expertise that is required for some activity. A *computer-mediated place* (e-community, *knowledge community*, or computer-mediated communication system) is a virtual space where people with questions or answers can gather. Finally, there is hope that one can collect people into *ad hoc groupings*: flexible arrangements of an organization's social network in order to solve specific, time-limited problems. We will cover each in turn.

Repositories

The original vision for a knowledge repository was relatively simple: A company should be able to remember what it has previously done. The idea, then, was to put that previous experience, or knowledge, in an information-base of some sort. These efforts ranged from data warehouse initiatives to new forms of text databases such as Lotus Notes.

Perhaps because repositories were the first technical augmentation in this knowledge wave, the issues surrounding it are perhaps best understood. This vision of knowledge repositories had four major weaknesses. First, proponents felt that they could construct one information-base for an entire company. This idea was quickly discarded, as the political realities and technical issues involved became apparent. Second, there was an assumption that all knowledge could be removed from individuals and placed into an information-base. This notion was also discarded, although more slowly, as proponents began to discover which information could be decontextualized appropriately or even made explicit. Third, it was assumed that people would share their knowledge spontaneously, and finally, that people would naturally understand what others had put into the information-base.

These problems result largely from not understanding the social and organizational dimensions, both social-relational and social-structural, of repositories. In an organization, information is not value-free. Nor is sharing free—it carries psychological costs, and the rewards may be unclear. If others plan to use the information, then it takes time and effort to properly write up the information. That is, the writer must go through the effort of properly decontexualizing the information, and then the reader must go through the effort of recontextualizing it to her or his own needs. Of course, storing knowledge "objects" is not the same as understanding the social processes that surround knowledge acquisition, dissemination, and understanding.

Expertise Locators
After people began to see the limitations of knowledge repositories, they started to explore how people might also provide information and knowledge to others. Accordingly, we have viewed expertise location as either finding the right person to answer the right question or finding a person to appropriately complement a team.

Many approaches to expertise location or expertise finding have been proposed. (For a summary of the various approaches and selected systems, see Ackerman, Pipek, and Wulf 2002.) The major difficulty with these approaches is keeping the finder engines stocked with up-to-date information about people. How to typify people, skills, and expertise is an open question. The dimensions of description are unclear, especially for social-relational issues, and it is hard to keep abreast of dynamically changing situations. Even though a company might have undertaken a skills inventory, new requirements arise and need to be included. For example, a company may have learned that Joe knows the C programming language well, but now needs to know who has learned the Perl and C# languages. We might also want to know how easy it is to interact with Joe or others (social-relational), or who else has overlapping skills if Joe is not available (social-structural). For expertise locators, then, not only must the engine's recommendations be accurate in its data, the data must also be correct, timely, and organizationally relevant (Ackerman et al. 2002).

Computer-Mediated Places
One way to find expertise currently residing in people is to find the people themselves. An alternative, however, is to have the people come to the problems. In other words, one would like to have an on-line place where people can go to have their questions answered. This requires an electronic community—also called an on-line community, a virtual community, or a community of practice—where people with expertise congregate and are able to help relative newcomers (and experts) along (Wenger 1998).

The major issue with this vision is that it assumes others will also want to join. Strong social ties, those existing between people with a history of social interactions, do lead to people spending time and energy. This could be fostered by having people spend time together, and since in a large, multinational corporation this might be impossible to do face-to-face, it would have to be done through on-line, computer-mediated places. Yet interesting people with expertise do not necessarily have the time to hang out together waiting for questions. Furthermore, not everyone is friendly or sociable. For example, Thomas Allen (1977) found that his gatekeepers—people who knew other people in an organization—were far more likely to be sociable (and productive) than were others in an organization.

Ad Hoc Groups
As we understand how to reconstitute the social network of an organization through new communication mechanisms (Sproull and Kiesler 1991), there arise the possibilities of creating ad hoc groupings or subnetworks. These groupings can come together quickly, actively work on a problem, and then disband after they have finished. Organizations have had crisis teams or tiger teams, but new technologies have offered the vision of supporting geographically distributed and extremely

short-duration teams. These teams are the virtual equivalent of the recent interest in the extreme programming teams (Teasley et al. 2000). This work is being attempted currently, but two major issues with this vision have already surfaced. First, it is hard to make up a team that can work together. Doing so entirely virtually is even more difficult. Ameliorating this difficulty, however, could be strong organizational cultures or earlier face-to-face work. The second issue is the same as for expertise locators—the great difficulty of finding the data to know who does what well. Moreover, those putting together the teams need to understand potential team members' work styles and other influences salient to accomplishing the work.

The above sections have attempted to indicate the current technical possibilities, and show how their potential use is deeply influenced by organizational and social issues. The following section contends that this influence is not a product of the problem per se. Sharing expertise requires an understanding of social and organizational concerns because it is deeply collaborative, and collaborative systems suffer from a so-called technical-social gap. The following section summarizes this gap, and shows why it cannot be avoided. It should be noted that many others have also argued for the existence of this gap; this section merely serves to show how sharing expertise is influenced heavily by this gap. After this brief overview, the next section will discuss some of our attempts to move around this gap in order to share expertise within organizations.

The Social-Technical Gap

Each of these technical mechanisms for sharing expertise has problems and issues, as the last section pointed out. These result from deep, underlying social concerns. Similar to much of CSCW (or groupware)—a research area that studies how groups or organizations use technology—knowledge management both gains and suffers from the interaction between technology and social phenomena. By concentrating merely on surface issues, knowledge management efforts have fallen below expectations. We shall argue below for a theoretical reason for this, and we shall show why seriously including social-structural and social-relational considerations (i.e., social capital) in the design of technology must be rewarding, but is also inherently difficult.

There is a set of well-known findings that are almost assumptions within CSCW. (A summary of these findings can be found in Ackerman 2000 and Ackerman 2001.) To a large degree, these are social findings that we know we need to deal with when building these systems.

This section discusses three of these social issues. The selection of these three is somewhat arbitrary in that many others could have been chosen. The three are: impression management, negotiating norms, and incentive structures. As will be discussed further below, each of these social issues is a well-known problem in constructing or using CSCW systems, but solutions remain illusive.

Impression Management
Humans are good at social interaction. People have very nuanced behavior concerning how and with whom they wish to share information. Erving Goffman (1961) noted that people present "faces" to one another: We present different information to fellow researchers, students, spouses, and even relatives. What we tell our mothers is not what we tell our best friends about ourselves. What we tell spouses about job prospects and career goals is not necessarily what we tell our managers. Goffman argued that a critical component of one's social psychology is the ability to do what he called "impression management." Indeed, people in face-to-face interactions find it very disconcerting to lose control of what they consider private information. The first author of this chapter remembers a colleague becoming almost violently angry because word of an award had leaked out before he could mention it— imagine what it would be like for truly private information. Goffman was fascinated by spies and thieves, but everyone does impression management and wishes to maintain control.

Computational systems, such as those for sharing expertise, are notoriously poor at helping people do impression management. Access-control systems often have quite simple models. Our social activity is fluid and nuanced (Garfinkel 1967; Strauss 1993; Suchman 1987), and we do impression management almost without thinking about it. In fact, in face-to-face interactions, we would find it strange to consciously consider each person, write down their category on a slip of paper, and then continue to communicate the details of our lives. Any access- or security-control system, on the other hand, tends to get in the way of, rather than foster, impression management and the underlying social interaction. We believe there is an inherent tension here—one either has the nuance of control or fluidity of interaction. Efforts to obtain both will necessitate the automatic inference of context; this is obviously difficult. As a result of this computational gap, sometimes it is easier and better to augment technical mechanisms with social ones to control, regulate, or encourage social behavior (Sproull and Kiesler 1991).

Norms of Use are Negotiated
People create norms to govern how they act in social settings. Norms have been considered as formal as rules or as informal as guidelines; for our purposes here, we use the term somewhat loosely to cover the entire spectrum (Feldman 1984). A collaborative system, such as a knowledge management system or an expertise location system, is no different than any other social setting: People still create and use norms to govern their social interactions. As with face-to-face or other social settings, the norms for using a collaborative system are often actively negotiated among users. That is, unless use is mandated and strictly controlled by a governing hierarchy (and perhaps even in those situations), the users themselves will have to work out "the rules of conduct." These norms cannot be merely looked up in a rule book of some sort—exceptions and new situations occur, people with new needs or abilities arrive, and formal rules are often too inflexible to get the actual work accomplished.

Accordingly, any norms of use are also subject to renegotiation (Strauss 1991). The people who use a system (or inhabit a social setting, however briefly) constantly interpret and reinterpret the norms of behavior, shaping them to the current inhabitants and their needs. Accordingly, collaborative systems often require some secondary mechanism or backchannel communication.

Incentives are Critical
Jonathan Grudin (1989) framed what is sometimes called the Grudin paradox: What may be in the managers' best interests may not be in the ordinary users' interests. In his analysis of group calendar systems, he notes that management would like to have employees' schedules so they

can be examined and managed. However, it is not in the interests of the employees to have their schedules open, if they achieve no other benefit from group calendar use. Grudin points out that the incentives for collaborative activities must be symmetrical; that is, there must be incentives and rewards for all users.

With expertise-sharing systems, there must be incentives for experts as well as the other users (Orlikowski 1992). For example, if users can ask questions of experts through a system, then clearly there is a benefit for the users. On the other hand, experts merely gain more overloaded inboxes and interruptions. The organizational reward system, culture, or work assignment therefore needs to be readjusted to provide a benefit to the experts as well.

A related problem is that the use of collaborative systems is often tedious. Additional data may need to be entered. For instance, users may need to enter access-control information to resolve who can read highly proprietary information. Many CSCW researchers try to use available data to reduce the cost of shared and collaborative work.

The Gap

The three sets of social findings above have been known, in one form or another, for many years. We do not know how to construct systems that meet these findings, however. (Ackerman 2000). Unfortunately, there is a gap between what we know how to do technically and what we know we have to do socially. For example, we cannot solve the impression management problem by adding more rules to a system; rules are brittle and require user intervention at inopportune times.

The gap, then, is between social requirements and our technical capabilities. This is not a new statement; it is merely a restatement of the difference between "technically working" and "organizationally workable." Given the current state of the art technically, there is an inherent tension between technically feasible systems and organizationally feasible ones. The history of CSCW demonstrates this tension, and CSCW is to be lauded for its understanding and interest in this tension.

As mentioned, many other researchers have argued for the existence of this gap. Our interest has been to allege that without new forms of technology, the gap is permanent (Ackerman 2001). Our second interest has been to consider how to edge around this gap. While the gap results from an inherent tension, there are clearly things that can be done to both acknowledge and ameliorate the social requirements. So what should one do? The rest of this chapter describes our efforts to grapple with this question for systems that share expertise.

Some Systems and Possibilities

This section looks at what we see as a combined organizational-technical or social-technical approach for expertise sharing. Through the development of a set of systems and an associated social studies, we have examined a number of possibilities for augmenting the location and sharing of expertise. These explorations incorporate both an understanding of the organizational or social realities as well as the technical possibilities. Below, we discuss three areas of research work, all of which attempt to find interesting points in the combination organizational-technical design space: tying together repositories with networks, self-feeding expertise locators, and lightweight social spaces.

Combining Repositories and Networks

In a series of studies, we examined combining information repositories with social networks. Our interest was in the iterative construction of information over time. In these systems, the user asks a question, and some expert answers it. The result, over time, is an information store.

In the original Answer Garden system (Ackerman 1993, 1994, 1996, 1998; Ackerman and Malone 1990), there was a set of commonly asked questions for some topic, a way to seek out the information if the answer was not in the information database, and most important, a way to grow and correct the database.

In the original Answer Garden, a user comes to the system to find an answer to some question. A user can browse through the information database by clicking through a set of diagnostic questions, browsing an outline or tree view, or through an information-retrieval engine. Figure 11.1 shows the interface for the original version in the X Window System; figure 11.2 shows a part of a Web version. (There exist

Sharing Expertise 283

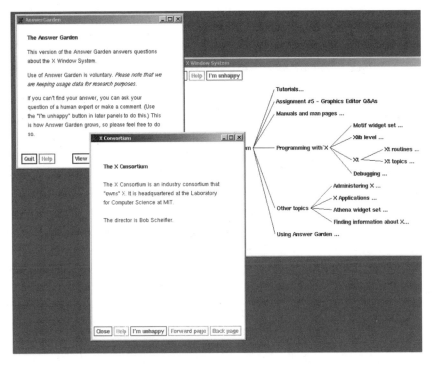

Figure 11.1
Answer Garden, the X Window System version.

alternative interfaces and Answer Garden implementations from third parties.)

If the user finds his or her answer, then he or she is finished. If the user does not find an answer, is confused with the answer or the navigation, or finds the answer to be incomplete, he or she can pop up a mailer (through the "I'm unhappy" button or link). The user asks his or her question, and the system routes the question to an appropriate human expert, who then answers. If the question and answer are common enough, the expert can insert the question-answer pair into the information store. This gave the system its name since the system grows over time where and when users demand extra information. Users get answers, experts get rid of their commonly asked questions, and the organization as a whole gets a collaborative memory or knowledge repository.

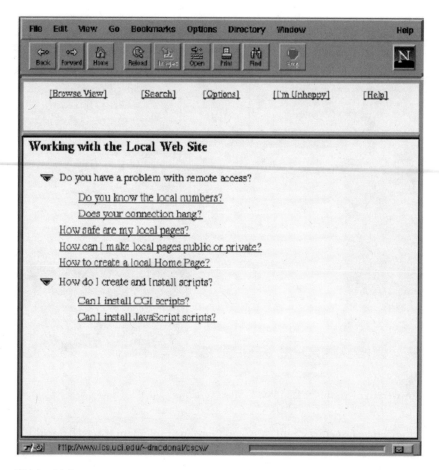

Figure 11.2
Answer Garden, the Web version.

Crucially, the Answer Garden system does not separate getting information from information repositories and people. In fact, users found this completely natural. When they could not find the answer in the information database, they were quite satisfied to have the ability to use the organization's social network.

Answer Garden 2 (Ackerman and McDonald 1996) added the ability to route questions to many forms of computer-mediated communication. Escalation agents, consisting of rules in our sample implementation,

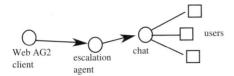

(a) The user's first attempt to get an answer goes to a chat channel.

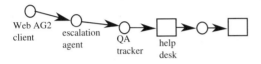

(b) The user's jth attempt to get an answer gets escalated to a help desk.

Figure 11.3
Answer Garden 2 functionality; two possible escalations for a question.

could "gracefully escalate" the question to a chat group (or now, an instant messenger list) of people nearby, then to a bulletin board, then perhaps to a help desk or consultant, and finally to an "expert" (figure 11.3). Answer Garden 2 corrected several issues in the original version. First, it eliminated the clear separation between experts and users. After all, many people have some level of expertise, and the true expert is a scarce and expensive resource in any organization. Second, the people nearby to the user are the most likely to understand the user's context. Since they know the user, they can also make the best judgments about how to present the answer. Of course, when no knowledgeable people are nearby, there is an organizational dysfunctionality, and Answer Garden 2 still provides for getting an answer.

These two systems augmented the repository model by not only making access to commonly needed information easier but also making access to people with the requisite knowledge easier. They tied people into the information system, while providing incentives for everyone involved. Their design deliberately rides the gap between social and technical approaches in an attempt to bridge the limitations of both.

Our approach to combining repositories with social networks has included both technology development and social studies. We believe that it is important to understand not only how the systems can be used (e.g., the adoption studies in Ackerman 1994, 1996; Ackerman and Palen 1996; Ackerman et al. 1997) and what the social requirements for such systems might be (e.g., the study of existing help systems in Ackerman and Palen 1996) but also how knowledge is stored and used in real systems.

We (Ackerman and Halverson 1998, 1999, 2000, 2002) examined the use of expertise in organizations, where expertise included both the use of documents and people. Our papers presented findings from an analysis of a human resources hotline. Hotlines are particularly interesting places to study expertise seeking because there is a constant, mostly repetitive flow of questions and information. The repetition makes the analysis easier, but there are enough varying questions to examine how nonroutine queries are processed. The particular hotline for these studies answered personnel questions in a large computer company.

In these hotline studies, we determined that:

- The memories used by the participants were *simultaneously embedded* within several organizational, group, and individual processes. In the site, information was complexly distributed and occasionally overlaid with multiple uses. Moreover, the memories had *mixed provenance*: sometimes the memory used was individual and private; sometimes it was group and public. For example, a call-tracking record is a digital record of a hotline call, which can later be accessed. The call-tracking records compile a variety of statistics used at different organizational levels. In this way, the call "memory" becomes part of the performance record for the person who handled the call, as well as information used at the group and organizational levels to plan. Thus, the same call record can have many different uses at different levels of organization.
- All of these memories must be used together seamlessly (or nearly so) to create an organizational product. The density and connectedness of memories used as resources can be remarkable. Within the organizational processes examined in the site, some information served as *boundary objects* (Star 1989). As mentioned earlier, while the representation is

the same and the information looks the same, their meanings change along with their users (Halverson 1995; Hutchins 1995). For example, when verifying that a person is an employee of the company, a hotline agent knows only the "facts" given in the payroll database. The telephone agent knows none of the details of the payroll record's creation or maintenance; almost the entire context has been lost. She or he will not know whether there are problems with the record, such as the database not showing longtime temporary workers, or with employee's employment, such as probation. A hotline agent can use the payroll record in a satisfactory manner, however, to determine the basic fact of employment. Removal of the detail and general agreement on a common-enough set of meanings enables the hotline agents to get their work done.

- To use information as a boundary object requires the loss of its contextualized information as it passes over the boundary. Those that need to use the information must expect this *de-contextualization*. To reuse information later, a user must then *recontextualize* that information (Ackerman 1993; Braudel 1980; Oakeshott 1983; Schmidt and Bannon 1992). The information, if not supplied by the same individual, must be reunderstood for the user's current purposes. This can be a difficult matter, although people do it every day in their work.

- Call records were not complete transcripts of everything said between hotline personnel and the caller. Instead, the agent who handled the call decided what would be the necessary information needed for subsequent reuse. To properly serve the reuser of information, the creator must adequately project the consequences of the memory's later use; that is, they must determine the information's trajectory. The *trajectory* (Hutchins 1995; Strauss 1993) describes the path of an event; in this case, we mean it to be the likely trajectory as anticipated. The incentives for keeping information for later reuse appear to follow the assumed trajectory and its *projected consequences*. In the hotline, if an agent assumed that a call was routine and would not be referenced again, she or he had little incentive to write many details of the call. If that record must be reused in the future, the future user must deal with the unanticipated consequences of the author's projecting the trajectory incorrectly.

Knowledge management largely restricts repositories to experience "objects" that are magically reusable, yet it is more fruitful to consider expertise sharing as both *object* and *process*. What is of interest is both a memory artifact that holds its state and an artifact that is simultaneously embedded in many organizational and individual processes. The container metaphor implied by objectifying memory is easier to consider computationally, yet it is extremely limited organizationally.

These field and technical studies have reinforced one another, enabling us to construct more organizationally viable systems. Throughout all of these studies, we have seen that by including the social-structural and social-relational aspects of an organization, we can foster more usable and useful knowledge management systems.

Finding Expertise

The Answer Garden series of systems assumed that considerable knowledge existed in the heads of people. Indeed, the back end of these systems required engines to find an expert or someone with suitable expertise. In this work, we have tried to augment the social-structural and social-relational aspects of an organization with systems, and our approach has been multipronged—combining repositories and networks. Again, our approach has been to determine how seeking expertise is done in natural settings through field-based studies, followed with how best to support and augment expertise seeking through the construction of experimental systems. The next section surveys these two approaches.

Field Studies of Expertise Finding

We have sought to understand expertise seeking through a set of social studies in natural settings. McDonald and Ackerman (1998) examined how people sought others' expertise in a medium-sized software company. McDonald and Ackerman called this software company MSC. MSC built a family of products to automate the back end of doctors' and dentists' offices. MSC's systems were long-lived transaction systems, several of which had been in production for over two decades. They used a propriety software infrastructure and tools.

McDonald and Ackerman found that expertise seeking in MSC could be analytically separated into three "phases": identification, selection, and escalation. These phases were often not separable or sequential in the everyday activity of the software engineers and others in MSC; however, they were separable enough to construct a system architecture based on them. Identification was the act of determining who might know the answer to a specific question. Selection was determining who was available or likely to provide the information. Escalation was the act of looking for additional people, perhaps crossing organizational boundaries or going to others that one might not normally consider. Each of these phases was necessary to obtain expertise. In McDonald and Ackerman's judgment, identification was the easiest to augment; selection and escalation could be augmented as well, but were more difficult. This analysis resulted in the system architecture of Expertise Recommender (ER), described below.

Recently, this line of research was extended to study an aircraft manufacturing hotline, available to help airline operators when they have significant repairs or problems with aircraft (Lutters 2001; Lutters and Ackerman 2002). The study examined how hotline engineers balance timeliness with safety and reliability. In addition to understanding the context of the work and the organizational structures that ensure safety, Wayne Lutters was able to determine the role of the information objects that are passed back and forth between the hotline and the airline engineers.

Systems for Finding Expertise

McDonald (2000; McDonald and Ackerman 2000) constructed a system based on the field study reported in McDonald and Ackerman 1998. The ER system was designed to help people in MSC find others with the suitable program expertise to answer specific program questions (see figure 11.4). ER's architecture assumed a number of general identification and selection heuristics (e.g., "find people nearby organizationally"), but also allowed site- and group-specific modules. In MSC, the programmers annotated their changes on line ten of a modified program; thus, one ER module for MSC searched who had most recently changed a program. While the architecture was designed to be general across organizations, the field study findings suggested that the methods used for identification and selection were local and contextualized. Therefore, ER's specific finding heuristics also had to be local and contextualized.

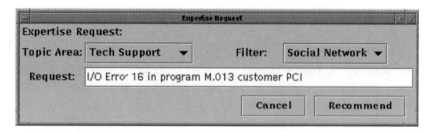

Figure 11.4
Expertise Recommender.

Finding Data

In addition to determining suitable systems architecture, we have also examined how to find the data to feed it. Our work has largely consisted of attempts to find what we call first-order approximations to measuring an organization's expertise network (Ackerman et al. 2002). Because fully measuring the network is too time-consuming and costly (and often cannot be maintained adequately), we have looked for discount methods. While we hope to largely use the digital traces of one's work and identity, these traces must be bootstrapped with some measurement.

Our efforts, then, have been to find quick measures of what is important to know within a group and who knows what. We found that we could get participants to construct "Trivial Pursuit" questions that would measure critical success factors for a group or organization. Not only did participants enjoy being measured in such a way, they were also surprisingly willing to guess how others would do. Indeed, we found that any seven randomly selected software engineers or three managers were able to estimate group members' performances nearly as well as administering the test to everyone. More work will be necessary to determine whether this will be an approximation to the approximation, and whether this straightforwardly extends to a large company. Nonetheless, these first results were encouraging.

Lightweight Social Spaces

One type of augmentation to an organization's sharing of expertise is to route queries to appropriate others with suitable expertise. Another type of augmentation is to create places or virtual spaces where people with suitable expertise congregate. Those with questions or those who wish to gain the expertise can then also go there.

There have been many successes within the use of computer-mediated communication systems where new forms of collaboration emerged (Sproull and Kiesler 1991). A study of the Zephyr messaging system at MIT (Ackerman and Palen 1996) showed that chat or instant messenger–like systems could effectively be used for providing help. Furthermore, this study (as well as Ackerman et al. 1997; Muramatsu and Ackerman 1998; Lutters and Ackerman 2002) revealed many of the role, reward, and norm structures important to socially maintaining the place over time.

Zephyr is a heavily used instant messenger–like system created at MIT (though it was in use before instant messenger). The Help Instance alone (one channel on Zephyr) receives over thirty thousand messages per semester. The system is over eleven years old and has only discretionary use (i.e., no one is paid to answer questions on the system). The following is an example of Zephyr use:

Time: 06:27:32 Date: Thu Oct 14 93
From: College life is vastly overrated, according to US News and World Report. ⟨elf⟩
Who wrote "Hallelujah!"? Or is the author unknown?

Time: 06:28:27 Date: Thu Oct 14 93
From: Mobeus was two-faced ⟨benjy⟩
If you're speaking of the Hallelujah chorus,
it is from Hayden's Messiah.

Time: 06:28:36 Date: Thu Oct 14 93
From: Kathy Talbott ⟨shilla⟩
Handel, not Hayden

One should note that the answer was obtained in slightly more than a minute (at 6:30 A.M.). Furthermore, the public and visible nature of the questions and responses makes it possible to obtain correct, authoritative answers. A number of other social mechanisms, including reinforcement for displaying expertise in the MIT environment, contribute to obtaining correct answers as well.

Two systems that were especially constructed to facilitate knowledge transfer were Babble and Loops (Erickson et al. 1999). Babble and Loops are two versions of a chatlike communication tool, originally designed to support small work groups. Babble is a client-server based application in which typed messages are transmitted across a TCP/IP network, stored on a server, and displayed to each client. Babble allows its users to engage in synchronous or asynchronous textual conversations, and provides visual feedback regarding who has recently participated in a conversation (see Erickson et al. 1999). Loops moves the Babble experience onto a Web-based platform, but maintains the features in Babble.

Babble and Loops look and feel like other forms of computer-mediated communication (CMC); yet neither is a bulletin board, a chat room system, a MUD, an e-mail system, or a newsgroup. Babble and Loops merge the persistence and sequencing found in asynchronous bulletin boards with the immediacy and informality of MUDs and chat rooms. Loops adds more explicit shared bulletin board aspects—creating a space that is public and semipersistent (i.e., changeable by anyone). These combined features result in a blended synchrony user experience, in which interactions can shift naturally between synchronous and asynchronous modes depending on who is around and how actively they are participating.

This is possible because a major component of both Babble and Loops is the social proxy: a minimalist, graphical representation of user activity. Seeing who is currently participating and the state they are in makes the social proxy a resource-governing social behavior (Bradner, Kellogg, and Erickson 1999).

Both Babble and Loops (figures 11.5 and 11.6) display most of the same information, just in a different organization. This includes a list of all connected users, the social proxy, a list of conversations, and the content of the current conversation (i.e., the text). They use slightly different metaphors, so that in Babble, conversations are referred to as topics, while in Loops they are places where conversations occur. (In figure 11.6, the pull-down in Loops that shows all the places is hidden.) Messages are appended to the bottom of the conversation pane and appear in the order posted.

Sharing Expertise 293

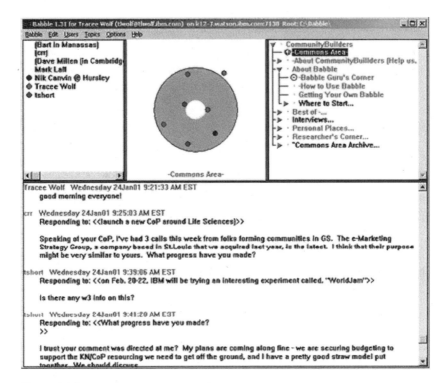

Figure 11.5
The Babble interface. From left to right, the top row of panes shows the user list, social proxy, and topic list. Below is the selected conversation.

Zephyr, Babble, and Loops are examples of technically created places in which people can gather to exchange information, ask particular types of questions, and share expertise with one another. Interestingly, Zephyr fosters relational ties as a second-order effect of the expertise search. In Babble, which foregrounds these social interactions, we have seen new instances of expertise searching and sharing behaviors (e.g., waylay, as seen in Bradner, Kellogg, and Erickson 1999). We have also seen situations with information sharing as a side effect of the social relations in these social spaces. As such, these lightweight social spaces begin to fold back into expertise finding, augmenting the social network of an organization.

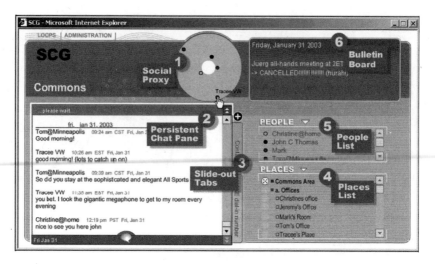

Figure 11.6
The Loops user interface, including a visualization of the presence and activity of participants; a chat area that supports synchronous or asynchronous conversation; public slide-out tabs that can hold editable text and URLs; a list of rooms; a list of people who are present; and a public bulletin board that can contain editable text and URLs. *Note*: The image has been edited to remove about a third of its height; the gray circles and rectangles are call-outs, and not part of the interface. The social proxy is on the far left, with the user list below. The current conversational place is in the middle, while the right side acts as a bulletin board.

Conclusions and Research Challenges

In this chapter, we have demonstrated some new possibilities for knowledge management and expertise sharing that attempt to combine organizational and technical feasibility. These efforts assume the possibility of technical augmentation to the way an organization shares expertise either through a social network or repository, but they try to do this augmentation in organizationally feasible ways.

This feasibility results from a consideration of the social-structural and social-relational aspects of an organization—critical dimensions of social capital. Not only are the organizational aspects described crucial for organizational adoption and use; we hope that properly constructed systems can also further foster and promote a sense of social cohesion—social capital in its best light.

Above we showed our work in:

- Augmenting repositories with social networks: Incorporating an organization's social network into a repository makes the repository more organizationally robust and responsive. Some of the work incorporates mechanisms to ask people nearby (however measured), assuming that people who were more tied to the information seeker were more likely to respond.
- Self-organizing expertise locators: This work directly augments the social network of an organization to promote expertise sharing. The work also assumes that relational aspects (e.g., trust) are critical to effective use.
- Communities that maintain and promote themselves as places: This work attempts to create new social structures for an organization to foster expertise sharing.

In all of this work, then, we have assumed that incorporating the social-structural and social-relational aspects of social capital were critical to effective knowledge management and expertise seeking. All of the systems include or reconfigure an organization's social network (social-structural), provide incentives and inculcate trust (social-relational), and lead to shared understanding and mental models (social-cognitive). While first generation knowledge management approaches were built on individual, cognitively based technologies (e.g., incorporating an intranet with an information-retrieval search), we believe that significant benefits will accrue only with understanding the need for social capital and incorporating its dimensions into all types of knowledge management technologies.

We are currently extending this work to consider how to form new connections among parts of a social network. Even-newer possibilities exist. One can imagine creating distributed coalitions, where social sub-networks of two organizations, institutions, or voluntary associations were joined, perhaps internationally. As well, one can imagine augmenting all of this social activity with agent clusters to find, broker, and reward those in the subnetworks.

In our current work, which extends Ackerman and Halverson 2002, we are considering how to design within a social-technical codesign space, and how to create new assemblages as resources for users

(Halverson and Ackerman 2003; Halverson 1995; Hutchins 1995). We expect this work to give us significant insights into designing new and organizationally feasible systems.

Acknowledgments

Babble and Loops are projects of IBM. Other projects described in this chapter have been funded, in part, by grants from the National Science Foundation (IRI–9702904 and IRI–0124878), the University of California at Irvine/NSF Industry/University Cooperative Research Center at the Center for Research on Information Technology and Organizations, NASA, the University of California MICRO program, Interval Research, Quality Systems, and the MIT/Project Oxygen partners.

This work has benefited from far too many conversations over the last decade to even mention. Nevertheless, we especially want to thank our collaborators, Wendy Kellogg and Tom Erickson. David McDonald, Wayne Lutters, and Jack Muramatsu were members of the team at the University of California, Irvine. We would also like to thank Volker Wulf and Marlene Huysman for their continued interest and support in this work.

References

Ackerman, M. S. 1993. Answer Garden: A tool for growing organizational memory. Ph.D. diss., Massachusetts Institute of Technology.

Ackerman, M. S. 1994. Augmenting the organizational memory: A field study of Answer Garden. In *Proceedings of the conference on computer-supported cooperative work*. New York: ACM Press.

Ackerman, M. S. 1996. Definitional and contextual issues in organizational and group memories. *Information Technology and People* 9, no. 1:10–24.

Ackerman, M. S. 1998. Augmenting organizational memory: A field study of Answer Garden. *ACM Transactions on Information Systems* 16, no. 3:203–224.

Ackerman, M. S. 2000. The intellectual challenge of CSCW: The gap between social requirements and technical feasibility. *Human-computer interaction* 15, nos. 2–3:179–204.

Ackerman, M. S. 2001. The intellectual challenge of CSCW: The gap between social requirements and technical feasibility. In *HCI in the new millennium*, edited by J. Carroll. New York: Addison-Wesley.

Ackerman, M. S., J. S. Boster, W. G. Lutters, and D. W. McDonald. 2002. Who's there? The knowledge mapping approximation project. In *Sharing expertise: Beyond knowledge management*, edited by M. S. Ackerman, V. Pipek, and V. Wulf. Cambridge: MIT Press.

Ackerman, M. S., and C. Halverson. 1998. Considering an organization's memory. In *Proceedings of the conference on computer-supported cooperative work*. New York: ACM Press.

Ackerman, M. S., and C. Halverson. 1999. Organizational memory: Processes, boundary objects, and trajectories. In *Proceedings of the IEEE Hawaii international conference of system sciences*. (HICSS–32). Los Alamitos, CA: IEEE Computer Society Press.

Ackerman, M. S., and C. Halverson. 2000. Re-examining organizational memory. *Communications of the ACM* 43, no. 1:58–63.

Ackerman, M. S., and C. Halverson. 2002. Organizational memory: Processes, boundary objects, and trajectories. Working paper.

Ackerman, M. S., D. Hindus, S. D. Mainwaring, and B. Starr. 1997. Hanging on the wire: A field study of an audio-only media space. *ACM Transactions on Computer-Human Interaction* 4, no. 1:39–66.

Ackerman, M. S., and T. W. Malone. 1990. Answer Garden: A tool for growing organizational memory. In *Proceedings of the ACM conference on office information systems*. New York: ACM Press.

Ackerman, M. S., and D. W. McDonald. 1996. Answer Garden 2: Merging organizational memory with collective help. In *Proceedings of the ACM conference on computer-supported cooperative work*. New York: ACM Press.

Ackerman, M. S., and L. Palen. 1996. The Zephyr help instance: Promoting ongoing activity in a CSCW system. In *Proceedings of the ACM conference on human factors in computing systems*. New York: ACM Press.

Ackerman, M. S., V. Pipek, and V. Wulf, eds. 2002. *Sharing expertise: Beyond knowledge management*. Cambridge: MIT Press.

Allen, T. 1977. *Managing the flow of technology*. Cambridge: MIT Press.

Bradner, E., W. Kellogg, and T. Erickson. 1999. The adoption and use of "Babble": A field study of chat in the workplace. In *Proceedings of the sixth European conference on computer-supported cooperative work*. New York: Kluwer.

Braudel, F. 1980. *On history*. Chicago: University of Chicago Press.

Cohen, D., and L. Prusak. 2001. *In good company: How social capital makes organizations work*. Boston: Harvard Business School Press.

Erickson, T., D. N. Smith, W. A. Kellogg, M. Laff, J. T. Richards, and E. Bradner. 1999. Socially translucent systems: Social proxies, persistent conversation, and the design of "Babble." In *Proceedings of the ACM conference on human factors in computing systems*. New York: ACM Press.

Feldman, D. C. 1984. The development and enforcement of group norms. *Academy of Management Review*, 9, no. 1:47–53.

Garfinkel, H. 1967. *Studies in ethnomethodology*. Englewood Cliffs, N.J.: Prentice-Hall.

Goffman, E. 1961. *The presentation of self in everyday life*. New York: Anchor-Doubleday.

Grudin, J. 1989. Why groupware applications fail: Problems in design and evaluation. *Office: Technology and People* 4, no. 3:245–264.

Halverson, C. A. 1995. Inside the cognitive workplace: New technology and air traffic control. Ph.D. diss., University of California at San Diego.

Halverson, C., and M. S. Ackerman. 2003. Yeah, the rush ain't here yet—take a break. In *Proceedings of the IEEE Hawaii international conference of system sciences*. Los Alamitos, CA: IEEE Computer Society Press.

Hutchins, E. 1995. *Cognition in the wild*. Cambridge: MIT Press.

Lesser, E. L., M. A. Fontaine, and J. A. Slusher. 2000. *Knowledge and communities*. Boston: Butterworth-Heinemann.

Lesser, E., and L. Prusak. 2000. Communities of practice, social capital, and organizational knowledge. In *Knowledge and communities*, edited by E. L. Lesser, M. A. Fontaine, and J. A. Slusher. Boston: Butterworth-Heinemann.

Lutters, W. G. 2001. Supporting reuse: IT and the role of archival boundary objects in collaborative problem solving. Ph.D. diss., University of California, Irvine.

Lutters, W. G., and M. S. Ackerman. 2002. Achieving safety: A field study of boundary objects in aircraft technical support. In *Proceedings of the ACM conference on computer-supported cooperative work*. New York: ACM Press.

McDonald, D. W. 2000. Supporting nuance in groupware design: Moving from naturalistic expertise location to expertise recommendation. Ph.D. diss., University of California, Irvine.

McDonald, D. W., and M. S. Ackerman. 1998. Just talk to me: A field study of expertise location. In *Proceedings of the ACM conference on computer-supported cooperative work*. New York: ACM Press.

McDonald, D. W., and M. S. Ackerman. 2000. Expertise Recommender: A flexible recommendation system architecture. In *Proceedings of the ACM conference on computer-supported cooperative work*. New York: ACM Press.

Muramatsu, J., and M. S. Ackerman. 1998. Computing, social activity, and entertainment: A field study of a game MUD. *Computer-Supported Cooperative Work: The Journal of Collaborative Computing* 7, no. 1:87–122.

Nahapiet, J., and S. Ghoshal. 1998. Social capital, intellectual capital, and the organizational advantage. *Academy of Management Review* 23, no. 2:242–266.

Oakeshott, M. 1983. *On history and other essays*. Totowa, N.J.: Barnes and Noble Books.

Orlikowski, W. J. 1992. Learning from notes: Organizational issues in groupware implementation. In *Proceedings of the ACM conference on computer-supported cooperative work*. New York: ACM Press.

Schmidt, K., and L. Bannon. 1992. Taking CSCW seriously: Supporting articulation work. *Computer-Supported Cooperative Work* 1, nos. 1–2:7–40.

Sproull, L., and S. Kiesler. 1991. *Connections: New ways of working in the networked organization*. Cambridge: MIT Press.

Star, S. L. 1989. The structure of ill-structured solutions: Boundary objects and heterogeneous distributed problem solving. In *Distributed artificial intelligence*, edited by L. Gasser and M. Huhns. San Mateo, Calif.: Morgan Kaufmann.

Stewart, T. A. 2001. *The wealth of knowledge: Intellectual capital and the twenty-first century organization*. New York: Doubleday.

Strauss, A. 1991. *Creating sociological awareness: Collective images and symbolic representations*. New Brunswick, N.J.: Transaction.

Strauss, A. 1993. *Continual permutations of action*. New York: Aldine de Gruyter.

Suchman, L. A. 1987. *Plans and situated actions: The problem of human-computer communication*. New York: Cambridge University Press.

Teasley, S., L. Covi, M. Krishnan, and J. Olson. 2000. How does radical collocation help a team succeed? In *Proceedings of the ACM conference on computer-supported cooperative work*. New York: ACM Press.

Wenger, E. 1998. *Communities of practice: Learning, meaning, and identity*. New York: Cambridge University Press.

12
Pearls of Wisdom: Social Capital Building in Informal Learning Environments

Robbin Chapman

What is social capital, and to what extent should it impact technology development for supporting communities? It is imperative to any meaningful discussion that we describe *social capital*, an overloaded term whose ambiguity stems from typical assumptions regarding the creation and generation of capital. A succinct definition of capital is "the time-dependent accumulation of subsequently accessible resources" (Putnam 2000). Thus, social capital is often thought of analogously. This viewpoint, however, lacks the requisite fidelity to depict social capital's complexity. The fundamental features of social capital are the networks, shared values, social trust, and norms of a community. In essence, social capital functions as the glue holding a community together. A central tenet of social capital is trust, both of the individual's and group's goodwill, the reciprocity of their relationships, and the quality of the connections between them.

The idea of social capital is not new or novel, but it has more recently taken shape as an important area of study following the work of James Coleman and Robert Putnam (Coleman 1990; Putnam 1993, 2000). The connections among people, rather than the properties of objects or individuals, has been illuminated as a key feature of social capital. What role can technology play in supporting the development of social capital? While the previous definition of social capital intuitively fits my notion of how and why people form coherent, productive groups, this same definition is problematic for technology developers and researchers. It often proves difficult to wrap our hands and minds around such an ethereal idea as social capital, especially when designing what I hope are nondisruptive systems that can support and augment it. Pearls of Wisdom

(PoW), a community-based, asynchronous knowledge-sharing system, provides an example of how social capital formation can motivate the design rationale of the software and activities design process. Pearls, computational artifacts with meaningful project design content, are the basic unit of the PoW system. Each Pearl contains basic how-to information in a variety of media formats along with personal reflections on the author's learning experience—for the exchange of ideas in the form of contributed comments and links to Pearl-inspired projects.[1]

Test Community: The Computer Clubhouse Network

This study will focus on a physical community, the Computer Clubhouse Network, a learning community that already relies heavily on social connections, trust, and norms for knowledge sharing. My study takes place as the Clubhouse Network expands from a few, local sites to a distributed, global network of clubhouses. The Computer Clubhouse (Resnick and Rusk 1996), founded in 1993, is a network of after-school technology centers where underserved youth participate in constructionist, project-based learning activities with the support of adult mentors. Seymour Papert coined the term *constructionism*, building on Jean Piaget's constructivist theory of learning, which maintains that people learn by constructing their own cognitive structures in the context of their previous knowledge and environment (Papert 1980; Piaget 1977). Constructionism takes this theory further by contending that people learn best when engaged in actively constructing external artifacts to share with and be critiqued by others (Papert 1993; Resnick and Rusk 1996). Papert argues that we learn through interacting with artifacts, and gain an understanding of the world by creating as well as experimenting with artifacts, and modifying them to work better. Being out in the world, these artifacts provide a point of reference for sharing ideas within a community. This emphasis on sharing casts constructionism as a social theory of learning, although the focus is primarily on creating environments where constructionist learning can flourish. Social constructionism (Shaw 1995) extends these ideas to include the development of social connections as well as artifacts, and stresses these social connections as a fulcrum for learning within a community. The Computer Clubhouse constructionist approach to learning also highlights the importance of interpersonal relationships and community in the learning process. The four guiding principles of the Computer Clubhouse model are:

1. a focus on constructionist activities, where youth become designers, inventors, and creators;
2. encouraging youth to work on projects related to their own interests;
3. creating a sense of community, where young people work together with one another with support and inspiration from adult mentors; and
4. providing resources and opportunities to those who would not otherwise have access to them.

A Computer Clubhouse typically has eighteen high-end computer workstations along with several dedicated computers for music and movie constructions. Members can come whenever they want and can stay as long as they wish. They have access to computers, professional graphic and multimedia development tools, the Internet, digital cameras, robot construction kits, and a sound studio. Adult mentors motivate members to develop their own projects—like a song, a Web site, or an animation—help members to get started with the different tools, and introduce them to other members who may have developed similar projects. Eventually, members start developing and pursuing their projects, and may be called on to offer support and encouragement to the rest of their Clubhouse community. The Computer Clubhouse community sees learning, technology, active participation, and collaboration as empowering tools. Clubhouse members not only have access to technology but also to new ways of thinking about how computation fits into their self-expression. Collaborations and social interactions are enhanced by, among other things, the layout of the space. Most Clubhouses have a large, oval table devoid of computers where members sit together and plan their projects. Clubhouse walls provide a venue for showcasing member projects, thereby inspiring new projects and collaborations.

The Computer Clubhouse environment serves as a microworld where members learn about and reflect on their place in the larger community (Chapman and Burd 2002; Pinkett 2001). For example, Clubhouse members participate in the Clubhouse Council, discussing issues relevant to the local Clubhouse organization. Mutual respect and sharing goes both ways. Clubhouse mentor and member relationships, for instance, contribute to a member's improved "sense of self" and an understanding of one's value to a community—something many youth seldom experience. Within the Clubhouse, there is fluidity between the roles of mentor and learner. In fact, role switching is a commonplace occurrence, fueling the learning experience through a cycle of learning by doing and learning by reflection. Learning by doing is embodied in the design process. Learning by reflection occurs via a process of teaching others a newly acquired skill as well as discussing their project and its development (Lave and Wenger 1991).

At the Computer Clubhouse, learners enter a culture that empowers them to control what they learn and how deeply they delve into any particular subject. This leads to learning opportunities situated in member experiences inside and outside the Clubhouse community. The result is a Clubhouse full of bright, eager learners expressing themselves and disseminating their unique viewpoints through the meaningful use of digital technologies. They also participate in their Computer Clubhouse council, a body that discusses and resolves issues related to local Computer Clubhouse organization.

Currently, there are approximately sixty-four Computer Clubhouse's worldwide, with another forty planned over the next two years. This increasing population of distributed Computer Clubhouses is connected via an intranet called the Computer Clubhouse Village, which facilitates communication throughout the network using e-mails, discussion forums, and project galleries. Here, we will look at two Clubhouses in Boston, Massachusetts.

The Role of Social Capital in Clubhouse Learning Experiences

In the broadest sense, a knowledge-building community is any group of individuals dedicated to sharing and advancing the knowledge of the collective (Scardamalia and Bereiter 1994). More specifically, expertise sharing has gone beyond the sharing of explicit knowledge to the building of those ties that give knowledge sharing a rich social context (Ackerman 1998; Ackerman, Pipek, and Wulf 2003). Much of the learning at the Computer Clubhouse is realized through the social interactions and connections among its members and mentors. The context for

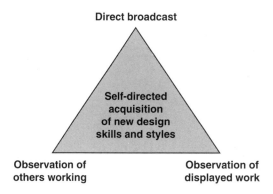

Figure 12.1
Computer Clubhouse learning triad. Individuals generally take one of three actions to obtain knowledge: they directly broadcast by making a general announcement or talking to a specific individual; they observe others working and ask specific how-to questions; or they observe work on the Clubhouse walls and try to locate the creator of a particular project for help with accomplishing their own similar tasks.

learning occurs on three levels, best described as the Clubhouse learning triad (figure 12.1).

Members observe the displayed work of others and attempt their own similar projects, learning the necessary software and hardware tools along the way. Members also learn by observing what others are currently working on, entering into conversational exchanges of knowledge. Finally, members broadcast to their friends, mentors, or anyone present, their desire to learn a particular skill. This may result in connections to people with the appropriate skills. Over time, within a local Clubhouse, experts gain a reputation, enjoy increased social status, and become a resource to others within their Clubhouse. A member relies on this store of social capital to support their learning.

Social capital can be thought of as the framework that supports the process of learning through interaction. The quality of the social processes and relationships within which learning interactions take place is especially influential on the quality of learning outcomes. This suggests that social capital plays an important role in fostering the social networks and information exchanges necessary for learning to happen. Figure 12.2 categorizes the types of social interactions that lead to the

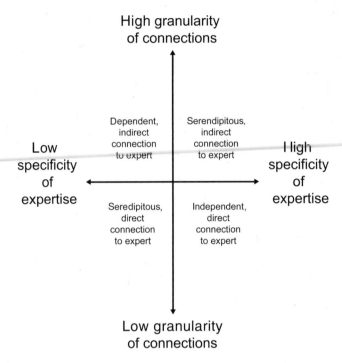

Figure 12.2
Connecting novices and experts at the Clubhouse. The dimensions of Computer Clubhouse expertise exchange include specificity and granularity. A high specificity means a greater chance of making connections to the exact expert to answer a query. A low granularity means that less intermediary contacts are necessary to locate the exact expert to query.

sharing of skills and ideas. The quality of these interactions can be viewed in relation to the specificity of the response to a member's query as well as the granularity (number of intermediary contacts) necessary to get applicable help. There are four of these categories:

1. *Serendipitous, direct*: by chance observation of another member working, and direct query of that member.

2. *Independent, direct*: by firsthand knowledge of the exact expert to query.

3. *Dependent, indirect*: by intervention of one or more community members knowledgeable of the exact expert to query.

4. *Serendipitous, indirect*: by verbal broadcast to members currently present in the Clubhouse. This can result in one or more intermediary contacts before the exact expert is found, or in no expert being identified.

The quality of social processes where learning interactions take place especially influences the quality of learning outcomes within the Clubhouse. A member who knows what skills other individuals possess can more easily establish connections to gain knowledge. Members who are not so versed must rely on chance encounters or must track down an expert. Often, in the latter case, frustration and project abandonment is the outcome.

The Tool: Pearls of Wisdom

Social capital provides a powerful lens through which to view sustainable, community-based technology development. For communities like the Computer Clubhouse, those connections, networks, and social norms must be supported by the technologies designed for that community. Any technology designed to promote expertise sharing must also provide for constructionist learning activities while encouraging activities amenable to the culture. For these technologies to be relevant and functional in the real world, the valuation and nurturing of social networks must be considered an integral part of the technology's design rationale.

As mentioned earlier, PoW, a community-based tool for knowledge sharing, will be the focal point of this exploration of IT's role in social capital building. PoW is a suite of software tools designed to support asynchronous knowledge sharing among community members. The system enables users to create Pearls, computational artifacts containing community-supplied expertise with creative design tools (i.e., graphic programs, robotics, music studio software, etc.). This expertise is then disseminated to the community in the form of interactive Web pages. Pearls also contain information about their creator, mechanisms for e-mail and community discussions, and links to projects. For example, a member may search the Pearls database for advice on how to use PhotoShop graphic design software to create an image that appears engulfed in flames. Another member can subsequently attach a finished project to the same Pearl to show off their use of this "fire effect." Yet another member may leave a comment about other ways of achieving the fire effect via a different set of operations. All of these actions emulate the type of person-to-person Clubhouse interactions familiar to that community.

Through their use of the PoW system, Pearl creators share their constructions (projects) and constructed knowledge (specific design skills) with the community. PoW allows the Pearl creator to be expressive of their thinking as part of this Pearl design process. For instance, the Pearl creator decides what examples to use, what medium or mediums will best convey information, and what visual attributes (color, font, etc.) allow for their own personal expression. These actions lead to a self-examination of the individual's learning process (Duckworth 2001; Hutchins 1995; Salomon and Perkins 1998). Through their construction experiences, learners gain fluency in articulating their problem-solving and learning styles. There is great value in reflecting on one's knowledge because it supports the metalearning process (Jonassen 1991; Salomon and Perkins 1998; Duckworth 2001). Pearl creation stimulates a learner's reflection on what one understands, how old knowledge links to new, and what aspects of that knowledge are crucial to shared understanding. The act of construction hones a learner's fluency in manipulating ideas by thinking about what one knows, deciding what is important to convey to others, and settling on the right way to formalize ideas for others to use (Wertsch 1985; Lave and Wenger 1991).

Objects-to-think-with act as carriers of powerful ideas and can help learners form new relationships with knowledge they already possess (Papert 1980). Pearls are objects-to-think-with, both during their construction and use, bridging the gap between personal knowledge and the symbolic representation of that knowledge. As a Pearl is created, it functions as a tool in the transformation of an individual's knowledge from the conceptual to concrete understanding. As a Pearl is used, it provides the scaffolding to support the individual in the self-directed exploration of new ideas.

Pearls of Wisdom 309

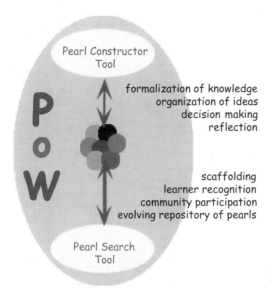

Figure 12.3
The PoW flow of support cycle. Pearl authoring can lead to rich learning opportunities. Pearl use can also provide necessary venues for participation, recognition, and learner scaffolding.

PoW Architecture

This section introduces the basic PoW system architecture. PoW currently resides on a MIT Media Lab server. Its future home will be on the Computer Clubhouse Network Village, an intranet developed by Intel to link all existing Clubhouses worldwide. We opted to utilize the same development resources as Intel Corporation to ensure uninterrupted user support and software upgrades. Table 12.1 shows a list of PoW development components.

Figure 12.4 describes program component interaction. In summary, servlets act as controllers to handle requests sent by Web pages and the application applet. JavaBeans implement the connection logic to and querying of the database. Results are returned to the servlets via JavaBeans, which call on the applet and Java server pages to display the results.

This design includes the benefits of logic reusability and modularity as well as faster page display.

Table 12.1
PoW development environment

Component	Description
Operating system	Microsoft Windows 2000 service pack
Web server	Microsoft Internet Information Server 5.0
Database management system	Microsoft SQL Server 2000
Web application server	JRun 3.1 Enterprise Edition
Programming language	JDK1.4x, Java Servlet 2.2, JSP1.1, JavaBean, and EJB 1.1

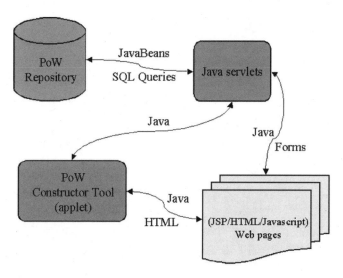

Figure 12.4
Schematic of PoW information flow.

PoW Design Rationale

The PoW interface is designed to motivate Pearl creation and use. The Pearl-Viewer and Pearl-Search interfaces are organized to reflect the practices around knowledge sharing that are currently part of the Clubhouse's culture. For example, when viewing a Pearl, the user may decide to link a similar project to the Pearl for other visitors to see. This sharing is also evident in Computer Clubhouses worldwide, as members hang their projects on the walls so others can see how they have mastered a particular design skill. A communication pathway is provided in the Pearl-Viewer design to facilitate Pearl creator recognition and a variety of community-to-individual interactions. In local Clubhouses, experts become known and are contacted often for conversations about their work or for help. The design rationale for PoW follows a two-tier approach, where functionality is provided: to support the current Clubhouse culture of knowledge sharing, and to address known hindrances to knowledge sharing at the Clubhouse and in a distributed environment. The current culture of knowledge sharing at the Computer Clubhouse is dependent on social capital for its functioning. The three tenets of social capital building—namely, networks, norms, and trust—can be seem in various aspects of Clubhouse life; therefore, PoW must support these already-established activities.

Networks
PoW must provide ways for members to connect to one another. It must also establish new connections among individuals and provide opportunities for chance connections.

Norms
Opportunities to display work as well as comment on the work of others are a cornerstone of Clubhouse culture. There needs to be mechanisms to allow this in the PoW software. There also needs to be a way to broadcast the need for information to the larger community. In Clubhouses today, this is done by making a general announcement, but this method lacks the efficiency of reaching people not present or too busy at the time to take notice. Finally, recognition of experts, by their work or through facial recognition, is an important motivator for members to help each other learn new things.

Trust
The community makes corrections to erroneous information or augments information that is present. PoW facilitates this through its discussion forum. Multiple Pearls on the same topic are allowed by the system, giving users a choice of which Pearl they will use. Trust in an individual's ability to convey correct information is acknowledged by

using PoW to access all Pearls that have been created by that individual. Also, with the visibility that a Pearl author receives also come accountability. An author's knowledge is rooted in the social context of the Clubhouse; hence, they have a stake in ensuring the accuracy and clarity of the instantiation of that knowledge.

Pearl-Constructor Tool
Face-to-face occurrences of knowledge sharing at the Clubhouse seem to form a familiar pattern. Other than conveying how-to information, there is disbursed throughout these exchanges example projects or comments on interesting, intermediary results. There is also usually some sharing of nontechnical information, such as personal anecdotes or comments.

The design rationale for the constructor was to support the richness of face-to-face interactions, which include conveying information, some personal sharing of anecdotal information, and viewing of the expert's project(s). The first page screen real estate is divided into three, resizable sections. The first section, labeled "What I Did," is where the creator can display the finished project as well as any intermediary states. The second section, "Why I Did It," is an area for reflection, personal anecdotes, and so on. The third, "How to Do It," contains instructions for the user to follow. Completion of this first screen is required for the Pearl to be accepted into the knowledge repository. The remaining pages consist of two sections: one for how-to information, and the other for reflection and comments.

Low Floor, High Ceiling
PoW design is driven by the "low-floor, high-ceiling" approach espoused by Seymour Papert when developing his Logo programming language (Papert 1980). Since the Computer Clubhouse boosts a membership of youth, ages ten to eighteen, the challenge was to develop a design tool that would make sense to the younger users and yet offer a richer palette of options for the older, more sophisticated ones. For young users, the only required skills are being able to use a mouse, to point and click, for selecting a media icon and placing it somewhere on the Pearl work area. A simple iconic interface allows users to select and place various media

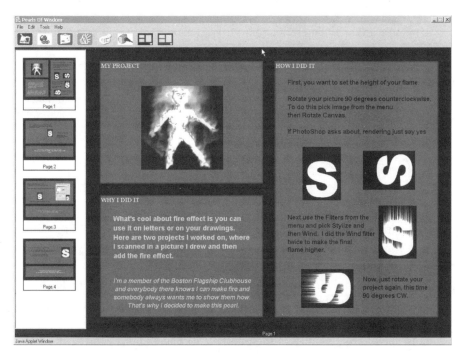

Figure 12.5
PhotoShop fire-effect Pearl under construction. The opening screen real estate is partitioned into three resizable areas that aid in organizing user ideas. There is an area for showing examples of the representative project, another for organizing how-to information, and a third for personal reflections and comments. Subsequent pages contain two resizable areas for how-to and reflective text.

types—including text, audio, video, image, and plug-ins—onto the work canvas. For example, creating objects to place within the Pearl is accomplished with two mouse clicks: first on the appropriate media icon, and second on the desired location for that media within the Pearl. More sophisticated users can further customize their Pearl layout (i.e., color, panels sizes, fonts, etc.) to obtain the desired design. For instance, a user may decide they want the bulk of their Pearl real estate to be devoted to audio instructions, interspersed with example images.

The Pearl-Constructor program, which includes a simple editing and multimedia-object selection toolbox, is designed to facilitate Pearl construction with minimal time investment. Time is important because early favorable results that are not time-consuming provide members with the

Table 12.2
Low- to high-threshold user options for Pearl authoring

User Type	Capacity
Novice user	Point-click insertion of media
	Point-drag media relocation within panel
Experienced user	Resizing of Pearl panels
	Set panel, font, and background colors
	Font characteristics
	Pearl page reordering
	Pearl page, panel, or object duplication

motivation and patience to go back and refine their work. This is also true for most Clubhouse projects. Members like to see early results for their efforts before committing longer periods of time to project completion. Projects without this quality are often abandoned. Pearls must therefore be low floor and easy to get some reasonable result, or unfinished Pearls will be abandoned. Once motivated to refine a Pearl, users must have the flexibility in the interface to make choices for their Pearl designs. For example, users can control how real estate is partitioned on any particular page of the Pearl. Pearl creators can resize the reflection and how-to areas to fit their design needs. The goal of the Pearl-Constructor tool is to provide a user interface that incorporates features to motivate Pearl creation and revision. Key design rationales include the following:

• *Low floor, high ceiling*: An iconic interface and simple editing toolbox flexible enough to create Pearls as simple or complex as the individual user desires. More advanced editing functions are also available, allowing for further customization (i.e., color, font, etc.).[2]

• *Quick "return on time investment"*: Rudimentary Pearls can be constructed and published quickly. Pearl construction minimally requires that the user complete the first page of the Pearl. This includes an example project, how-to information, reflection, and a Pearl title.

• *Edit after publishing*: Editing can occur anytime, from Pearl inception to after Pearl publishing. This allows for an iterative Pearl design cycle, either self-motivated or as a result of feedback from others.

- *User control*: The Pearl creator decides when to publish. Once published, Pearls may be edited, but they cannot be "unpublished."

Users also may edit their Pearls as often as necessary; Pearls are not searchable by the community until the Pearl designer "publishes" it, thereby giving the creator control over when the Pearl becomes public domain. This allows for Pearl "polishing" over several editing cycles. Likewise, Pearls can also be edited after publishing. The hope is that Pearl creation becomes another type of project seen at the Clubhouse.

Pearl Categorization

Pearls need to be organized to allow for easy browsing and Pearl retrieval. This involves, first, a relevant categorization schema. Pearls are categorized along three dimensions, closely following the natural categorization that I observed during conversations among various Clubhouse members around knowledge sharing. The following table summarizes these dimensions.

Currently, the number of categories that can be assigned to a Pearl are restricted to two.

Once a Pearl is ready for publication, the Pearl creator selects "publish" from the Pearl-Constructor menu. Additional information is

Table 12.3
Pearl category classes

Category Type	Description
Design tool	The design tool may consist of software or hardware. Some projects use multiple design tools. Currently, there is no way to look up additional tools, although this is being considered for the future.
Skill genre	The skill genre describes either some expertise or some known aspect of the project (i.e., special effects, robotics, Web page design, etc.). Up to two skill genres are permitted.
User-specified category	Users are asked to provide additional keywords to categorize their Pearl. This provides additional categorical Pearl descriptive.

captured automatically from the Clubhouse Village database, including the member's name and Clubhouse, which is included with the Pearl content, and their age and gender, which is logged and made available for later analysis by researchers. The user is prompted to assign Pearl categories, as mentioned earlier in this section.

Pearl-Search Tool

The PoW search interface serves as the main program screen. From here, users can elect to make new Pearls, browse Pearls, search, and a myriad of other actions. Just as when face-to-face connections are made at the Clubhouse, searching the PoW knowledge base offers subtle and complex opportunities for social capital building. For example, in the process of finding someone at the Clubhouse to help them, users will peruse member projects mounted on the walls or ask other members (see figure 12.6). The PoW search interface is organized into three sections with personal, community, and search functions to provide similar opportunities for seeing the work of other members. Key design elements include the following:

- *Personal area*: The user can create new Pearls, edit existing Pearls, or view a String-of-Pearls, which is a visual representation of one's interaction with the system, including all Pearls created or visited. It also serves as a future reference of project ideas of interest to the user. This is a valuable reflection activity, as the member can gain insight into how many skills have been learned or shared. Currently, members are not aware of the range of new skills they have acquired; they simply use these skills to develop their projects.
- *Search area*: Look for Pearls by category, perform targeted searches, or make a request to create a Pearl.
- *Community area*: View featured and new Pearls.

For example, the personal section provides user access to one's personal portfolio, which contains all the Pearls one has created. The community section offers a slide show of featured (randomly selected from the PoW database) and new Pearls. Currently, there is no ranking of the featured Pearls (say, according to how often they are used), just as all projects posted to Clubhouse walls are viewable by all, regardless of their level of sophistication. This provides inspiration for new project ideas (see figure 12.6). As the PoW database grows, along with the number of Clubhouses in the network, the issue of ranking will be revisited. The search section offers a standard keyword search capability and also a mechanism to broadcast requests for community members to create Pearls on particular topics. These requests are broadcast to the entire community via e-mail and are stored for later retrieval by potential Pearl creators.

The search algorithm scans through the Pearl metadata (design tool, skill genre, and user specified) as described in the Pearl categorization section. This search consists of a three-tier schema. First, there is a search of Pearl metadata, such as category and user-supplied keywords, design tool name, and the user-supplied Pearl description. If no results occur, a list of Pearls for the design tool category is displayed. The results are then ranked by their geographic location, from local to global, with respect to the local user. The ranks include:

- local (find those experts geographically close);
- regional (find those within a predetermined Clubhouse region, which may consist of a section of a country or a group of several countries; the Clubhouse regions have been configured by the Computer Clubhouse Network staff to group together Clubhouses with similar languages, cultures, etc.); and
- global (sample the entire Computer Clubhouse Network).

This ranking allows the user to see Pearls first from local experts, then from others with the same language or other similarities, and finally from anyone across the network. The results are displayed as a list of pearl titles with short user-supplied descriptions and project images.

Currently, there is no mechanism for dealing with the multilingual and intercultural differences that come with a global network of Clubhouses. As a result, all initial studies will take place at Massachusetts Clubhouses in the United States. The software architecture has been organized to accommodate multilingualism in future versions. At present, however, my main interest is to observe what type of pressures are put on the system by this smaller set of local users and where these pressures occur.

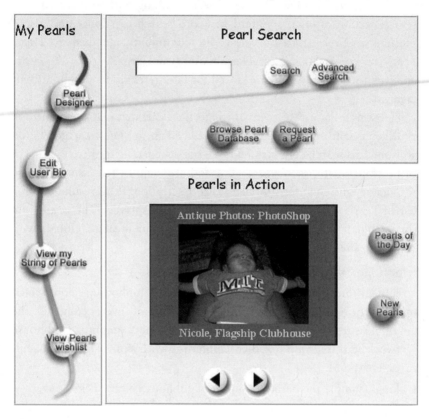

Figure 12.6
The search interface provides three levels of social interaction: personal, search, and community. The personal section contains user-specific Pearl management tools. The community section showcases new Pearls and also displays randomly selected Pearls from the PoW database. The search section provides search functions and a mechanism for broadcasting new Pearl requests.

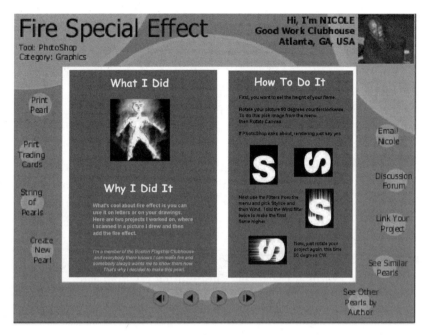

Figure 12.7
PhotoShop fire-effect Pearl, page 1. Besides Pearl content, the Pearl-Viewer interface provides options for community participation, recognition of the Pearl creator by the community, and output.

Pearl-Viewer

Besides displaying Pearl content, the Pearl-Viewer offers a variety of mechanisms for exchanging comments, viewpoints, projects, and Pearls. Its real estate is organized into sections dedicated to community connections, output functions, and Pearl creator recognition, all surrounding an author-created, content-filled information area.

Community functions include a discussion forum, an e-mail link to the Pearl author, an option for the Pearl user to link their Pearl-inspired project to the Pearl, and the ability to browse other creator or topical Pearls. The discussion forum allows for the exchange of ideas about the Pearl content, genre, or alternative tools or methods for achieving the same results. At Computer Clubhouses, members may post their projects on the walls. Others learn similar design skills as those seen in the initial project. Over time, Clubhouse walls reflect this propagation of ideas and expertise as more projects with the particular design skill are displayed. The forum offers a similar space for idea propagation as Pearl users may opt to list their own completed projects (with skills learned from using the Pearl) within the forum. Thus, we begin to see the spread of ideas and knowledge via Pearl use.

Included with the content is the Pearl creator's name, image, or logo, and Clubhouse information. This serves as a form of member recognition. An output functions area allows for a printer-friendly version of the Pearl, a trading-card version of the Pearl, and access to the user's personal "string of pearls." In summary, this interface design incorporates features to motivate Pearl creation, use, and community participation derived from similar, relevant motivators for face-to-face, local Clubhouse interactions.

PoW in Use: Learning Experiences in Informal Environments

An early PoW prototype, developed by the author, was field-tested in two Boston-area Clubhouses over a period of three months. The prototype did not include the community or output functions, as mentioned earlier in this chapter. I provided ten Pearls, which were used at two Boston Clubhouses. A total of thirty-three Clubhouse members attended a series of eight MicroWorlds Pro (LCSI 1990) workshops. The goal of the workshops was to use MicroWorlds Pro software (and its Logo programming language) to make simple games. The ten Pearls covered various basic, but widely applicable programming skills ranging from communicating with software objects, to creating procedures and variables, to creating special effects (i.e., music, vocals, etc.). The Pearls were introduced at the beginning of the workshop, and participants were encouraged to look them over. Modeling Pearl use was also an important aspect of its introduction to the participants. For example, when working with an individual, workshop leaders using that Pearl as a reference answered questions covered by an applicable Pearl. With the exception of one participant, repeating this process resulted in the participants checking Pearls first for information they needed. Part of their motivation for doing so was expressed in the workshop's exit questionnaire. Most found it quicker to get the information from the Pearl, rather than go through the effort of getting the attention of my assistant or myself. A smaller number of the participants, less than half, also mentioned they browsed the Pearls for ideas on what to do next in their project. While the participants were not able to make their own Pearls during the workshops, they were asked in the wrap-up session what if any, Pearls they thought would be nice to have. The majority wanted some way to have Pearls for the techniques they learned during their one-on-one workshop interactions. The workshop concluded with the participants presenting their various projects to the group. Only two participants had previously used MicroWorlds before, and neither had any significant Logo programming experience. Nine participants had some HTML programming experience. Some project examples provided included horse race, maze, and soccer games. After an introductory session on how to use MicroWorlds's interface and create programmable objects, each participant decided what game they wanted to make. All were given access to the ten Pearls: eight of them covered programming skills used in the example projects, and two contained other advanced programming content.

Research Methods

I used a participatory research framework in the spirit of Gerald Sussman and Richard Winter in the course of the workshops (Sussman 1983; Winter 1989). All observations and data were analyzed, and the research problem was reassessed, with this cycle repeated after each workshop. During the course of the workshops, the researchers informally discussed design experiences with members and took notes on salient points from these conversations. At the end of each workshop session, the participants answered a questionnaire that probed their comfort with using Pearls, the perceived usefulness and inadequacies of the tool, and whether they would recommend the tool's use to other members. The documentation included photos, researcher notes, member projects, and participant questionnaires.

Empirical Findings

The participants were eager to share their projects with other Clubhouse members (especially because MicroWorlds was rarely used prior to the workshop); indeed, they enjoyed being considered the local experts using the program.

The workshop participants made several characteristic uses of the ten Pearls:

- All ten Pearls were used with no bias toward the eight that applied specifically to the examples I showed them. Pearl use was driven primarily by what the members wanted to accomplish with their projects, and for some members, the two advanced-topic Pearls served that purpose.
- The members were vocal about what they felt was missing from a particular Pearl (i.e., not enough detail, too much detail, etc.). This openness is part of the Clubhouse culture and happens in face-to-face interactions as well.
- At the conclusion of the workshops, the participants were eager to show their completed projects to the researchers and other Clubhouse members.
- The members still roamed the physical space to see what others were working on and to share their ideas. This is also a normal part of Clubhouse culture.

The majority of the members entered the following cycle in interaction with their Pearls:

- They first decided their project's genre (i.e., making a game, animation, or story).
- They then browsed the ten Pearls to see which would be useful, and abandoned inapplicable Pearls.
- Next, they asked for help from researchers and mentors to explain parts of a Pearl or to get help that Pearls couldn't provide. Requesting help from mentors is a normal Clubhouse activity.

During the workshop, face-to-face collaboration was fueled by a participant's need to see what others were doing or talk about how to use the Pearl. The former is typical of Clubhouse interaction. The discussions about expertise are not generally seen at the Clubhouse. These discussions centered on the Pearl's content, and its shortcomings or usefulness. This is a significant finding for several reasons. First, the participants were able to pace their acquisition of knowledge. This appeared to be a real confidence builder. Second, the participants indicated they felt less dependent on someone else and someone else's schedule to get what they wanted out of their project. While getting help from others works well at the Clubhouse, there is value in learning to pursue information independently. Third, programmers enjoy maximum status at Clubhouses since few individuals do programming. There are also only a few mentors who can support members in learning to program. Previous to the workshop, the concept of programming was a mysterious and frightening prospect. After their workshop participation, a handful of the participants were eager to learn more programming skills—motivated in part by their desire to design games that their friends could play. They indicated in conversations during the workshop and/or on the exit questionnaire that they could program now and couldn't wait until everyone else found out. There was a real sense of the social status they could gain by improving their programming expertise and showing others how to do the same. Although most programs involved simple object manipulation, with a few also incorporating more a complex use of variables, all the participants felt they had accomplished something they had no previous knowledge of. A chief complaint was the lack of a sufficient number of Pearls, covering a wider breadth of programming topics. This complaint came up often and speaks to the importance of seeding an adequate number of Pearls into PoW before it is introduced to the entire community. Mentors and previously established expert members will do this with early submissions.

Clearly, being able to have high visibility around their work was crucial to members. Also, they valued the independence of more self-directed learning (i.e., having Pearls available to use on demand). These early prototype Pearls remain in use at the South Boston Clubhouse. Although how-to content is made available, community or output functions are not. In spite of this lack of Pearl functionality, many community interactions (the displaying of work and showing others how to use the software) were observed during the workshops, probably because those interactions are already part of the Clubhouse culture. I am continuing to observe use of the full PoW system to determine if community functions are utilized on-line and how this may differ from local, in-person interaction. I am interested in what kinds of interactions are stimulated, sustained, or hindered by the current PoW system, with its full complement of community and output support functions.

Discussion

Benefits of Knowledge Sharing at the Computer Clubhouse

For the learner, there is real power in the journey from tacit, localized knowledge to explicit, distributed knowledge. In the case of the Clubhouse, members commonly don't have a sense of what they have learned and the value of that knowledge. Tacit knowledge is the implicit knowledge an individual uses to accomplish and make sense of some goal (Polanyi 1958; Nonaka and Takeuchi 1995). By comparison, explicit knowledge has a tangible dimension, and must be codified in such a way that is easily accessed by and communicated to others. Where the tacit and explicit intersect is where knowledge transformation takes place for the learner. There are four, interconnected modes in this cycle of knowledge transformation: socialization, externalization, combination, and internalization (Nonaka and Takeuchi 1995). I propose this transformation can also serve as a strong learning indicator, evidenced by Clubhouse members' ability to recognize when they know something that can be useful to others, making that knowledge available to others, and making others' knowledge their own (figure 12.8). With the creation of Pearls and their use by others, a Clubhouse members' success at learning new things at the Clubhouse is made visible and becomes a resource to the rest of the community. This culminates in a life cycle of knowledge creation and sharing.

Another interesting outcome may be the blurring of the current barriers to knowledge sharing—such as time, distance, and social inequities—that without technology as a mediator, can prevent the continued spread of new ideas and knowledge. For example, at the Clubhouse, age and gender have typically played a role in limiting one-on-one knowledge sharing. This is merely a reflection of similar barriers seen in the larger society. Changing how information is exchanged in the Clubhouse community may result in exchanges that circumvent these accepted norms of interaction. Also, we must remain cognizant of how a system's function contributes to the building of social capital. Indeed,

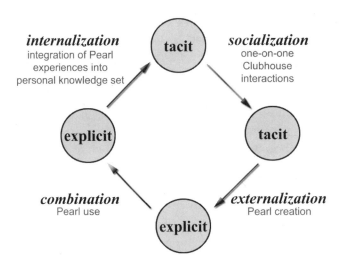

Figure 12.8
Modes of knowledge transformation through the creation and use of Pearls (adapted from Nonaka and Takeuchi 1995).

this is a key factor in the sustainability of the technology within a community. There is already a strong social capital dynamic in place around how information is shared within a local Computer Clubhouse. Member trust and network building are critical to the realization of project goals. A major tenet of the Computer Clubhouse philosophy is that individual development is derived from the interactions among individuals and the quality of the members' connections to each other. Clubhouse members learn from as well as share and work with each other to realize their project goals. The act of explaining and interacting with others often results in valuable learning opportunities.

Challenges to Building Social Capital
There are limitations to knowledge sharing in a distributed setting versus sharing face-to-face that can limit social capital building and result in adverse effects on the community (Portes 1998). One downside is the potential inability of users to recognize when they have something worth sharing, and in the case of many Clubhouse youth, a reluctance to display poor writing skills. Just as important is the level of communication efficacy that a community requires. Does the technology provide the requi-

site levels of interactions necessary to support the building of trust? Yet trust often takes time to cultivate. Maish Nichani (2001) points out that there is no such thing as instant trust; instead, time and (in this case virtual) space is required for trust to develop. It grows as people are exposed to one another and share their experiences. Even in lightweight social spaces such as those described by Mark Ackerman and Leysia Palen (1996), with almost complete anonymity among participants, trust that community members will self-correct such visible responses to questions is key to data integrity. Likewise, PoW, with its visible Pearls and mechanisms for public discourse, should experience similar self-correction by the Clubhouse community. A significant feature of Computer Clubhouse interactions has been the characteristic of synchronous information sharing. PoW affords asynchronous interactions, which are only meaningful if the community is amenable to this limitation. While the technology does allow the user to find information on demand from across a distributed membership, it does not offer the richness of personal interactions, which plays a role in the sustainability of Clubhouse culture. This challenges the developer to consider what value the community places on the use of the technology and to question if the advantages realized through the technology are a sufficient motivation for a community to adopt that technology.

The current PoW design generates a number of salient questions regarding interactions between communities and technologies. For example, what barriers are crossed and how are new interactions realized through the use of this technology? In the Computer Clubhouse, common barriers to information sharing include gender and age differences. Male members rarely ask female members for help. Older members rarely ask younger ones for help. These are perceived social barriers within this community, but these barriers depend on face-to-face connections for reinforcement. How less important do these differences become when gender and age, say, are not so easily identifiable?

Limitations of the Technology

For the Computer Clubhouse Network, globalization carries the opportunity for the rapid propagation of ideas throughout the network. As Gerry Stahl (1999) notes, people become aware of the world through these social interactions, bringing their own observations and experiences into their activities. The Computer Clubhouse Network is not a classic community of practice (Wenger 1998); instead, it has variations across the network that cannot be ignored, including language, culture, and standards of motivation as well as relevance. The increasing number of Computer Clubhouses across diverse cultures offers a unique opportunity for distributed knowledge sharing of design perspectives throughout the network. There is also the danger, however, of these cultural and language differences hindering the ability of users to make viable connections on the system. Central to the process of creating trust in this variety of cultures is the necessity of relating experiences. Richard Schwier (2001) observes that the active participation of a distributed community's members is critical to community sustainability and continuity. PoW uses the "Why I Did It" section for personal reflections and stories—the foundation that helps build relationships and trust. Unfortunately, language differences pose an insurmountable barrier to such sharing. One possible solution is to promote a kind of cultural symmetry in the search results. For example, if you are a member of the South Boston Clubhouse, then order search results with Pearls from Massachusetts, then the United States, and finally from any other Clubhouse locations. These "preorganized" search results are then returned to the user, increasing the likelihood that language and other cultural differences are minimized, while still leaving more diverse Pearls accessible.

It is because a culture of sharing knowledge is part of the Clubhouse community that it was chosen as the test site for the PoW system. As the use of the system increases, the balance between growing community involvement and providing adequate trust and reciprocity support will speak to the sustainability of this technology. Trust runs along two dimensions: trust in the source of information, and an individual's belief that community feedback will be affirming or constructive. An obvious PoW advantage is the realization of previously impossible connections and fewer constraints on access to information than interactions without the technology. There are also mechanisms embedded in the technology to recognize experts, either by using their Pearls, contacting them directly, engaging in on-line community discussions, and utilizing trading cards.

Many of the projects developed at the Clubhouse do not involve much writing. Although one could say that writing is difficult for some members, or that writing is not as socially attractive as making multimedia animations, it is important to note that the great majority of tools available at the Clubhouse do not emphasize the power of expressing ideas in words. PoW is a tool that includes writing as a possible means of expression. This practice will have to be accepted and adopted by the community.

Conclusion

The design rationale for community-based learning technologies must incorporate support for the sharing of values, development of personal networks, and establishment of identity within that community. In the case of technologies for knowledge sharing, software design must incorporate motivations for individual participation and the creation of knowledge. While social capital does provide a powerful lens through which to view sustainable, community-based technology development, careful attention must be paid to not affording agency to the technology alone. It is imperative to go beyond a mere technology-centered approach in order to create a culture that encourages and rewards the sharing of knowledge. This is a formidable challenge and will require a deeper understanding of community dynamics than previously thought. One of the ideas being put forth with this and similar works is the expectation that communities of all kinds have the ability to build their own knowledge-sharing structures and to determine what level of codification is acceptable. Finally, in the case of the Computer Clubhouse, technologies such as PoW are bound to change the cultural norms of how information and knowledge are shared. There is no way to guarantee that these changes will be positive ones.

While the Pearls used in the Clubhouse workshops did not contain any community-support mechanisms, the members fell back on traditional methods of seeking and sharing ideas as well as information with each other. The current PoW system, with includes community supports, will test whether there is any shift from face-to-face interactions, and if so, how that may impact the social capital dynamics of trust, accepted norms, and networking we've seen in the absence of this technology. The workshop participants did not generate the content for the ten Pearls made available for their use. What roles will ownership, in both the creation and social reinforcement of member-created Pearls, play in the adoption of this technology for knowledge sharing into Clubhouse culture? The Clubhouse learning model reinforces the notion that for social development and learning, everything comes down to people and their connections. People inspire each other, they learn by explaining and interacting with others with different perspectives and backgrounds, and community members leverage each other by working together.

Technology's role is to serve as a medium for personal expression and a support for the development of social connections. Throughout this chapter, I have endeavored to outline the significance of social capital building as an integral component of community systems design. We must continually focus on the importance of adjusting our design sensibility to incorporate social capital building mechanisms specific to the community for which these ITs are being developed. Poorly utilized knowledge-sharing tools will be the result if we ignore the importance of social capital within any community for which ITs are being developed. Without such a coherent design rationale, the acceptance and integration of the technology into the culture becomes problematic. My goal is to continue to investigate and address those obstacles that interfere with effective social capital building and knowledge sharing.

Notes

1. Media types include any combination of text, video, audio, image, or plug-in.
2. MacLean et al. 1990 advocates a similar design approach for tailoring environments, and suggests offering a "gentle slope" of increased complexity to encourage tailoring abilities.

References

Ackerman, M. 1998. Augmenting organizational memory: A field study of Answer Garden. *Association for Computing Machinery (ACM) Transactions on Information Systems* 16, no. 3:203–224.

Ackerman, M., and L. Palen. 1996. The Zephyr help instance: Promoting ongoing activity in a CSCW system. In *Proceedings of Association for Computing Machinery (ACM) conference on human factors in computing systems.*

Ackerman, M., V. Pipek, and V. Wulf. 2003. *Sharing expertise: Beyond knowledge management.* Cambridge: MIT Press.

Chapman, R. 2001. Redefining equity: Meaningful uses of technology in learning environments. Paper presented at IEEE International Conference on Advanced Learning Technologies, Madison, Wisconsin.

Chapman, R., and L. Burd. 2002. Beyond access: A comparison of community technology initiatives. In *Proceedings informatica 2002: The Latin American and Caribbean symposium on education, science, and culture in the information society—SimpLAC 2002.* La Havana, Cuba: UNESCO.

Coleman, J. S. 1990. *Foundations of social theory.* Cambridge, Mass.: Belknap Press.

Duckworth, E. 2001. *"Tell me more": Listening to learners explain.* New York: Teachers College Press.

Hutchins, E. 1995. *Cognition in the wild.* Cambridge: MIT Press.

Jonassen, D. H. 1991. Evaluating constructivist learning. *Educational Technology* (September): 28–33.

Lave, J., and E. Wenger. 1991. *Situated learning: Legitimate peripheral participation.* Cambridge: Cambridge University Press.

LCSI. 1990. MicroWorlds Pro. Software on CD. Highgate Springs, VT.: LCSI.

MacLean, A., K. Carter, L. Lovstrand, and T. Moran. 1990. User-tailorable systems: Pressing the issues with buttons. In *Human factors in computing systems, Proceedings of the conference on Computer Human Interaction*, edited by J. Whiteside. New York: ACM Press, pp. 175–182.

Nichani, M. 2001. Communities of practice at the core: The e-learning post corporate learning, community building, instructional design, knowledge Management, personalization, and more. Accessed 2 December 2002, <http://www.elearningpost.com>.

Nonaka, I., and H. Takeuchi. 1995. *The knowledge creating company: How Japanese companies create the dynamics of innovation.* New York: Oxford University Press.

Papert, S. 1980. *Mindstorms: Children, computers, and powerful ideas.* New York: Basic Books.

Papert, S. 1993. *The children's machine.* New York: Basic Books.

Piaget, J. 1977. *The essential Piaget.* Edited by H. E. Gruber and J. J. Voneche. New York: Basic Books.

Pinkett, R. 2001. Creating community connections: Socio-cultural constructionism and an asset-based approach to community technology and community building in a low-income community. Ph.D. diss., Massachusetts Institute of Technology.

Polanyi, M. 1966. *The tacit dimension*. London: Routledge and Kegan Paul.

Portes, A. 1998. Social capital: Its origin and application in modern sociology. Vol. 24 of *Annual Review of Sociology*.

Putnam, R. 1993. The prosperous community: Social capital and public life. *American Prospect* 13 (spring 1993): 35–42.

Putnam, R. 2000. *Bowling alone: The collapse and revival of American community*. New York: Simon and Schuster.

Resnick, M., and N. Rusk. 1996. Access is not enough: Computer Clubhouses in the inner city. *American Prospect* 27:60–68.

Salomon, G., and D. N. Perkins. 1998. Individual and social aspects of learning. In vol. 23 of *Review of research in education*, edited by P. Pearson and A. Iran-Nejad. Washington, D.C.: American Educational Research Association.

Scardamalia, M., and C. Bereiter. 1994. Computer support for knowledge-building communities. *Journal of the Learning Sciences* 3, no. 3:265–283.

Schwier, R. A. 2001. Catalysts, emphases, and elements of virtual learning communities: Implications for research and practice. *Quarterly Review of Distance Education* 2, no. 1:5–18.

Shaw, A. 1995. Social constructionism and the inner city. Ph.D. diss., Massachusetts Institute of Technology.

Stahl, G. 1999. Reflections on WebGuide: Seven issues for the next generation of collaborative knowledge-building environments. In *Proceedings of CSCL '99: the third international conference on computer support for collaborative learning*, edited by C. Hoadley and J. Roschelle. Mahwah, NJ: Lawrence Erlbaum.

Sussman, G. 1983. *Action research: A socio-technical systems perspective*. London: Sage Publications.

Wenger, E. 1998. *Communities of practice: Learning, meaning, and identity*. New York: Cambridge University Press.

Wertsch. J. V. 1985. *Vygotsky and the social formation of mind*. Cambridge: Harvard University Press.

Winter, R. 1989. *Learning from experience: Principles and practice in action-research*. Philadelphia: Falmer Press.

13

Expertise Finding: Approaches to Foster Social Capital

Andreas Becks, Tim Reichling, and Volker Wulf

The term *social capital* has gained importance in the scientific discussion of different disciplines. Pierre Bourdieu provides an early definition of the concept: "Social capital is the aggregate of the actual or potential resources which are linked to possessions of a durable network of more or less institutionalized relationships of mutual acquaintance and recognition" (1985, 248). Ronald S. Burt (1992, 9) understands social capital as the friends, colleagues, and other personal relationships that offer the opportunities to use one's personal or financial capital. Robert Putnam (1993, 2000) applies the concept of social capital even to cities, regions, and whole nations. He understands social capital as the set of properties of a social entity (e.g., norms, level of trust, or social networks) that enables joint activities and cooperation for mutual benefit.

All of these definitions of social capital have a point in common: the creation of social networks requires efforts (investments) and allows their purposeful use later on. Like financial capital invested in machinery or personal capital gained within educational institutions, social capital increases the productivity of labor. Therefore, the concept has considerable economic relevance.

Yet the findings concerning an appropriate structure of the social networks diverge. The mainstream assumes that closely knit social networks are advantageous. On the other side, Burt (1992) argues that rather loosely coupled networks, containing structural holes, are best suited to provide the appropriate resources. These networks allow human actors to get diverse, nonredundant information. Putnam (2000) distinguishes between two types of social capital: bonding social capital relates to social networks within an actors own community, while bridging social capital reaches beyond the community boundaries.

In this chapter, we want to focus on building social capital to foster collaborative learning processes. Don Cohen and Laurence Prusak (2001) argue that social capital offers an interesting new perspective with which to look at knowledge management. While earlier approaches focused on storing and retrieving explicit knowledge represented in documents, new work deals with implicit knowledge ("knowing," according to Michael Polanyi [1958]) as well. As such, research has to be centered around the problem-solving capabilities of individual actors and social entities—for example, communities (Ackerman, Pipek, and Wulf 2003). Contrary to knowledge management, theories of learning have concentrated on those institutional settings whose primary purpose is knowledge transfer (e.g., schools or universities). Within the field of learning theories, sociocultural approaches that focus on knowledge acquisition within communities of practice gain in importance (Lave and Wenger 1991; Wenger 1998). They complement or even replace approaches that stress individual learning (e.g., behaviorism or cognitivism). From the point of view of management science, Janine Nahapiet and Sumantra Ghoshal (1998) have tried to link the two lines of thought: communities of practice and social capital. Following Joseph A. Schumpeter's (1934) model of knowledge creation, they contend that social capital fosters the generic processes of combining and exchanging knowledge.

Given the importance of social capital for knowledge management and learning, applied computer science needs to take this perspective into account. One way to do this is to investigate how computer applications may contribute to increasing social capital. One can think of different roles that computer applications may play in increasing social capital, for example:

• Analysis of existing social capital: Algorithms may be capable of detecting interpersonal relations (e.g., analysis of the frequency of mail exchange). Appropriate visualizations ease the mutual understanding of the current state of a social entities' social capital. These visualizations may also be the starting point of interventions to improve the social capital of a social aggregate.

- Finding of (unknown) actors: To encourage bridging social capital, algorithms may make actors who have similar or complementary backgrounds, interests, or needs aware of each other. Therefore, personal profiles have to be created and updated either manually by the actors or automatically by appropriate algorithms.
- Communication among actors: While actors are often dislocated or need to communicate asynchronously, appropriate applications for computer-mediated communication are needed. This especially refers to the link between communication channels and the artifacts that the communication refers to.
- Building of trust within social entities: To establish and maintain social relations, trust plays an essential role. A computer application may open up additional channels among the actors to encourage trust building.
- Maintaining social relations: Bridging social capital is often characterized by rather infrequent personal relationships (e.g., among school or university friends, or among former colleagues). Within highly dynamic environments, there is an ongoing danger that these relationships may fade away (e.g., new addresses after changing jobs). Computer applications may help actors to stay informed about news concerning their old friends (e.g., address changes).

With regard to each of the different roles that computer applications may play, one has to reflect critically on the appropriate mixture between technologically mediated and technologically nonmediated activities.

This chapter concentrates mainly on the role of finding actors, that is, to make potentially fitting actors aware of each other in a virtual or real space. So these functionalities offer opportunities to introduce actors to each other by matching or visualizing aspects of their behavior, background, qualifications, expertise, or interests. Therefore, these functionalities need to grasp, model, and evaluate relevant personal data. These data can be either put in manually by the user, automatically grasped, or imported from other applications.

After reviewing the current state of the art, we will present an application that is supposed to foster social capital within an e-learning platform. E-learning platforms allow users to access content in a structured way typically by means of a navigation hierarchy. When grasping personal data automatically, the hierarchical content structure of the e-learning application eases the adding of semantics.

State of the Art in Expertise Finding

Research in the field of computer-supported cooperative work (CSCW) and artificial intelligence (AI) has created applications that can be understood as technical support for building social capital. Traditionally, CSCW research focused on the support of small working groups already containing a high level of social capital. Some of the techniques developed in this context, however, can also be used to support the creation of bridging social capital. Recently, CSCW research has also focused on the support of less-well-connected communities. Social capital can provide an interesting perspective on building applications for such communities. The AI community can contribute to the fostering of social capital because it creates algorithms that allow one to detect a pattern of similarity within unstructured data. These similarities can be used to match actors or make them aware of each other.

Here, we want to discuss five research directions in more detail that can contribute to encouraging the building of social capital:

1. expertise-profiling systems;
2. topic-oriented communication channels;
3. discussion and annotation systems;
4. collaborative recommender systems; and
5. mutual awareness.

The core question in the field of expertise-profiling systems is how to make explicit and implicit knowledge held by individuals visible and accessible to others. In the standard approach to personal-profiling systems, the actors are asked to input the data describing their expertise or interests by themselves (e.g., yellow pages). The creation and maintenance of personal profiles suffer from a couple of difficult problems, though. First, a common understanding of the different attributes of a personal profile has to be given (Ehrlich 2003). If the profiles are created and updated manually, the different human actors need to have a joint understanding of each attribute. Only in this case can their input be matched automatically. Second, the actors need to be motivated to input and update their personal profiles. The ongoing necessity to update these profiles threatens their validity (Pipek, Hinrichs, and Wulf 2003). Therefore, these data may be complemented by automatically generated data—derived, for instance, from analyzing an actors' home page or mail traffic. Still, automatically generated profiles aggregate data whose semantics are not clear. So it is doubtful whether these data really represent the actors' competencies and interests.

The core question in topic-oriented communication channels is how do electronic media change communication and social interaction among the actors? There are different approaches to realize topic-oriented communication channels such as newsgroups, mailing lists, MUDs, or MOOS. While the first ones are based on a purely content-oriented structure, the last ones apply a spatial metaphor to structure communication. Experiences demonstrate that topic-oriented communication channels are able to create virtual communities of mutual support (cf. Rheingold 2000; Hafner 2001). In cases where the actors communicate by revealing their personal identity, social relationships—even beyond the virtual space—may be established. In the domain of Web-based training, Martin Wessner and Hans-Rüdig Pfister (2001) have proposed that learning platforms may be supplemented with topic-oriented communication channels. To structure and focus users' contributions in learning environments, the authors introduce the concept of intended points of cooperation (IPoCs)—that is, starting points for communication in a learning unit that are defined by the authors of the unit. Specific communication channels (e.g. chat, video conferencing, shared whiteboards) are used to support different types of communication. In the sketched learning environment, the critical process of group formation can be performed manually by a tutor (supported by a tool that displays course- and class-related information) or automatically by matching learners who participate in the same course and have not yet completed common IPoCs. So the learners are matched without their direct involvement.

The core question in discussion and annotation systems is how to support the development or refinement of a mutual understanding on a certain topic by means of computer-mediated discourses, typically in textual form. There are many approaches that combine the presentation of content with integrated functionality to annotate or discuss (e.g., Buckingham Shum 1997; Pipek and Won 2002; Stahl 2003). In cases in which the different contributions to the discussion can be attributed to individual actors, such applications can support the building of social capital. Active participation in computer-mediated discourses is required to catch other actors' attention. Yet this may not always be an appropriate approach. The discourses are typically restricted to a rather specific issue, which makes it difficult to transfer a competency demonstrated in a specific discourse to other topics.

The core question in recommender systems is how to support actors in selecting an item from a set of rather similar items. Several recommender systems are relevant here because they have been designed to support the finding of human actors (cf. Yiman-Seid and Kobsa 2003). Systems like Who Knows (Streeter and Lochman 1988), the Referral Web (Kautz, Selman, and Shak 1997a, 1997b), Yenta (Foner 1997), or MII Expert Finder and XperNet (Maybury, D'Amore, and House 2003) extract personal data about human interests automatically from documents that are created by the actors. Adriana Vivacque and Henry Lieberman (2000) have developed a system that extracts personal data concerning a programmer's skill from the Java code that the programmer has produced. Based on these personal data, the system allows one to pose queries or match actors. These systems, however, have hitherto dealt with specific matching algorithms for one type of personal data. David W. McDonald, alone (2000) and with Mark Ackerman (2000), developed a framework for an expertise recommendation system that finds people who are likely to have expertise in a specific area. Contrary to the general approaches to expertise matching mentioned above, the framework allows specific heuristics to be developed that are tailored to the individual organizational context. Thus, it does not focus on an automatic evaluation of many different documents or programs written by an actor but rather on a context-specific heuristic. This heuristic needs to be revealed by a preceding ethnographic study in the application field. If found, such a heuristic is probably better suited than an automatic algorithm. As in the approaches mentioned above, the heuristic matches experts with people looking for support.

The core question in the field of mutual awareness is how to make the activities of distributed actors visible to each other. With their study on the importance of mutual awareness for cooperation, Christian Heath and Paul Luff (1991) have motivated a whole series of design approaches. These approaches tried to capture selected activities of individual actors and make them visible to their cooperation partners (e.g., Rodden 1996; Sandor, Bogdan, and Bower 1997; Fuchs 1997, 1999; Fitzpatrick et al. 2002). With regard to the data captured, one can distinguish between structured and unstructured ones. Structured data record the use of a system's functionality, and unstructured data typically consists of video streams. The visualization of these data is supposed to compensate for a lack of visibility of individuals' activities and their context in a distributed setting. Awareness features are typically built for groups that contain a high level of social capital and cooperate intensely. Nevertheless, awareness data and the resulting histories of interaction can also be applied to match people who are not yet well known to each other. For instance, recent approaches try to apply structured awareness data to make individuals aware of others who access the same Web site. The Social Web Cockpit provides awareness data that informs users about the presence of other users at a site of interest. Moreover, it allows for collaborative content rating and recommendation functionalities (Gräther and Prinz 2001). Thus, communities can be set up in a self-organized way based on common interests. The Social Web Cockpit, however, has to be installed as an additional application on each users' computer before benefiting from the matching functionality.

Matching Personal Data with Algorithms

As the discussion above has shown, identifying, collecting, and maintaining appropriate personal data is difficult. The information kept in a profile may stem from many different sources (e.g., manually created interest statements, or professional or personal histories). A great deal of semantic background knowledge may also be necessary to realize an algorithmic matchmaking of experts based on personal data.

With regard to the creation of personal profiles in learning platforms, we are in a rather advantageous position. The structure of the content represented in the platform provides an ontology of the knowledge domain. This ontology can be used to add semantics to automatically recorded data. Specific features of a learning platform, like the results of tests, may allow one to update personal profiles automatically. So, learning platforms provide semantics and specific data that ease the automatic identification of expertise. It is therefore easier to match expertise in learning platforms than in other applications.

In this chapter, we want to use histories of interaction and awareness data concerning the production and use of the platform's content to create and update personal profiles. Due to the fact that the content is prestructured, the automatic capturing and processing of these data seems to be promising. In the following, we want to show which data are relevant and how to gain semantic information from these data:

• Data concerning the production of learning material; actors who have produced specific content for the platform may be experts in this domain.

• Data concerning the update of learning material; actors who have updated or refined specific content for the platform may be experts in this domain.

• Data concerning tutoring responsibilities; actors who are doing or have done tutoring tasks concerning specific content of the platform may be experts in this domain.

• Data concerning test results; actors who have passed tests concerning specific content of the platform may be knowledgeable in this domain.

• Data concerning the actual use of certain material; actors who are navigating through specific content of the platform may be interested in this domain.

• Data concerning the history of interaction with certain material; actors who were navigating through specific content of the platform may be interested or even knowledgeable in this domain.

Further aspects of the user's profile can be imported from sources outside the learning platform. Keyword vectors or higher-order structures derived from an actor's mail (incoming or outgoing) or document production (letters, papers, slides) can be automatically captured as well as those derived from an actor's home page (Foner 1997; Streeter and

Lochman 1988). Further data may be extracted from an automatic evaluation of aspects of actors' task performance (e.g., elements of a program ming language used; cf. Vivacque and Lieberman 2000). These automatically captured data can be supplemented by profile data entered by the user concerning their personal background, interests, or competences. A cross-check between manually and automatically created profile data may reveal inconsistencies. These inconsistencies can be indicated at the user interface to initiate an update of the personal profile.

Now the question arises of how to apply these data in matching learners, tutors, and content providers. The matching algorithms make use of the ontology given by the hierarchical structure of the content. Whenever a learner looks for support, that support can be located by referring to the specific learning unit that the learner browses. With regard to this learning unit, the system can retrieve data about the production history of the content. The creator of a unit as well as the actor who did the last update can be presented to the learner along with the one responsible for tutoring. In a similar way, histories of passed tests can be applied to match learners with those who have already demonstrated capabilities within a certain time span. Finally, the matching algorithms allow one to identify those actors who are actually browsing the same learning unit or have done so within a certain period of time.

Prior to presenting our approach to expert finding, it is necessary to discuss some further requirements for the matching framework. First, we sketch a direct consequence of the discussion of relevant information sources for user profiles (e.g., information on the professional training status, information gained from produced documents, user context and history of interaction, etc.): Quite clearly, each source of information requires a specific method of matching. The expert matching framework should thus consist of modules (with well-defined technical interfaces) that encapsulate a certain information type and then contribute to a global matching result by calculating a degree of matching based on that particular type of data (e.g., similarity in learning or project history, interest profile, etc.). Note that this kind of modularity is a prerequisite for adapting an expert-finding component to different application contexts—for example, different learning platforms or knowledge management environments: Specific matching modules can be exchanged or adapted according to the relevant requirements.

Second, matching expertise affects privacy issues: Learners—or more generally, users of any kind of platform that includes expert-finding functionalities—might not be willing to make available any kind of personal information to the public. In order to protect the users' right of informational self-determination, each user must know and be able to define which of their personal data is used for matching or publication, respectively.

An important problem of matching personal data is the question of "data quality": Histories of interaction, for example, are collected successively and tend to become more expressive with each history item collected. The completeness of personal data may also vary from user to user due to individual privacy decisions. Yet the matching quality will depend on the "completeness" of the information available. The more complete a specific type of personal data, the more reliable one can expect the matching result to be. As a consequence for the algorithmic framework, a "degree of completeness" of the different types of personal data should be measured where possible and be taken into account when calculating a matching degree.

Finally, given that a user agrees to use certain types of personal data for the matching process, they should also be able to adapt the expert-matching algorithm to a certain degree: The perception of which modules—that is, types of personal data—contribute to a good profile matching may vary for different users in different contexts. In order to let the user decide which "factors" contribute to the expert matching, and to what degree, we propose to incorporate a factor that weighs the impact a certain module has on the overall matching result into the matching framework.

A Modular and Adaptable Expert-Matching Approach

This section presents an algorithmic framework for expert matching that takes into account the requirements discussed above. We make use here of the following terminology: *Expert finding* means matching a prototype set of personal data (i.e., profile of a certain user in the application environment or a query profile) against a collection of other actors' personal profiles in order to determine a ranking of fitting actors. We distinguish between two modes of using an expert-finding component:

1. *Filter functionality:* In this mode, a user applies the expert-finder system in order to find other users with personal data that are similar to their own (or relevant parts of their own data, respectively). This functionality is not only relevant in a learning environment where a learner wants to find other learners with similar backgrounds, interests, and knowledge in order to build a learning group. It also applies to organizational knowledge management scenarios. Consider an enterprise environment where an expert in a certain field wants to set up an expert network of people with a similar project background in order to share experiences. Alternatively, the user can pose a query (in terms of a user-defined profile) to the expert-finder system in order to find people that match *explicitly defined* needs of the user. As an example, consider an enterprise environment where an employee needs to find an expert in a certain field who is able to solve a specific problem.

2. *Cluster functionality:* Here, the expert-finder system is used to cluster the profiles of all users in order to present a "landscape of expertise" for analysis or exploration purposes. Consider an enterprise environment where a project manager tries to identify the expertise of those members of the staff who could potentially take over a certain subtask.

The term *personal data* is deliberately kept very general. As discussed earlier, data subsets may include an actor's professional and training status, interest statements, self-assessment of abilities, certain kinds of history information (learning history in a learning environment, project history in a company's expert database, etc.), and similar data that describes the user's expertise depending on the application context.

We now describe the algorithmic matching framework of our expert-finder system in a more formal way. We point out the elements for making matches of profiles and define their constraints. Let P denote the set of all possible personal data (i.e., user or query profiles). These data consist of the full set of information available for each user or the query, respectively, in a certain application environment (i.e., a certain learning platform, an enterprise expert database, etc.).

As discussed above, there are different kinds of data that help to determine the degree of expertise. According to the nature of the data subsets, different algorithms for matching subsets (e.g., self-assessment, history) have to be applied. In the following, this is expressed by the notion of modules: Each module contains functions for matching the respective relevant subsets of personal data (which contain the data items used for matching). Intuitively, each module realizes a criterion for expert matching (e.g., one module for matching the histories of interaction, another for matching the training status of users).

Formally, each module M_i consists of a matching function $m_i: P \times P \to [0,1]$, which determines the degree of similarity for each pair (p_a, p_b) of personal data collections (e.g., the similarity of learning histories),[1] and a completeness function $c_i: P \to [0,1]$, which measures the degree to which relevant data for M_i is available in a profile p (e.g., the amount of interaction history in the profile).

The matching function is realized for each module. The realization of this function has to meet the following requirements: The more similar two profiles are, the higher the value calculated by m_i (with 1 representing a "perfect" match). Furthermore, matching identical profiles should produce a perfect match (which is quite intuitive)—that is, $m_i(p_a, p_a) = 1$ for all $p_a \in P$. In cases where the clustering functionality is used, we also assume m_i to be symmetrical—that is, $m_i(p_a, p_b) = m_i(p_b, p_a)$ for all $p_a, p_b \in P$.

The completeness function takes into account the quality of data available for module M_i's matching process, where $c_i(p) = 0$ means that no data is available and $c_i(p) = 1$ corresponds to the maximal degree of completeness. For example, assume a module M_i that matches profiles based on certain history data of a user. If the user is new to the platform in which the expert finder is embedded, only few (or even no) history data may be available in p. In this case, the matching function m_i might yield a high matching value based on the considered pair of profiles, but this matching would be based on sparse data. In such a case, $c_i(p)$ should yield a low value.[2]

In order to allow the user to adapt the matching process, the user can adjust the influence that different matching criteria have on the overall matching result. Formally, this is done by assigning weights to the single modules: For each module M_i, a weight $w_i \in [0,1]$ is given, where $w_i = 0$ means that module M_i is switched off and $w_i = 1$ means that module M_i has full influence. All values of w_i between 0 and 1 correspond to a more or less strong influence of module M_i.

We also need to take care of privacy issues: If users do not want the system to use certain parts of their profile for matchmaking, they should be able to switch off the corresponding modules. Formally, we introduce flags $priv_{ai}$, $priv_{bi} \in \{0,1\}$ for all users a and b and each module M_i that indicate whether the respective user wants module M_i to be used or not.

Now we can define the overall matching result for profiles p_a and p_b from users a and b:

$$m(p_a, p_b) =_{def} \frac{1}{k} \sum_{i=1}^{k} priv_{ai} \cdot priv_{bi} \cdot w_i \cdot \min(c_i(p_a), c_i(p_b)) \cdot m_i(p_a, p_b)$$

Summing up, the overall matching value of two profiles p_a and p_b is based on the matching degree of each individual module. A module can only contribute to the overall result if both user a and user b agree that their profile may be used for matchmaking by setting the respective flag to 1. Furthermore, the contribution of each module depends on the completeness (and thus as we claim, trustworthiness) of the respective profile data and the user-defined weighting for that module.

Realization of the Expert Finder

Based on the methodical approach presented in this chapter, we built an expert-finder system for the Fraunhofer e-Qualification framework—an e-learning environment that offers extensive technical, methodical, and didactic support for both authors of Web-based trainings and learners. In this section we sketch the generic architecture of our system and present a user interface for the e-Qualification platform that mainly focuses on the filter functionality of our framework. Moreover, we sketch another application of the framework where the actors' personal data are input for a cluster functionality and an advanced visual interface for exploring a landscape of expertise is applied.

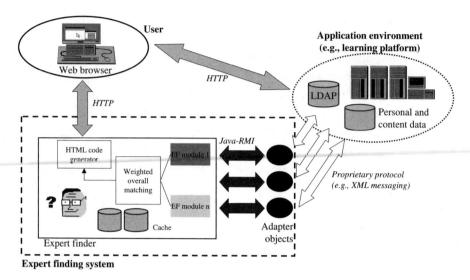

Figure 13.1
Architecture of the expert-finding system.

Architecture

The architecture (cf. figure 13.1) we have chosen for our expert-finding system has been designed in order to keep the system flexible and easily adaptable to different application platforms. The current architecture requires that the application platform provide a Web-based interface. The expert-finding system consists of three major parts: The *expert finder* itself (which realizes the algorithmic matching framework described above), the *connection to the application environment* (e.g., a learning platform), and the *connection to the client* (i.e., Web browser). The expert-finding system contains internal databases where information about the user and the content of the application environment is "cached." This is done to have quick access to those data that are frequently used by the expert finder, and to store additional information about users and content that is not kept in the main databases of the application environment. The connection to the learning environment is done by "adapter objects" that translate personal data and content information from the application environment's proprietary format into the data structures used by the expert finder. The connection to the client is

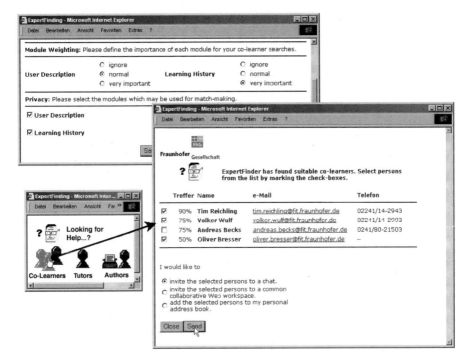

Figure 13.2
The expert-finder interface of the e-Qualification platform: the main window of the expert finder (bottom left), the result list of a colearner search (bottom right), and the options window for configuring modules for a colearner search (top left).

realized by a Java™ servlet that forwards requests to the expert finder and generates HTML codes from the computed matching results.

User Interface

Figure 13.2 depicts the user interface of the expert-finder within the Fraunhofer e-Qualification platform: Users of the platform who are logged in as learners can use the expert-finder system in order to contact authors, tutors, and suitable colearners. For learners, the main window of the expert finder is available during the complete training session (either as a separate window or integrated in the training Web pages as a frame layout). Learners can seek advice from the author of their whole session or specifically assigned tutors by clicking on the respective symbol in the main window. In both cases, the result of the user request is the name, e-mail address, and telephone number of the author(s) or tutor(s), respectively.

When the learner looks for potential colearners, they can click on the corresponding icon in the main window (or expert-finder frame). As a result of this request, the learner directly receives a list of suitable colearners, ranked by the matching degree of that learner's own profile and the profile of each candidate colearner (thus realizing the filter functionality of our framework). The user can then select potential colearners from the list, and either invite them to a chat (as a means of synchronous communication), ask them to join a common collaborative work space within the learning platform (as a means of asynchronous communication), or just add them to their personal address book for later reference.

In a specific options window, the user can define a relative weighting of the modules used to assess the suitability of other users of the platform as colearners, thus determining the influence each module has on the matching result. In this case, the modules "learning history" and "user description" are available. The latter contains, among other items, the educational status and interest statements of each learner subscribed to the platform. The options window also allows a learner to adjust their individual privacy flags. The privacy flags are used globally—that is, as soon as a learner deactivates a flag, the corresponding profile data will not be used by the expert finder at all. The module weights, on the other hand, are used locally: If a learner selects "ignore" for one of the modules, this only affects their colearner searches. For the searches of other learners, the respective personal data are still available.

Expert Maps: Clustering Profiles for Exploration
In this section, we sketch another application of our expert-finder framework where the calculated similarity information of personal profiles is used to automatically calculate a graphic landscape of expertise: In larger, traditional enterprises or virtual organizations, similar or complementary expertise is usually spread over different departments or sites, making it difficult to get an overview of the areas of competence available in the organization. We conducted informal interviews in several

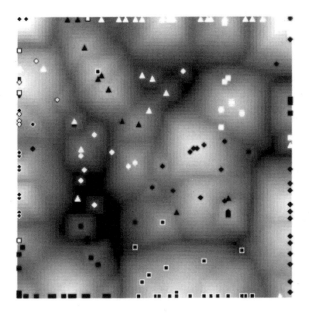

Figure 13.3
An expert map that displays the similarity structure of all pairs of personal profiles can be used for navigating through and exploring an expert landscape.

medium-size companies that were interested in advanced knowledge management technology. The study revealed that a graphic presentation of an expertise landscape would be of high value. We therefore propose to combine the clustering functionality of our framework with a technique for visualizing the structure of complex information spaces, originally developed for data and document mining (Becks, Sklorz, and Jarke 2000). Given a degree of similarity for each pair of actors (e.g., the similarity of each pair of personal profiles), the visualization approach automatically generates an overview map (cf. figure 13.3, which intuitively displays the relationships of actors, represented as points on the map; similar collections of personal data are grouped as neighbored points—the more similar, the closer they appear on the display. The shades of gray indicate additional similarity information: Groups of similar actors can be found in the brightly shaded areas. Dark borders separate these groups. The darker the border, the more dissimilar are the respective profile groups.

The map in figure 13.3 shows how a landscape of expertise based on the personal data of employees can look like in a medium-size company. The color and shape of the profile points on the map represent the department or site that the respective employee (who is represented by the personal profile) works in. Applications of such expert maps (as discussed with our industry partners) would include the identification of competence areas and their distribution within the company or the identification of only sparsely covered areas of expertise. In other applications, project managers would use the map for identifying suitable team members by browsing through the competence landscape, and innovation managers would apply the map for identifying groups of people with related or complementary personal profiles in order to set up networks of experts for experience exchange and creative brainstorming sessions. Initial discussions of the concept with our industry partners yielded positive feedback.

Conclusion

The concept of social capital has recently gained importance. Social capital seems to have an impact on different aspects of the performance of social entities. The application of IT is likely to influence the level of social capital. In this chapter, we have looked at computer applications to encourage human learning. We assume that a high level of social capital among the learners will have a positive effect on their performance. Given the fact that actors within e-learning platforms will not know each other in general, we have developed a framework for matchmaking. Contrary to face-to-face learning environments, e-learning environments do not allow users to meet in person. So many of the real-world clues that enable the establishment of initial forms of social relationships among learners are missing (e.g., behavior, gesture, facial expressions, or eye contact). The question comes up of whether there can be virtual substitutes that may ease the establishment of social relationships. Here, we have explored ways to match the learners' personal data. We have developed a framework that allows one to match different subsets of personal data in a weighted manner. Depending on the subset of personal data, the matching algorithms may favor similarity or complementarity.

Many research issues discussed here need further investigation. First of all, an empirical evaluation of our approach is required. In particular, a key question concerns the value added by automatic matchmaking. This question has to be addressed in a number of smaller steps. First, quite clearly, the quality of the matching needs to be assessed by the target users. For this, we will apply the matchmaking strategy in an in-house study based on profiles of our colleagues. A following series of interviews will be conducted in order to find out whether the suggested partners are regarded to be an appropriate choice. Second, given a sufficient quality, an essential question is whether users are willing to establish social contacts based on the recommendations of computer algorithms. This can only be assessed in current learning environments. We will therefore integrate our expert-finding system into several learning environments and collect data. Possible information can be extracted from log files, which show the communication activities (e.g., chat, e-mail) started in response to the system's recommendations. More important, empirical investigations with platform users need to be conducted in order to learn from the users' reactions. Refinements of the different matching algorithms are also required. We have to find out which are the right subsets of personal data and which algorithms are best suited to match a certain subset of data. Empirical research like the study presented by Cross and Borgatti (in this volume) can inform the design of these algorithms. As the application context will have a strong influence on the choice of the appropriate algorithms, the expert-finding components should be highly tailorable. Appropriate concepts to tailor these applications have to be developed.

Further research is required to find out how to support the building of relationships after having matched the actors. Questions arise such as: Which private data should become visible to the matched partners? What are the appropriate media for communication? Which role may face-to-face meetings play in such an environment? Another research issue refers to the preconditions necessary to successfully match actors. Probably a somehow common language and culture are crucial prerequisites for establishing contact among the actors. These factors cannot be grasped easily by means of personal data.

We have approached the problem of how to support the establishment of social relationships in the virtual world, where many traditional clues for self-selection among the actors are missing. Our approach to expertise matching in an e-learning environment needs empirical evaluation and further refinements. As the study indicates, however, there is the potential to support the building of social capital by technical media. A thorough investigation of the possible risks as well as opportunities is required.

Notes

1. Note that dissimilarity ("distance") measures can be used as well since their results can be converted to similarity values.

2. Of course, certain matching functions may take sparse data into account and yield corresponding matching values. In such a case, consider $ci(p) = 1$.

References

Ackerman, M. S., V. Pipek, and V. Wulf, eds. 2003. *Expertise sharing: Beyond knowledge management.* Cambridge: MIT Press.

Becks, A., S. Sklorz, and M. Jarke. 2000. Exploring the semantic structure of technical document collections: A cooperative systems approach. Paper presented at the fifth IFCIS international conference on cooperative information systems, 6–8 September, (CoopIS'2000), Eilat, Israel.

Bourdieu, P. 1985. The forms of capital. In *Handbook for theory and research for the sociology of education*, edited by J. G. Richardson. New York: Greenwood Press.

Buckingham Shum, S. 1997. Negotiating the construction and reconstruction of organisational memories. *Journal of Universal Computer Science* 3, no. 8:899–928.

Burt, R. S. 1992. *Structural holes: The social structure of competition.* Cambridge: Harvard University Press.

Cohen, D., and L. Prusak. 2001. *In good company: How social capital makes organizations work.* Boston: Harvard Business School Press.

Ehrlich, K. 2003. Locating expertise: Design issues for an expertise locator system. In *Expertise sharing: Beyond knowledge management*, edited by M. S. Ackerman, V. Pipek, and V. Wulf. Cambridge: MIT Press.

Fitzpatrick, G., S. Kaplan, T. Mansfield, D. Arnold, and B. Segall. 2002. Supporting public availability and accessibility with Elvin experiences and reflexions in computer-supported cooperative work. *Journal of Collaborative Computing* 11, nos. 3–4:447–474.

Foner, L. N. 1997. Yenta: A multi-agent, referral-based matchmaking system. In *First international conference on autonomous agents*. New York: ACM Press.

Fuchs, L. 1997. Situationsorientierte Unterstützung von Gruppenwahrnehmung in CSCW-Systemen. Ph.D. diss., University of Essen.

Fuchs, L. 1999. AREA: A cross application notification service for groupware. In *Proceedings of the sixth European conference on computer-supported cooperative work*. Dordrecht, Netherlands: Kluwer.

Gräther, W., and W. Prinz. 2001. The social web cockpit: Support for virtual communities. In *Proceedings of GROUP 2001, ACM 2001 International Conference on Supporting Group Work*, September 30–October 3, 2001. Boulder: ACM Press.

Hafner, K. 2001. *The well*. New York: Carroll and Graf.

Heath, C., and P. Luff. 1991. Collaborative activity and technological design: Task coordination in London underground control rooms. In *Proceedings of the third European conference on computer-supported cooperative work*. Dordrecht, Netherlands: Kluwer.

Kautz, H. A., B. Selman, and M. Shak. 1997a. The hidden web. *AI Magazine* (summer): 27–36.

Kautz, H. A., B. Selman, and M. Shak. 1997b. Referral web: Combining social networks and collaborative filtering. *Communications of the ACM* 40, no. 3:63–65.

Lave, J., and E. Wenger. 1991. *Situated learning: Legitimate peripheral participation*. Cambridge: Cambridge University Press.

Maybury, M., R. D'Amore, and D. House. 2003. Automated discovery and mapping of expertise. In *Expertise sharing: Beyond knowledge management*, edited by M. S. Ackerman, V. Pipek, and V. Wulf. Cambridge: MIT Press.

McDonald, D. W. 2000. Supporting nuance in groupware design: Moving from naturalistic expertise location to expertise recommendation. Ph.D. diss., University of California of Irvine.

McDonald, D. W., and M. S. Ackerman. 2000. Expertise recommender: A flexible recommendation system and architecture. In *Proceedings of the international conference on CSCW*. New York: ACM Press.

Nahapiet, J., and S. Ghoshal. 1998. Social capital, intellectual capital, and the organizational advantage. *Academy of Management Review* 23, no. 2:242–266.

Pipek, V., J. Hinrichs, and V. Wulf. 2003. Sharing expertise: Challenges for technical support. In *Expertise sharing: Beyond knowledge management*, edited by M. S. Ackerman, V. Pipek, and V. Wulf. Cambridge: MIT Press.

Pipek, V., and M. Won. 2002. Communication-oriented Computer Support for Knowledge Management. *Informatik/Informatique—Magazine of the Swiss Informatics Societies* 1:39–43.

Polanyi, M. 1958. *Personal knowledge: Towards a post-critical philosophy.* London: Routledge and Kegan.

Putnam, R. 1993. The prosperous community: Social capital and public life. *American Prospect* 13:35–42.

Putnam, R. 2000. *Bowling alone: The collapse and revival of American community.* New York: Simon and Schuster.

Rodden, T. 1996. Populating the application: A model of awareness for cooperative applications. In *Proceedings of the international conference on computer-supported cooperative work.* CSCW' 96, Boston. New York: ACM Press.

Rheingold, H. 2000. *The virtual community.* Cambridge: MIT Press.

Sandor, O., C. Bogdan, and J. Bowers. 1997. AETHER, An awareness engine for CSCW. In *Proceedings of the fifth European conference on computer-supported cooperative work.* Dordrecht, Netherlands: Kluwer.

Schumpeter, J. A. 1934. *The theory of economic development: An inquiry into profits, capital, credit, interest, and the business cycle.* Cambridge: Harvard University Press.

Stahl, G. 2003. Building collaborative knowing: Elements of a social theory of learning. In *What we know about CSCL in higher education,* edited by J.-W. Strijbos, P. Kirschner, and R. Martens. Dordrecht, Netherlands: Kluwer.

Streeter, L. A., and K. A. Lochman. 1988. An expert/expert location system based on an automatic representation of semantic structure. In *Proceedings of the fourth conference on artificial intelligence applications.* New York: ACM Press.

Vivacque, A., and H. Lieberman. 2000. Agents to assist in finding help. In *Proceedings of the conference on computer-human interaction.* New York: ACM Press.

Wenger, E. 1998. *Communities of practice.* Cambridge: Cambridge University Press.

Wessner, M., and H.-R. Pfister. 2001. Group formation in computer-supported collaborative learning. In *Proceedings of the 2001 international ACM SIGGROUP conference on supporting group work,* Boulder, Colo., New York: ACM Press.

Yiman-Seid, D., and A. Kobsa. 2003. Expert finding systems for organizations: Problem and domain analysis and the DEMOIR approach. In *Expertise sharing: Beyond knowledge management,* edited by M. S. Ackerman, V. Pipek, and V. Wulf, Cambridge: MIT Press.

14

Fostering Social Creativity by Increasing Social Capital

Gerhard Fischer, Eric Scharff, and Yunwen Ye

Complex design problems require more knowledge than any single person can possess, and the knowledge relevant to a problem is often distributed among all stakeholders who have different perspectives and background knowledge, thus providing the foundation for *social creativity* (Fischer 2000). Bringing together different points of view and trying to create a shared understanding among all stakeholders can lead to new insights, ideas, and artifacts. Social creativity can be supported by innovative computer systems that allow all stakeholders to contribute to framing and solving these problems collaboratively. Such systems need to be designed from a metadesign perspective by creating environments in which stakeholders can act as active contributors and be more than just consumers (Fischer 1998).

To foster social creativity, people need to be motivated to be active contributors—for example, to add content to organizational memories, share tools that they have developed, bring their unique knowledge into a discussion, or formulate their ideas in a way that other stakeholders can understand. We are particularly interested in those sociotechnical environments in which these activities are not performed as a work assignment (i.e., they are not required, nor are they explicitly rewarded) but in which people are motivated to accumulate social capital. The incentive is to be a good colleague, contribute and receive information as a member of the community, acquire a reputation of being a good citizen, and jointly construct artifacts that could not be developed individually.

Social Creativity and Social Capital: A Conceptual Framework

The Western belief in individualism romanticizes the perception of a solitary process in creative activities, but the reality is that scientific and artistic forms emerge from the joint thinking, passionate conversations, and shared struggles common in meaningful relationships (John-Steiner 2000). The power of the unaided individual mind is highly overrated (Salomon 1993). Although creative individuals are often thought of as working in isolation, intelligent and creative results come mainly from interaction and collaboration with other individuals. Creative activity grows out of the relationship between an individual and the world of his or her work, and the ties between an individual and other human beings. Much human creativity arises from activities that take place in a social context in which interaction with other people and the artifacts that embody group knowledge is an important contributor to the process. Creativity (Florida 2002) does not happen solely inside a person's head but in the interaction between a person's thoughts and a sociocultural context (Engeström 2001). Situations that support social creativity need to be sufficiently open-ended and complex that users will encounter breakdowns (Schön 1983). As any professional designer knows, breakdowns—although at times costly and painful—offer unique opportunities for reflection and learning.

The concept of social capital provides an anchor to analyze the nontechnical and nonmanagerial aspects of social creativity. Social capital creates the resources on which a person can draw to obtain knowledge, cooperation, and help from others.

Social Creativity

To make social creativity (Fischer 2000) a reality, we have explored new forms of knowledge creation, externalization, integration, and dissemination based on the observation that the scarce resource in the information age is not information; rather, it is the human resource to attend to this information. One aspect of supporting social creativity is the externalization of an individual's and a group's tacit knowledge (Polanyi 1966). Individual tacit knowledge means intuition, judgment, and common sense: the capacity to do something without necessarily being able to explain it. Group tacit knowledge means knowledge existing in the distinct practices and relationships that emerge from working together over time. Externalizations (Bruner 1996) support social creativity in the following ways:

- they cause us to move from vague mental conceptualizations of an idea to a more concrete representation of it;
- they provide a means for others to interact with, react to, negotiate around, and build on an idea;
- they allow more voices from other stakeholders to be brought in; and
- they create a common language of understanding (including boundary objects that are understandable across different domains) (Arias and Fischer 2000).

Creating these externalizations requires active contributors and not just consumers. Externalizations of individual knowledge make it possible to accumulate the knowledge held by a group or community. An important challenge for social creativity is to capture a significant portion of the knowledge generated by work done within a community. Experiences with organizational memories and collaborative work have exposed two barriers to capturing information. First, individuals must perceive a value in contributing to an organizational memory that is large enough to outweigh the effort (Grudin 1994). Second, the effort required to contribute to organizational memory must be minimal so that it will not interfere with performing the work at hand (Carroll and Rosson 1987). Since human beings often try to maximize utility in the decision-making process (Reisberg 1997), increasing the value and decreasing the effort of knowledge externalization and sharing are essential. No objective evaluation of value and effort exists; evaluation is subject to each person's perception and background.

Many factors play a role in the evaluation of the perceived value; among these are the three major dimensions of economic, intellectual, and social values. Economic capital includes the monetary rewards that a person can receive; intellectual capital refers to the intellectual satisfaction that a person derives from making a contribution, such as the boost of confidence, self-assurance of individual capability, and acquisition of knowledge and skills; and social capital refers to the influence brought about by individuals' behaviors on the relations with their sociocultural environment.

Social creativity entails taking a new perspective on how we design the supporting technological, social, and organizational environments. Without this perspective, technology to support working and learning is often designed in ways that fail to support social creativity. Organizational aspects (e.g., course structures and curricula in educational settings) are often concerned more with transmitting facts and basic skills and do not adapt well to open-ended problem solving and collaborative learning.

An important prerequisite to bring social creativity alive is that media and environments be available to support metadesign (Fischer and Scharff 2000). The perspective of metadesign characterizes objectives, techniques, and processes that allow users to act as designers and be creative (Henderson and Kyng 1991). The need for metadesign is founded on the observation that design in the real world requires open systems that users can tailor, customize, modify, and evolve (Fischer and Girgensohn 1990; Mackay 1990; MacLean et al. 1990; Oppermann 1994; Lieberman 2001). Because problems cannot be completely anticipated at design time (when the system is developed), users at use time will discover mismatches between their problems and the support that a system provides.

Social Capital
Social creativity is essential to support collaborative design through the active participation of all stakeholders. Although previous research in knowledge management, organizational memory, and groupware systems (Preece 2002) has recognized the important role that social context plays (such as motivation and trust), a unified conceptual framework does not exist to analyze such a role. *Social capital* characterizes the interpersonal relationships that an individual has with other members in a surrounding community, and it provides the basis for analyzing the sense of community and the degree to which the individual is connected with others in the community.

The concept of social capital first appeared in the writing of Lyda Judson Hanifan, who used it to describe those tangible substances, such as goodwill, fellowship, sympathy, and social interactions, that "count for most in the daily lives of a people" living in a rural community (1916, 130). Recently, this concept has been widely applied to a variety of disciplines, including political science (Putnam 1995), sociology (Coleman 1988), education (Bourdieu 1983), business management (Meyerson 1994), organizational theory (Cohen and Prusak 2001), and collaborative software construction (Raymond and Young 2001). Many definitions of social capital exist, including: "Features of social organization, such as networks, norms, and trust, that facilitate coordination and cooperation for mutual benefit" (Putnam 1995, 67); and "An instantiated informal norm that promotes cooperation between two or more individuals" (Fukuyama 2001, 7).

A useful framework for understanding the role of social capital in fostering social creativity as well as approaches to increasing social capital is proposed by Janine Nahapiet and Sumantra Ghoshal, who define social capital as "the sum of the actual and potential resources embedded within, available through, and derived from the network of relationships possessed by an individual or social unit" (1998, 243). They identify the three operational dimensions in social capital:

1. structural: a relationship network that connects people and helps individuals to find people for assistance or cooperation;

2. relational: the sense of trust that individuals have toward each other along with connections; and

3. cognitive: the bonding force, such as shared understanding, interest, or problems, that holds the group together.

This definition indicates that social capital exists at both the individual and group level (social unit). Most studies on social capital have focused on the group level—namely, the active connections among the group members such as trust, mutual understanding, and shared values and behaviors. These connections generate better knowledge sharing and transfer within a group due to established trust, shared language and goals, and informal ties. Lower transaction costs due to trust and cooperation also make less necessary any costly, formal coordination mechanisms such as contracts and the like. The social capital of a group depends on the social capital held by its individual members, although without a group that provides the social context, an individual could not have any social capital. Within a given social context, an individual's social capital is the actual and potential resources that the individual could draw on for acquiring cooperation from other members.

In systems that require user participation and collaboration, social capital, by its definition, is a significant factor to consider. We use the term *social capital–sensitive system* to refer to systems whose success is influenced by social capital and that are designed to take issues of social capital, explicitly or implicitly, into consideration.

The Seeding, Evolutionary Growth, and Reseeding Process Model

The development and deployment of computer systems that support social creativity require the continual participation and contribution of developers and users not only at the design time (when the original system is developed) but also at the use time (when people use and evolve the systems in their activities). System development techniques have typically emphasized a strong distinction between the activities of designing a software system and the use of that system. Even efforts such as participatory design (Schuler and Namioka 1993), which has improved user involvement in software design, have stressed system development and use separations. In order to accommodate open problems and continuously changing requirements, it is necessary to avoid the separation of design and use, and to create models in which continuous change can take place.

Seeding, Evolutionary Growth, and Reseeding (SER) is a process model that aids in the understanding and development of systems that evolve over time (Fischer et al. 2001). This model has been developed based on experience with analyzing and building systems supporting collaborative design. The SER model helps in understanding social creativity by identifying critical phases in the evolution of systems supporting collaborative construction and evolution. Each of the phases of the SER model raises questions that systems supporting social creativity must face.

Social creativity in the SER model is a continuous process that encourages users to act as designers. For a system that relies on the participation of a community to succeed, the social and technical aspects of the system must both be designed to leverage and increase the social capital of a community. As mentioned above, we call such systems social capital–sensitive systems because their successful operation is strongly affected by the social capital in the environment in which they are deployed.

The SER model postulates three phases of development. The *seeding* process involves the creation of an initial system that can grow over time. During *evolutionary growth*, users make changes to and extend the initial seed through their own work. When the incremental changes in evolutionary growth are no longer practical, a *reseeding* phase takes place in which the incremental changes are organized, reformulated, generalized, and incorporated into the initial seed. Cycles of evolution and reseeding continue as long as the community is actively developing and using the software.

The first goal of providing a seed is to offer immediate benefits to users to attract them to the system. The second equally important goal of seeding is to engage in metadesign by creating environments that allow users to become active designers through extending, refining, and augmenting the existing system. An initial system open to user participation is the prerequisite to support social creativity in the context of collaborative knowledge construction and design.

For the seeding phase, the following questions relevant to a social capital perspective are crucial:

- How can a seed be perceived as providing increased value to members of a community so that people will feel motivated to participate?
- What aspects of a seed can reduce the effort needed to use and modify that seed?
- Who must participate in the creation of the seed?
- How does the seed balance the goals of the initial developers and the needs of the community at large?

During the evolutionary growth phase, the users must incrementally change a system in the context of their work, generating the following questions and challenges for social capital–sensitive systems:

- What kinds of extension mechanisms are necessary?
- What motivates people to contribute?

- Is there a collaborative process that encourages people to make and share contributions?
- Do participants receive individual benefits and social rewards that increase the utility and justify the effort of contribution?

Managing incremental changes is difficult, and the reorganization of the reseeding phase is an attempt to synthesize these incremental changes and create a new stable system on which new changes can be created. Reseeding requires major modifications and it raises the following issues:

- When is reseeding necessary?
- If reseeding requires a great deal of effort, who perceives the value in doing it?
- How can reseeding efforts integrate the modifications of members of the community and continue to acknowledge the importance of individual contributors?

The SER model characterizes the typical life cycle of computer systems that support social creativity in the context of collaborative design. The success of such systems requires social capital that motivates users to actively participate, and facilitates collaboration and coordination.

Analysis of Existing Success Models

In order to develop a conceptual framework for social capital–sensitive systems, we have analyzed open source developments (a methodology to engage the "talent pool of the whole world"), and the Experts Exchange (an environment in which IT experts compete and collaborate with each other to provide specific solutions to specific problems, and in which the experts receive recognition by the questioners). These analyses were based on numerous sources, including firsthand participation in open source and Experts Exchange efforts, long-term evaluation of the communication and evolving artifact in open source communities, and comparisons with other ongoing open source studies (Scharff 2002b).

Open Source

Open source development (Raymond and Young 2001) is an activity in which a community of software developers collaboratively constructs systems to help solve problems of shared interest and for mutual benefit. Contemporary perspectives on open source tend to emphasize the technical aspects of projects, the projects themselves (DiBona, Ockman, and Stone 1999; Mockus, Fielding, and Herbsleb 2000), the licenses under which the software is shared (Perens 1999), or the use of software in a particular domain (Davis et al. 2000; Claypool, Finkel, and Wills 2001). Although these are important aspects of open source software, they do not acknowledge the collaborative processes involved in producing the software. Understanding open source as a process of collaborative design and construction highlights the sociotechnical aspects of open source software as social capital–sensitive systems in which the participation of a community is vital to its success, and the final product emerges from the contributions of the entire community. Development in the open source community has been characterized by principles related to social capital, such as "(1) in gift cultures, social status is determined not by what you control but by what you give away, (2) prestige is a good way to attract attention and cooperation from others, and (3) utilization is the sincerest form of flattery" (Raymond and Young 2001).

Open source provides evidence of social capital both as a noun (the resources that a community has that facilitate group interaction and sustainability) and a verb (the processes that sustain ongoing community activity). Seen from a collaborative design perspective, the ingredients that make up an open source community include a group of participants willing to contribute to the community, a concrete software artifact to which explicit extensions are made, and a set of collaborative technologies that help the community coordinate the objects that they are producing and communicate.

The collaborative processes used to coordinate activity in the project form the basis of joint activity. A defining feature of open source software is that the source code for that software can be obtained and modified by anyone who wishes to do so. The ability to change the source code is an enabling condition for the collaborative construction of software because it allows software developers to make changes to the behavior of the software. This simple condition changes software from a fixed entity that is produced and controlled by a closed group of designers to an open effort that allows a community to collaboratively design based on personal desires following the framework provided by the SER process model. Open source invites passive consumers to become active contributors. Of course, modifying software requires the time, inclination, and technical aptitude to do so—conditions that do not exist in all communities.

Studies and analyses of open source software (Scharff 2002b) underscore the significance of the three dimensions of social capital noted earlier.

The Structural Dimension

- *Coordinating distributed work*: Collaborative technology helps to provide the structure through which collaboration in the community can take place. Open source projects rely heavily on collaborative technology to coordinate the software itself (using technical tools such as source code management systems [Fogel 2000] as well as to communicate about the project (using tools such as chat, e-mail, and threaded discussions). Asynchronous tools are favored due to the involvement of large, geographically distributed communities, and face-to-face meetings are rare. Therefore, maintaining social cohesion requires the active use of collaborative technology. Collaborative technology helps to create a sense of community and supports work. Different communities use technology differently, but the need for focused technical discussion, general discussion, and off-topic social discussion is commonplace.

- *Role distribution*: The social capital in an open source community is not based on a set of homogeneous individuals who could merely be replaced by other competent participants. Rather, the capital is based on the complementary skills and interests of participants. In a healthy open source project, the skill sets of different participants will complement each other (Campbell 1969). Different participants take on different roles that facilitate the rapid change and creation of stable releases, including testing, contributing new changes, coordinating releases, and maintaining documentation. A key aspect of the collaborative approach is to help individuals find tasks in which their talents would be best applied.

The Relational Dimension

- *Supportive process*: The process of adopting changes proposed by the community creates interesting conflicts and trade-offs. On the one hand, people must be given the feeling that their contributions are worthwhile and that they will benefit the group; this facilitates ongoing participation. On the other hand, not all changes are universally useful, and conflicting extensions must somehow be mitigated. The most visible part of this process manifests itself as a single project leader who has the ultimate (dictatorial) control over what should be included in the system. Yet the change adoption process (as an essential component of reseeding) is far more complex, and relies heavily on discussion and comparison of changes via on-line discussions. Even if the changes are eventually not adopted, community members tend to be actively supportive of all contributions, trying to make contributors realize that their input is worthwhile. When this process breaks down, open source projects tend to split into separate projects and bifurcate the community, although this situation is extremely rare (Raymond 1998).
- *Peer pressure*: Public code deliverables encourage developers to write good code. Developers take pride in their contributions, and low-quality contributions are detrimental because they lower the contributors' reputation.
- *Extrinsic motivation*: Without active participation, open source projects would simply be static pieces of software rather than ongoing community activities. Similar to community participation in other areas, participation in open source communities is a mix of intrinsic and extrinsic motivation. Extrinsic motivation exists in forms of reputation, respect, and trust among peer members, and provides a positive feedback loop by which a group of core participants who are well respected and trusted for their superior skills excite and motivate other developers to join a project.

The Cognitive Dimension

- *Public deliverables*: Public code deliverables provide a concrete object to which all members of the community contribute. This collaboratively constructed artifact includes not only the source code (the core of the software system) but also documentation, tests, examples, community discussion, and so on. It is the construction of a software system that supplies the bonding force. The group has a shared goal to make a system, the features and aspects of which come from the contributions to that system. Discussion does not exist in the abstract but in the context of creating a shared artifact.

• *Intrinsic motivation*: Open source projects are developed by a group of people who work together because of their shared interest in the projects. The shared interest comes not only from the common need for the specific functionality offered by the open source software but also the need for sharing knowledge about the software, and with it the excitement of engaging in intellectually stimulating and personally enriching activities. Participating in an open source project is a way to learn about a new technology that may be useful to the further professional development of technically inclined participants. It can also be an investment in the future, perhaps by gaining socioeconomic status through a better job or by enhancing professional relationships through meeting people.

Experts Exchange
The Experts Exchange (Experts Exchange 2002; see figure 14.1) is a Web-based, knowledge-sharing environment that exemplifies many of the crucial aspects of social creativity and social capital. Users of the Experts Exchange are IT professionals who share knowledge through a question-answering process. Based on firsthand participation in and observation of the system, it appears that there are thousands of experts who regularly participate in answering questions posed by the community. (The Experts Exchange boasts that it has nearly 100,000 experts, and although it is impossible to determine how much each expert participates, the environment clearly has a large and vibrant community.) Unlike some systems that separate novices and experts, anyone can pose a question or become an expert. The unique characteristic of the Experts Exchange is a point system to motivate active participation. When users join, they are given a number of points. They can offer points to others in exchange for answering questions, and can gain points by answering questions posed by other users. Questions are organized into categories related to topics important in IT, such as Linux and Java programming.

Fostering Social Creativity by Increasing Social Capital 367

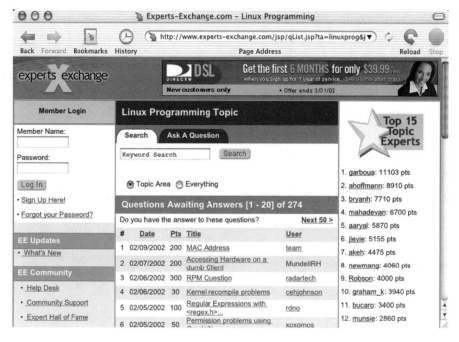

Figure 14.1
A screen image of the Expert Exchange.

The categories are designed by the Experts Exchange staff (the application domain of the Experts Exchange is IT-related subjects), but the users may post any questions within these subject areas. Persons asking questions have the ability to judge the quality of the answers posed to their questions. Given this system of demands, recognition, and rewards, the Experts Exchange has developed an active user community of questioners and answerers.

Examining the Experts Exchange as a social capital–sensitive system using the SER model illustrates the importance of social creativity and social capital at both the individual and group level. The seed for the Experts Exchange is a Web site that allows people to post questions, respond to already posted questions, and evaluate these responses. Incentives include giving new users initial points (so that they can ask questions). The effort in posting new questions is low, and the potential value is high if there are experts who can and will answer the question. The system evolves through the addition of new questions and answers to posed questions. When someone submits a candidate answer to a question, the user who posted the question can evaluate the quality of the answer and determine if the points should be awarded for that answer. The Experts Exchange site itself has evolved, including: a redesign of the user interface, and the addition of new features such as finding jobs for users who demonstrate a certain expertise; a reseeding of the content by placing previously answered questions and answers into the "IT Knowledge Base" (providing a collection of questions and answers available as a service for commercial subscription); and support for linking questioners with potential experts, providing subject-based categories for questions, and notifying people when new questions are posted. The reorganization and repackaging of information captured in questions and answers has been a primary business model for the Experts Exchange. The Experts Exchange staff filters and organizes the answered questions, and users can pay for access to high-quality resources.

The organization and features of the Experts Exchange support the structural dimension of social capital. The system offers a forum that connects IT professionals with each other. Domain-specific categories of questions help people find experts who are capable of answering their questions. These categories also reinforce the cognitive dimension so that there is mutual interest in the questions being discussed. Since not every IT professional is an expert in every area, users self-select the topics of interest to themselves. Perhaps most important to the Experts Exchange is the mix of external rewards and internal motivation that is exemplified by the point system. The point system has served as a metric for the activity and aptitude of the members in this community. Over time, some users have amassed large numbers of points due to the quality of their answers, highlighting significant individual and social aspects of creating social capital, including the following:

- *Explicit recognition and reputation enhancement*: Points provide a mechanism not only to reward individuals for participating but to provide explicit public recognition of an individual's contribution to the community. The top users with the most points in a given forum are presented in a list prominently displayed on entering the forum. Thus, if users wish to demonstrate their skills (such as being a Java programming expert), they can answer questions posed in the Java programming forum. This has encouraged informal competition for recognition to be the expert in a given area. Experts interested in enhancing their reputation tend to frequently check for unanswered questions in the topic area of their expertise, providing an informal social mechanism (rather than a technical feature) that ensures that people get answers to their questions. The public display of points and recognition for valuable contributions gives contributors an easily quantified form of satisfaction. The trust and reputation enhancement facilitated by the point system are critical elements of the relational dimension of social capital.

Due to the large number of questioners, answerers, and posed questions, it is difficult to get a sense of the activity of members in the community. In order to match experts with questions, users may browse unanswered questions or may request notification about questions that they may be able to answer. The organization of the system into topics along with the search and browsing capabilities form the structural and cognitive dimensions for sharing in this community. Unlike a threaded discussion or ongoing coordinated activity, users will usually not be aware of work outside the relatively small numbers of questions they have asked or answered. The public display of members' points is an indicator of the activity of community members, making users aware of contributors who are currently involved in answering questions as well as people who have been involved in the past (Erickson and Kellogg 2001). Point sharing in the Experts Exchange adds to the relational aspects of social capital by establishing a sense of trust and a positive atmosphere for contribution. These aspects include:

• *Positive reinforcement*: The point system emphasizes the positive aspects of contributing valuable answers while minimizing potentially socially detrimental behavior. Good answers are publicly praised and rewarded, but weak answers are not presented for public ridicule, and users are not criticized for the questions they were not able to answer well. This creates an environment that encourages a competition to succeed and assists others without emphasizing a "zero-sum game" in which winning, by definition, involves losing.

- *Encouragement of information exchange*: Points give potential contributors an explicit reason to contribute, based on the personal benefits described above. The points that contributors have are one indication of the members' social capital in the community. The members can use the points to obtain help from others by asking questions, thus leading to a social process of sharing. Users may earn points by answering questions or may purchase additional points. Hence, even if a user feels unable to answer questions, by purchasing points and offering them to other users, experts become encouraged to answer more questions.

- *Measurable positive outcome*: The point mechanism provides a measurable positive community aspect, and it is a way of quantifying the social capital existing in the community. Not only do points indicate that individuals have received answers to their questions (one explicit measurable outcome), but the social system as a whole can obtain a measurable positive outcome by indicating that the community as a whole is serving its own question-answering needs. Social capital in the community increases because the number of questions without answers remains low, and the satisfaction of questioners who obtain answers remains high. The more points that are exchanged in the system, the greater the indication that a community's needs are being met.

Examples of Sociotechnical Developments Promoting and Relying on Social Capital

Studying existing examples of social capital–sensitive systems in conjunction with the initial conceptual framework laid out at the beginning of this chapter has provided the basis for our own system developments to foster social creativity with social capital. This section further illustrates this perspective by discussing how the understanding of social capital has influenced our system-building efforts, and in return, how our systems have enriched our understanding of social creativity and social capital.

CodeBroker: Evolutionary Construction of Reusable Software Component Repositories

CodeBroker (Ye 2001) is a system empowering software developers to share and reuse source codes. Although the general belief is that software reuse improves both the quality and productivity of software development, systematic reuse has not yet met its expected success due to two basic difficulties: creating and maintaining a reuse repository; and enabling software developers to build new software systems with components from the reuse repository and then encouraging them to contribute to the repository. These two issues are in a deadlock: if software developers are unable to reuse, the investments on reuse cannot be justified; conversely, if companies are unwilling to invest in reuse, software developers have little to reuse.

We have taken a social-technical approach (Ehn 1988; Greenbaum and Kyng 1991; Grudin 1994) to address both issues simultaneously by fostering a reuse culture from the bottom up by encouraging software developers to reuse components from a repository that may not be of high quality in its initial state but can be evolved through the continual contributions of software developers during the ongoing use of the repository. The concept of social capital helps us realize the important role of informal social networks in supporting reuse. To increase social capital to facilitate reuse and the evolutionary construction of reuse repositories, we are creating reuse repository systems that not only provide information on reusable components but enable developer communities to form, develop, and maintain a sense of shared identity around reuse repositories (Brown and Duguid 2000).

Our approach is guided by the SER model described earlier. In the seeding process, an initial reuse repository is created by clustering functionally related software components through the analysis of existing software systems, indexing the components, and providing an interface for users to access the components. This initial repository need not be of the highest quality but must supply opportunities for evolution. The initially seeded repository is not only the first incarnation of the intellectual capital owned by the software development organization but also an instantiated form of social capital of the organization because it creates the awareness for the expertise existing in the organization as well as a road map to the expertise distributed among software developers. Mediated by reusable components, knowledge sharing and indirect collaboration takes place, enabling software developers to create new systems by building on each other's subsystems.

Software developers who attempt to reuse have to locate task-relevant components from an initially seeded reuse repository and comprehend the located components. If no components that completely fit are located, or if the located components have some bugs to be fixed, software developers need to modify the components, and the modified components need to be contributed back to the repository through a sharing process, resulting in the evolutionary growth of the reuse repository.

A reseeding process is needed when the growth makes the repository too chaotic to grow further. During reseeding, the repository is reorganized, and its components are refactored and generalized, based on their use and the information added by software developers during the evolutionary growth.

Tools that support the easy location of reusable components are an essential factor to the success of our proposed approach to reuse due to two reasons: First, the goal of creating a reuse repository is to get components reused. Second, to involve software developers in the evolution of the reuse repository, they must first be made aware of the opportunities and benefits of reusing existing components in the repository because only when they enjoy certain benefits from the existing repository will they reciprocate with their own contributions. Most current reuse repository systems (Mili, Mili, and Mittermeir 1998) are created as systems independent of a software development environment. They are costly to use because software developers have to switch between development environments and reuse repository systems. Furthermore, because they support user-initiated information access mechanisms only, software developers who are not aware of the existence of reusable components end up reinventing the wheel. As the size of reuse repositories grows as a result of contributions made by software developers, it becomes increasingly difficult for software developers to anticipate the existence of such components.

Figure 14.2 shows a screen image of CodeBroker and its system architecture. CodeBroker consists of four software agents: Listener, Fetcher, Presenter, and Illustrator. Listener infers the needs for reusable components from the partially written program in the editing space (the top buffer) and creates a reuse query based on the inferred needs. The reuse query is passed to Fetcher, which searches the reuse repository to return

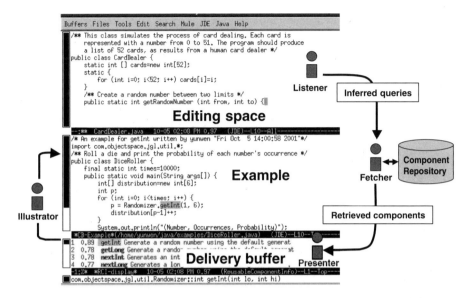

Figure 14.2
A screen image of CodeBroker and its architecture.

a set of components that match the query. Presenter delivers the matching components to the developer in the delivery buffer (the low buffer). Presenter does not deliver all retrieved components; it uses discourse models to filter out task-irrelevant components and user models to filter out user-known components (Fischer and Ye 2001). Discourse models that specify which part of the component repository is not of interest in the current development session are created by developers during their previous interactions with the system to limit the search space. User models that contain those components that the developers know already are created by analyzing the programs that developers have written before.

CodeBroker helps developers utilize the existing social capital to seek help from other developers. When developers have difficulties in comprehending and applying the component of interest in their current task, they can invoke Illustrator, which searches a list of predetermined directories (such as those directories that contain programs written by peer developers) for a program that uses the component. Such an example provides the context for understanding how to use the component (Fischer, Henninger, and Redmiles 1991; Wulf 1999).

Furthermore, the first line of the Example buffer reveals the author of the found example, which enables the creation of a knowledge transfer channel between the current developer who is experiencing difficulty and the developer who wrote the example. For such knowledge transfer to take place, the advanced developer must be willing to share their precious time and expertise, and the learning developer must have trust in the advanced developer. Both of the above issues rely on social capital, and the system creates a connection between the two developers that otherwise might not be associated. A successful collaboration between these two increases their sense of trust and makes later collaboration easier—important elements of social creativity and social capital.

CodeBroker increases the value of the reuse repository as a whole because it widens the reuse possibility of its components. At the same time, it increases the perceived value of components contributed by software developers through the sharing process because they know, with the help of CodeBroker, that their contribution will have a wider reach. Despite these motivating factors, a process of sharing that requires too much extra effort will inhibit software developers from doing so. We are currently developing a new tool called Enhancer to reduce the cost of contributions by providing the following functionality:

- Enhancer automates as much as possible the process of packaging a component that existing software developers decide to share with others for reuse through seamless integration with existing popular development environments. With the assistance of Enhancer, software developers are able to contribute a component by invoking a high-level command such as "share it" within their normal development instead of carrying out detailed operations.
- Enhancer provides mechanisms to help software developers modify an existing component and then contribute the modified component back to the repository. The modification is made in the context of normal development so that software developers do not need to do too much extra work, and Enhancer automatically establishes a relationship between the newly modified component and the original one. Software

developers who have contributed to the improvement of the component are acknowledged as a positive reinforcement to their reputations both as subject matter experts and individuals who are willing to contribute.

• Enhancer captures the usage of reusable components by recording the developers who reuse a component to provide circumstantial evidence on the quality of the component, create a list of subject matter experts that can potentially be tapped for help, and explicitly recognize the contributions made by the original developer of the component and other contributors. Such data will give clear and measurable indications of how well knowledge sharing is taking place as reinforcement mechanisms for increasing social capital.

The ongoing enhancement of CodeBroker with tools such as Enhancer is driven by our conceptual framework and our assessment of earlier versions of CodeBroker (Ye 2001) to increase its potential to become a truly social capital–sensitive system.

The Envisionment and Discovery Collaboratory (EDC)

The EDC (Arias et al. 2000) is an environment in which participants collaboratively solve problems of mutual interest. The problem contexts explored in the EDC, such as urban transportation planning, flood mitigation, and building design, are all examples of open-ended social problems. In these contexts, "optimal" solutions to problems do not exist, and the solutions depend on the participation of diverse stakeholders (Rittel 1984). Solving problems in the EDC therefore requires development of social capital, and the technical and social features of the EDC are designed to enhance social capital. The EDC empowers users to act as designers in situated learning and collaborative problem-solving activities. For most design problems, the knowledge to understand, frame, and solve them does not already exist but is constructed and evolved during the solution process, exploiting the power of the "symmetry of ignorance" and "breakdowns." The EDC is an environment in which social creativity can come alive (Fischer 2000).

Figure 14.3 shows the EDC in use, illustrating some of its features. The EDC supports face-to-face problem-solving activities by bringing

Figure 14.3
The Envisionment and Discovery Collaboratory.

together individuals who share a common problem. The problem is discussed and explored by providing participants with a shared construction space in which users interact with physical objects that are used to represent the situation currently being discussed. As users manipulate physical objects, a corresponding computational representation is updated by using technologies that recognize the placement and manipulation of physical objects. Computer-generated information is projected back onto the horizontal physical construction area, creating an augmented reality environment. This physical construction is coupled with information relevant to the problem currently being discussed.

The EDC is designed to foster social creativity by increasing social capital along the following dimensions:

- *Putting owners of problems in charge*: The collaborative design activities supported by the EDC rely on the contribution and active participation of all involved stakeholders. Design domains consist of ill-defined and wicked problems (Schön 1983) in which there are no correct answers and the framing of problems is a major aspect of solving them. Collaboration of stakeholders is an inherent aspect of these domains. For example, the citizens, urban planners, and transportation experts who solve or are influenced by urban transportation issues are themselves an integral part of the problem context. The goal of the EDC is to bring these stakeholders—each of them or each group owning one part of the problem—together to solve problems of mutual interest and take active roles in addressing the problems that shape their lives.

- *Recognition and awareness of other participants*: The face-to-face nature of the EDC promotes recognition and awareness of other participants. By bringing people together around a table to solve a shared problem, the EDC attempts to acknowledge the importance of the various participants in the problem-solving context. Providing participants with a physical language to describe problems reduces the effort required to make the tacit knowledge of the participants explicit. This construction serves as a common reference point that allows participants to communicate through a shared representation, creating a neutral representation through which (potentially heated) discussion can take place (Arias 1988). Computer recognition of physical representations is designed to allow for the computer to reduce the effort of capturing and formalizing problem information. Face-to-face discussions without some capture mechanism may be rich interactions, but only participants around the table benefit, and when the discussion is over, the interaction is lost. Interaction with physical objects (and corresponding computational objects and models) offers a natural way to capture discussion by using formal structures that may be useful for indexing, history mechanisms, comparing multiple solutions, and helping to integrate diverse communication technologies.

- *Supporting communities of interest*: Providing multiple avenues for participation is critical because participants in the EDC may not share common backgrounds. They represent a community of interest (Fischer 2001), bringing together stakeholders from different domains who have different background knowledge and different things to contribute. The exchange of information is encouraged by supplying stakeholders with the tools to express their own opinions, and this requires an open system that can accommodate and evolve based on new information. For example, city planners may bring formal information (such as the detailed planning data found in Geographic Information Systems), whereas citizens may use less formal techniques (such as sketching) to describe a situation from their points of view.

Courses-as-Seeds

Courses-as-seeds (dePaula, Fischer, and Ostwald 2001) is an educational model that explores social creativity and social capital in the context of fundamentally changing the nature of the courses taught at educational institutions, primarily at the university level. Its goal is to create a culture of informed participation (Fischer and Ostwald 2002) that is situated in the context of university courses and yet extends beyond the temporal boundaries of semester-based classes. Traditionally, the content of a course is defined by the resources provided by instructors, such as lectures, readings, and assignments. By asking students to play an active role in helping to construct their courses, courses rely not only on the intellectual capital supplied by instructors but on the social capital offered by students. Courses are conceptualized as seeds, rather than as finished products, and students are viewed as informed participants who take an active part in defining the problems they investigate (Rogoff, Matsuov, and White 1998). The output of each course contributes to an evolving information space that is collaboratively designed by all course participants, past and present.

An essential aspect of courses-as-seeds is the transformation of traditional classroom roles. Students act as active contributors—active not only in the assignments that are given to them but also in the design of the courses themselves. Instructors' roles are likewise transformed from a "sage on the stage" to a "coach on the side." The relationship network connecting students becomes important in fostering collaborative construction. We have developed and used social capital–sensitive collaborative technologies to provide opportunities to bring social creativity alive by relying on our experience with open source projects (discussed earlier; see also Scharff 2002a). Students chose their own projects and formed teams based on personal interests. These projects were linked together by the project leader (in our case, one of the instructors). The projects involved frequent concrete public deliverables, not in the form of progress updates and final reports that are detached from the actual work of the project but to make the work on the project itself visible to team members. The students collaboratively constructed a course information environment using the Swiki collaborative technology (Rick et al. 2002). Swiki is a collaborative website in which anybody can easily create, edit, or link between the pages.

Courses-as-seeds is grounded in the SER model (described earlier). The seed for a course (embedded in the course information environment) is a set of themes, concepts, readings, and other materials initially identified by the instructors. Throughout the semester, students participating in the course (which includes face-to-face class meetings involving presentations and group discussion) evolve the course information environment. Reseeding involves the reorganization of the course information environment to become more useful for future generations of students, and to form the seed for future classes. In order to not overburden students with too much engagement (requiring unrealistic levels of social capital), we defined different roles on a rotating basis for specific activities. For example, some students answered homework assignments, whereas other students read these submissions, summarized them, reorganized them, and created summaries that represent a single distilled document for future readers (rather than requiring that students read all previously posted assignments). These analyses formed a small kind of reseeding. In course projects, teams created public "home pages" (similar to the home pages found in open source projects) for potential use by future students.

Transforming courses from traditional models of information transmission to collaborative construction requires a transformation of mindsets for students and instructors (Fischer 1999). In an educational culture in which collaboration is stifled and often considered as cheating (Norman 2001), the standard expectations of students often hinder the development of social capital. We encouraged group work through participation in the course information environments, by creating group projects, and by rewarding students for building on others' work. Nevertheless, many students resisted this collaborative construction, which creates a formidable challenge when students with "negative" social capital are asked to participate. We have found that self-selection is key for course success, discouraging students who did not wish to collaborate from participating. A crucial question remains: Is it more desirable to attempt to "force" collaboration (potentially creating social capital where none existed before) or to cultivate existing social capital?

Public deliverables have played a major role in establishing social capital in our courses. Encouraging students to send their contributions to each other can be a source of social creativity. Students reported that public deliverables are a good motivator to make higher-quality contributions because poor contributions make them look bad in the eyes of their peers. This is similar to findings in open source, in which developers claim that public deliverables encourage higher-quality components because anyone can observe (and evaluate) the contribution. Students who contributed frequently and with more substance were recognized as leaders by other students. In peer evaluations, students seem to react strongly to the activity of their team members. Because every change to the course information environment is recorded along with the user making the change, it is possible for students to see who is contributing what. Public deliverables encouraged students to share, which is rare in team projects, but is common with open-source-like course projects. When one team created a home page design that other teams liked, the other teams quickly adapted their sites to match the model. This increased the quality of their own sites and can be seen as a source of social creativity. The members of the original team were flattered by (and were glad that they were credited for) the adoption of their ideas by other teams.

Assessment

The analysis of existing systems (disscussed earlier) coupled with experiences derived from our own systems (also described above) has led to a greater understanding of the challenges involved in fostering social creativity with social capital–sensitive systems in collaborative design. Table 14.1 provides a summary statement of our findings.

Table 14.1
Summary of social creativity and social capital aspects of the environments discussed

	Brief description	Social creativity	Social capital
Open source	Approach and methodology for collaborative design and construction of software	Source code availability breaks down barriers between users and designers; decentralized model encourages contributions by many	Open process recognizes contributors and motivates them; public deliverables create objects for discussion and encourage high-quality deliverables
Experts Exchange	Web-based knowledge-sharing environment	High-quality answers encouraged by points and competition; Web site is collaboratively evolved by both questioners and answerers	Point system rewards participation; awareness of who does what; everyone wins: folks get answers, experts get recognition
CodeBroker	Reuse environment	Reuse of components leads to better software; environment encourages sharing and building on stable subcomponents	Trust in developers leads to trust in code; willingness to help; reciprocity
EDC	Face-to-face support for collaborative design	Contribution of stakeholders having different perspectives and knowledge; solutions to problems emerge from discussions and individual design activities; support for the owners of problems	Low barriers to contribute based on integrated and domain-oriented environment; owner of problems can directly engage in personally meaningful activities
Courses-as-seeds	New methodology and computational support for learning and teaching	Learning from each other is supported; instructors learn from students	Encouraged group work to foster social capital; public deliverables in the class increase peer pressure and peer recognition

Addressing the Adoption Barriers of Collaborative Technologies

Building a system that supports collaborative design is merely an enabling condition for fostering social creativity by increasing social capital. All too often when systems are created, the expected use never matches the reality, which can be characterized by "build it—and they will not come." Analyses of the failures of the adoption of groupware applications (Grudin 1994) point to many complex factors, from the institutional buy-in of proposed technologies, to the involvement of system users during development. We will analyze these barriers using the equation "utility = value / effort," meaning that people will decide on the worthiness of doing something (utility) by relating the (perceived) value of an activity to the (perceived) effort of doing it.

Applying this equation, one technique for increasing utility is to decrease the effort necessary to contribute to a system. Capturing interactions automatically (Hill et al. 1992; Adachi 1998) is a technique that decreases the effort needed to contribute. Giving participants an extensive seed rather than a completely empty open system helps provide immediate value; it also diminishes effort by helping to ground new contributions by following the example set by the initial organization of the seed. Incremental formalization (Shipman and McCall 1994) can reduce effort by allowing unstructured data to be added quickly, and then permitting the more time-consuming formalization of the data to be added gradually. This allows users to strike a balance between the added utility of more formalized (and machine-interpretable) structures (such as increased search capabilities) and the cost of adding the information necessary for searching. Techniques such as latent semantic analysis (Landauer and Dumais 1997) can be added to incremental formalization by helping to create new associations without the additional efforts of formalization. By measuring the similarity between different texts, latent semantic analysis can create associations between unstructured text documents without the human effort of adding structure or metadata to that text. In the EDC, techniques such as physical interaction decrease the participation effort by making the interaction more intuitive or natural than similar actions performed using the traditional computer input methods of mice and keyboards. In CodeBroker, analyzing the formal structure of the current problem (such as code comments and program structure) provides usage data that greatly reduces the difficulty in formulating complex queries. In addition to these technical interventions, the social aspects of communities can help to reduce the perceived effort of contributions. As contributions increase, familiarity with systems and the community will also increase, and users will see evidence that contributing may not be too difficult. A sense of peer pressure will be created: If others can find time to contribute, then an individual can as well. For this reason, the awareness of the contributions of others and the positive acknowledgment of their accomplishments are critical.

Increasing perceived value requires an understanding of what an individual and community will find valuable. Timely information that will help solve immediate problems will encourage people to associate contributions to this effort with something from which they can see an immediate benefit. Possible financial rewards from such information is one possibility (Andrews 2002), but not all valuable efforts will lead to them. In our courses-as-seeds model, students' participation (as well as any tangible outcome stored in on-line repositories) and the grades they receive are directly connected, so each contribution potentially adds value through the improvement of students' grades. In open source, the realization that problems are intellectually interesting or fun helps to contribute to the perception of value in participating. Other community members are another important source of value: if someone can interact with a large number of people who can help to solve a problem (see the Expert Exchange discussed earlier), then the community will create value. Similarly, if participation gives access to a group of people with whom someone would like to work, their presence will create value for that individual.

An interesting question raised by studying social capital is whether the notions of perceived utility and effort exist solely in the heads of individual participants, or whether group activity creates an emergent and collective conceptualization of these ideas. As we have seen, social interactions play a role in individual perceptions. Is there a similar way that groups as a whole have notions of effort and utility? Systems such as the Experts Exchange indicate that this may be the case, as the participation of a large community of experts lowers the effort of any single individual and increases value due to the desire for a whole group of experts to collaborate.

Motivation

When an initially seeded system is adopted by its users, the further development and evolution of the system through the thriving of social creativity depends mainly on the motivation of users to contribute and participate, as well as the trust that users place in the quality of contributions made by other people. The relational dimension of the social capital definition provides a framework to assess the roles that motivation and trust play in fostering social creativity. Nahapiet and Ghoshal (1998) argue that the obligations of generalized reciprocity, shared norms and values of the community, mutual trust among members, and identification with the community are the four critical components of the relational dimension of social capital. Without active contribution and participation from motivated users, the process of collaborative knowledge construction will not succeed. Factors that affect users' motivation to contribute and participate are both cognitive (intrinsic) and sociotechnical (extrinsic).

The precondition for motivating users to contribute to systems such as evolving reuse repositories or to participate in open source systems is that they must derive an intrinsic satisfaction from accomplishing their tasks and goals. In other words, the activity of contribution and participation itself must be perceived as engaging, satisfying, and pleasurable enough to be worth its associated efforts. The satisfaction may come from the intellectual challenge inherent in the task, the feeling of accomplishment and fulfillment, the joy of mastery of new knowledge, or the desire for social acceptance by identifying with some communities and associating with their members.

Social factors that positively reinforce intrinsic motivation include generalized reciprocity, social recognition, and rewards within a community.

Generalized Reciprocity

In a community with a good stock of social capital, generalized reciprocity (Putnam 1995) is considered a social norm. This encourages members to do their own share of contributions in the expectation of receiving reciprocated help from other members. In our evolutionary reuse repository project, the CodeBroker system is designed to create an opportunity for users to obtain immediate benefits from the existing reuse repository through its delivery mechanisms. To abide by the social norm of generalized reciprocity, users are more likely to feel obliged to return the favor by reciprocating with their own work for the benefit of others (Nahapiet and Ghoshal 1998). In open source communities, software developers receive assistance in testing, debugging, and functionality enhancing from other developers. In the EDC, participating users learn new knowledge by receiving constructive critiques from other knowledgeable participants. Achieving social creativity requires the active participation of all stakeholders who have different perspectives and knowledge. The norm of generalized reciprocity motivates each stakeholder to contribute their own knowledge and understanding relevant to the common problem faced by all members, in the expectation of receiving reciprocated contributions from others who have different expertise and insights. As a result, the combined contributions can solve a problem that no individual member can solve alone. Recognizing the necessity for collaboration motivates some members to contribute at first. As members start to contribute, the sense of peer pressure causes other members to contribute reciprocally, gradually leading to the establishment of generalized reciprocity in the community.

Social Recognition
Contributors' motivations increase when the community explicitly recognizes the effort and especially the value of user contributions. For example, listing the most active contributors in the Experts Exchange produces a quantified form of satisfaction and creates an informal competition for recognition as a subject matter expert. In open source communities, the existence of a large number of users provides psychological and social support for voluntary developers similar to the support an audience gives to theatrical performers (Ye and Kishida 2003). Such explicit recognition mechanisms create community-wide awareness of the existence of experts along with the distribution of expertise that facilitates potential information exchange and cooperation, therefore increasing the social capital of the community.

Rewards

Rewards for user contributions may come from both inside and outside the community. Active contributing users increase their own stock of social capital within the community by increasing their name recognition and reputation as subject matter experts as well as willing helpers and collaborators (or good citizens). In return, they command respect and trust from other members, and have the capability of executing larger influence in the community. Peer respect and trust make it easier for them to attract collaborators to advance their personal agenda in the future. We have found a strong correlation between user contributions and influential powers in open source communities. Actively contributing developers gradually move toward the center of the community and play increasingly important roles in making decisions about further development (Nakakoji et al. 2002). If a user is recognized as an expert within an open source community, this reputation can spread to society at large, earning the user socioeconomic benefits such as a higher salary or social status. Some active open source developers have become internationally renowned for their strong technical skills and great insights, and some have benefited from the reputation they have built in open source communities to obtain highly paid jobs or offer consulting services to companies for profit. Because such external recognition depends on the influence the community as a whole could have on society, members have indirect and potential personal interests to behave as good citizens to sustain the community.

All the above factors are components of social capital. Therefore, by taking social capital into consideration during the design of collaborative computer systems that facilitate social creativity, the utility of contributions increases due to the increase of perceived value, leading to higher motivation for active contribution and participation.

Technology can either promote or thwart a user's motivation to contribute and participate because it can either increase or decrease the effort of contribution. In terms of motivation to active contribution or participation, individual differences exist; some participants are more motivated than others. The technological barriers to contributing might thwart those less motivated. If the technology difficulty is extremely high, even the most motivated members might not be able to contribute. For example, it is impossible for extremely motivated and well-skilled users to contribute to the development of closed systems. If the technology requires no extra effort to contribute, no motivation is required to participate—users become active contributors indirectly through the work that they perform. In order to avoid making unreasonable demands on social capital, it is critical to design collaborative computer systems that are consistent with the normal work process of users so that little extra work is required to contribute, allowing the population of motivated contributors to increase.

The decreased difficulty enables more people to participate, enhancing the social capital of the community by increasing the number of participants and enriching the connectedness of the members. The increased number of participating users means that each contribution can benefit more people through the enriched connectedness, and hence, the value of contribution increases, leading to higher motivation for contributing.

Trust
As a key benefit of social capital, mutual trust among members of a community establishes better knowledge-sharing conduits, lowers transaction costs, resolves conflicts more easily, and creates greater coherence in terms of collective actions. Because "trust is the expectation that arrives within a community of regular, honest and cooperative behavior, based on commonly shared norms on the part of other members of that community" (Fukuyama 1995, 26), it develops when there is a history of favorable past interactions among those members (Preece 2002). Trust leads community members to expect positive future interactions, while also encouraging further interactions and participation.

In the context of collaborative computer systems that support social creativity, it is important that the initially seeded system have relatively coherent, stable, and predictable behaviors for users to develop a sense of trust in the system. By being able to manipulate familiar physical objects and receive immediate reactions from the system through simulation, users of the EDC transfer their trust in real-world objects into the system. Having a ready-to-run system in open source is the first step for its users to determine the trustworthiness of the system through firsthand experience by observing its behaviors. The ability to fully access and freely change source code reinforces the sense of trust by assuring the smooth resolution of future uncertainties because users have control over the system when it behaves unexpectedly or unsatisfactorily. Impersonal trust (McKnight, Cummings, and Chervany 1998) in the system is transferred into interpersonal trust in the system developer or developers, which again is transferred to impersonal trust in the subsequent systems built by them. We are currently replicating this trust model in our attempt to create evolutionary reuse repositories through collaboration.

During the evolutionary growth phase, the initial trust develops or decays, depending on whether the continuing and repeated interactions between users and the system as well as among users match the expectation of the interacting partner, either the system or another user. In the EDC, interpersonal trust among users develops naturally when they are engaged in face-to-face meetings, relying on their everyday life experiences. In virtual communities created in open source and reuse repositories, where face-to-face meetings rarely happen, the system needs to provide extra support to address the trust issue. Because users do not directly interact with each other, trust among them has to be mediated through common objects such as source code and reusable components that are in common use. The original developers of open source systems are generally trusted by other members due to the reasons discussed in the previous paragraph. Interpersonal trust in members of an open source community is transferred from the impersonal trust in the quality of their code contributions made over a relatively long period of participation in the community. The quality of code contributions is judged by the execution and inspection of the code. This model, however, does not work well in the case of reuse repositories for two reasons. First, most reusable components are not immediately executed and their quality cannot be easily judged. Second, careful inspection of the code requires a lot of time, which conflicts with one of the primary benefits of reuse: to speed up the development. Our approach to address the trust issue is to look at the social context of components (Brown and Duguid 2000): who produces it, how many times it has been reused, who has reused it, and what people who have reused it say about it. This approach aims to develop a trust model that offers circumstantial evidence of the quality of a component based on the interpersonal trust among members in the community. Although the software developer may encounter the reusable component for the first time, the history of favorable past interactions between other trusted members and the component can help that developer trust it. The interpersonal trust and impersonal trust are mutually dependent. If the impersonal trust in a component based on interpersonal trust generates positive interactions, the interpersonal trust is positively reinforced; otherwise, the interpersonal trust decreases.

Evolution by Designers and Evolution by Users

To support social creativity, we need a new conceptual model to understand the relationship between information repositories that are the place to store information and knowledge and their designers and users. Figure 14.4 describes two general models of creating and using an information repository. Model (a) requires a thick, good input filter, which can be applied to select important and reliable information or a few dedicated information producers of high caliber, resulting in a relatively small information repository that contains only information of good quality, but often misses potentially useful information. Model (a) supports information consumers, who are mostly passive, to locate and choose what they need from such an information repository. It is suited for static information repositories and is effective in domains in which changes are slow. It is not suitable, though, for evolutionary domains that require social creativity.

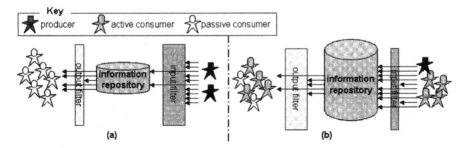

Figure 14.4
Two general models of creating and using an information repository.

Model (b) describes the collaborative construction of information repositories. It has a thin input filter that allows not only dedicated producers but also active consumers (or local developers (Nardi 1993)) who are able and willing to contribute to putting information into the repository, resulting in a large information repository. This model requires a good, thick output filter that can provide information contextualized to the task at hand and the background knowledge of individual users (Ye 2001). Information repositories created in model (b) support social creativity based on the following reasons:

- they are information spaces owned by the people and communities who use them to do work, not by management or an IT department;
- they support the collaborative and evolutionary design of complex systems by allowing users to become active contributors;
- they are open and evolvable systems, serving not only as repositories of information but also as mediums of communication and innovation;
- they are information spaces that can be evolved through many small contributions made by many people, rather than through large contributions made by few people—as has been the case for knowledge-based systems of the past following model (a).

A purely self-organized (decentralized) evolution of complex artifacts and information repositories is a myth, however. For example, although open source software is constructed in a bazaar style—any user can change its source code—the main version for mass distribution is still controlled by a core set of "project leaders" who have the final say on the course of the evolution of a project. These people perform centralized integration of information (during the reseeding phase) that is contributed by others in a decentralized manner (during the evolutionary growth phase). Contributors are explicitly acknowledged and often assume responsibility for the evolution of their subsystem. Open source projects have many varieties of control structures, but each project has some centralized responsibility—none of them relies totally on decentralized evolution.

The evolution of living information repositories must have elements of both decentralized evolution and centralized integration. The mix of these modes—as well as the means of selection of individuals to assume roles of responsibility—takes many forms. The goal of making systems modifiable by users does not imply transferring the responsibility of good system design to the user. In general, modifications that normal users make will not be of the quality of those a system designer would make. Users are not concerned with the system per se but with doing their work. On the other hand, users are concerned with the adequacy of the system as a tool to do their work, and as such, they experience the fit, or misfit, between their needs and the capabilities of the tool. This is knowledge that the designer lacks because the designer does not use the tool to do work. Such systems do not decrease the responsibility or importance of the system designers but shift the design emphasis on a finished system at design time to a system that can be adapted and modified at use time, by both users and designers.

Sustaining the usefulness and usability of evolving information repositories over time involves challenges and trade-offs, as summarized in table 14.2 (Fischer and Ostwald 2001). Such factors depend on whether these information repositories are evolved by specialists, who do not actually use the systems to do work, or in the working context by knowledge workers, who are the owners of problems and evolve the environments in the context of their work.

Self-Application

Building tools to support communities is not restricted to settings in which designers facilitate the work of other people. We have attempted to design our own community and tools to be social capital–sensitive. The advantage of this approach is greater than simply having an easily available community of users. Self-application allows us to see if ideas and tools work for us—if we are not willing to adopt them, then why should others? Another advantage is breaking down the barriers between designers of interactive experiences and end users subjected to that experience. We wish to create a community in which all can help shape the experience, and the value gained by the individual to contribute is greater than the effort expended. Ideally, the community as a whole will strengthen the ties that bind it together and create social capital.

Although it is difficult to measure our group's social cohesion, emergent participation, and facilitation of participants directly, it is possible

Table 14.2
Information repositories evolved by specialists versus evolved in the working context

	Evolved by designers	Evolved by users in the working context
Examples	Digital library of ACM	Reference libraries and Web sites of research groups, Eureka (Xerox copier technicians) (Brown and Duguid 2000)
Nature of individual entries	Database-like entries	Narratives, stories, best practices, cases
Economics	Requires substantial additional resources (e.g., clerical support)	Puts an additional burden on the domain workers
Cognitive dimensions	Possible in domains in which objects are well defined and tasks can be delegated	Problem owners need to do it because the objects are emerging products of work
Motivation	Work assignment	Social capital

to understand some of these issues indirectly through our experience with collaborative technologies that we have both designed and used. The adoption challenges with organizational memories and collaborative work discussed above have been observed in our own work as well. Our key finding from this experience has been that collaborative sociotechnical environments require more than new technologies. Collaborative technologies themselves will not change people's perceptions of technological barriers: people still must perceive a direct benefit for adopting new technologies, and they must gain more from contribution than what is required. People must not only focus their efforts on the immediate need to complete tasks but on the longer-term effects to act collectively (Resnick 2001). To create this change in behavior, the effort to contribute to a system must be minimal so that it will not interfere too much with getting the "real work" done. Social structures and work expectations have to change as well in a collaborative work environment. As one of our students observed, "Collaborative systems will not work in a noncollaborative society." In our courses, if students do not want to contribute, or do not have a preconception that collaborative activity is expected or desirable, no amount of technology is going to change their lack of participation. Mindsets cannot be easily altered (Fischer 1999). Work practices must change so that they are perceived as collaborative, involving a need for people to share authority and power, and breaking down traditional barriers (as well as personal ones) (Zuboff 1988). Not only do we rely on each other more than before (increasing the dependencies required to complete tasks); we do not always clearly communicate the expectations that are necessary from other people. Breaking down barriers of authority and control leaves individuals unclear about their precise roles, which are more obvious and defined in a closed system. As roles are transformed, so too must we transform both our understanding of our roles along with our ability to effectively negotiate who will do what.

Conclusions

Solving complex design problems requires the collaboration of numerous individuals with complementary skills. The resources necessary to solve problems are distributed among members of the community, and creative solutions emerge out of collaborative work. Collaborative systems need to be designed to facilitate social creativity, relying on the participation of members of the community and encouraging collaborative construction. The potential rewards of facilitating social creativity are great: encouraging participants to contribute to problems that shape their lives; distributing work so that individual contributions can be small, but combined output can be substantial; leveraging the unique skills of individuals to create solutions that overlap mutual, complementary competencies; and helping to create new ideas that would not have been created in isolation. Creating socially creative activities requires social capital–sensitive environments. All the resources of individuals and social groups—including the networks that connect people, the motivation and trust among members of a community, and the shared understanding and interests that bind communities together—must be both utilized and cultivated.

Future generations of collaborative systems that wish to support social creativity can benefit from an increased understanding of the many aspects that affect social capital. A sociotechnical perspective is important because new media, technology, approaches to collaboration, and communities must coevolve in order to foster a richer form of collaboration. Just as technologies and documents provide structure, which influences contribution, so too must new technologies be designed to better support the kinds of contribution of a particular community. Designing for and increasing social capital involves a process of continuous evolution. The SER process model that encourages ongoing change highlights the different challenges faced in creating initial seeds for collaboration, handling incremental changes, and periodically reorganizing and revising the contributions.

Increasing social capital requires an understanding not only of a group's internal cohesiveness but what role individuals play in the formation of that group. Individual participation is a necessary precondition for group work, and that participation relies on the balance between the perceived benefit of a contribution and the perceived effort of becoming involved. Personal motivations, individual relationships, expectations from participation, recognition by peers, and tangible metrics for success are crucial for participants. Technology can be used not only to decrease the potential effort made by individuals but to bring people together, and by doing so, can increase the perceived value of participation. The collective activity and sharing culture of a community can be amplified by technologies that encourage individuals to participate in a group effort. A critical challenge for the next generation of collaborative systems is to acknowledge the individual and collective forces required in collaborative design, and to foster social creativity by increasing social capital.

Acknowledgments

The authors thank the members of the Center for LifeLong Learning and Design at the University of Colorado; they made major contributions to the conceptual framework described in this chapter. The research was supported by the National Science Foundation, grants REC–0106976 ("Social Creativity and Meta-Design in Lifelong Learning Communities") and CCR-0204277 ("A Social-Technical Approach to the Evolutionary Construction of Reusable Software Component Repositories"); SRA Key Technology Laboratory, Inc., Tokyo, Japan; and the Coleman Initiative, San Jose, California.

References

Adachi, T. 1998. *Utilization of usage data to improve organizational memory.* Master's thesis, University of Colorado at Boulder.

Andrews, D. C. 2002. Audience-specific online community design. *Communications of the ACM* 45, no. 4:64–68.

Arias, E. G. 1988. Bottom-up neighborhood revitalization: A language approach for participatory decision support. *Urban Studies* 33, no. 10:1831–1848.

Arias, E. G., H. Eden, G. Fischer, A. Gorman, and E. Scharff. 2000. Transcending the individual human mind: Creating shared understanding through collaborative design. *ACM Transactions on Computer-Human Interaction* 7, no. 1:84–113.

Arias, E. G., and G. Fischer. 2000. Boundary objects: Their role in articulating the task at hand and making information relevant to it. In *International ICSC symposium on interactive and collaborative computing.* Wetaskiwin, Canada: ICSC Academic Press.

Bourdieu, P. 1983. Forms of capital. In *Handbook for theory and research for the sociology of education*, edited by J. G. Richardson. New York: Greenwood Press.

Brown, J. S., and P. Duguid. 2000. *The social life of information.* Boston: Harvard Business School Press.

Bruner, J. 1996. *The culture of education.* Cambridge: Harvard University Press.

Campbell, D. T. 1969. Ethnocentrism of disciplines and the fish-scale model of omniscience. In *Interdisciplinary relationships in the social sciences*, edited by M. Sherif and C. W. Sherif. Chicago: Aldine Publishing Company.

Carroll, J. M., and M. B. Rosson. 1987. Paradox of the active user. In *Interfacing thought: Cognitive aspects of human-computer interaction*, edited by J. M. Carroll. Cambridge: MIT Press.

Claypool, M., D. Finkel, and C. Wills. 2001. An open source laboratory for operating systems projects. In *Sixth annual conference on innovation and technology in computer science education.* New York: ACM Press.

Cohen, D., and L. Prusak. 2001. *In good company: How social capital makes organizations work.* Boston: Harvard Business School Press.

Coleman, J. C. 1988. Social capital in the creation of human capital. *American Journal of Sociology* 94:S95–S120.

Davis, M., W. O'Donovan, J. Fritz, and C. Childress. 2000. Linux and open source in the academic enterprise. In *Conference on user services: Building the future.* New York: ACM Press.

dePaula, R., G. Fischer, and J. Ostwald. 2001. Courses as seeds: Expectations and realities. In *Proceedings of the second European conference on computer-supported collaborative learning.* Maastricht, Netherlands: Maastricht McLuhan Institute.

DiBona, C., S. Ockman, and M. Stone, eds. 1999. *Open sources: Voices from the open source revolution.* Sebastopol, Calif.: O'Reilly and Associates.

Ehn, P. 1988. *Work-oriented design of computer artifacts.* Stockholm: Arbetslivscentrum.

Engeström, Y. 2001. Expansive learning at work: Toward an activity theoretical reconceptualization. *Journal of Education and Work* 14, no. 1:133–156.

Erickson, T., and W. A. Kellogg. 2001. Social translucence: Designing systems that support social processes. In *Human-computer interaction in the new millennium,* edited by J. M. Carroll. New York: ACM Press.

Experts Exchange. 2002. The experts exchange. Accessed on 5 November, <http://www.experts-exchange.com>.

Fischer, G. 1998. Beyond "couch potatoes": From consumers to designers. In *Proceedings of the* Asian Pacific Computer-Human Interaction Conference, Los Alamitos, CA: IEEE Press.

Fischer, G. 1999. Lifelong learning: Changing mindsets. In *Seventh international conference on computers in education on "new human abilities for the networked society,"* edited by G. Cumming, T. Okamoto, and L. Gomez. Omaha, Nebr.: IOS Press.

Fischer, G. 2000. Social creativity, symmetry of ignorance, and meta-design. *Knowledge-Based Systems Journal* 13, nos. 7–8:527–537.

Fischer, G. 2001. Communities of interest: Learning through the interaction of multiple knowledge systems. In *Twenty-fourth annual information systems research seminar in scandinavia.* Bergen, Norway: Department of Information Science.

Fischer, G., and A. Girgensohn. 1990. End-user modifiability in design *environments.* In *Proceedings of 1990 conference on human factors in computing systems CHI'90,* Seattle, WA. New York: ACM Press.

Fischer, G., J. Grudin, R. McCall, J. Ostwald, D. Redmiles, B. Reeves, and F. Shipman. 2001. Seeding, evolutionary growth, and reseeding: The incremental development of collaborative design environments. In *Coordination theory and collaboration technology,* edited by G. M. Olson, T. W. Malone, and J. B. Smith. Mahwah, N.J.: Lawrence Erlbaum Associates.

Fischer, G., S. R. Henninger, and D. F. Redmiles. 1991. Cognitive tools for locating and comprehending software objects for reuse. In *Proceedings of thirteenth international conference on software engineering* ICSE'91, (Austin, TX). Los Alamitos, CA: IEEE Computer Society Press.

Fischer, G., and J. Ostwald. 2001. Knowledge management: Problems, promises, realities, and challenges. *IEEE Intelligent Systems* 16, no. 1:60–72.

Fischer, G., and J. Ostwald. 2002. Seeding, evolutionary growth, and reseeding: Enriching participatory design with informed participation. In *Proceedings of the participatory design conference*. Palo Alto, CA: Computer Professionals for Social Responsibility.

Fischer, G., and E. Scharff. 2000. Meta-design: Design for designers. In *Third international conference on designing interactive systems*. New York: ACM Press.

Fischer, G., and Y. Ye. 2001. Personalizing delivered information in a software reuse environment. In *Proceedings of user modeling 2001 (eighth international conference, UM 2001), Sonthofen, Germany*, edited by M. Bauer, P. Gmytrasiewicz, and J. Vassileva. Heidelberg: Springer Verlag.

Florida, R. 2002. *The rise of the creative class: And how it's transforming work, leisure, community, and everyday life*. New York: Basic Books.

Fogel, K. F. 2000. *Open source development with CVS*. Scottsdale, Ariz.: Coriolis Group.

Fukuyama, F. 1995. *Trust: The social virtues and the creation of prosperity*. New York: Free Press.

Fukuyama, F. 2001. Social capital, civil society and development. *Third World Quarterly* 22, no. 1:7–20.

Greenbaum, J., and M. Kyng, eds. 1991. *Design at work: Cooperative design of computer systems*. Hillsdale, N.J.: Lawrence Erlbaum Associates.

Grudin, J. 1994. Groupware and social dynamics: Eight challenges for developers. *Communications of the ACM* 37, no. 1:92–105.

Hanifan, L. J. 1916. The rural school community center. *Annals of the American Academy of Political and Social Science* 67:130–138.

Henderson, A., and M. Kyng. 1991. There's no place like home: Continuing design in use. In *Design at work: Cooperative design of computer systems*, edited by J. Greenbaum and M. Kyng. Hillsdale, N.J.: Lawrence Erlbaum Associates.

Hill, W. C., J. D. Hollan, D. Wroblewski, and T. McCandless. 1992. Edit wear and read wear. In *Proceedings of CHI'92 conference on human factors in computing systems*, edited by P. Bauersfeld, J. Bennett, and G. Lynch. New York: ACM Press.

John-Steiner, V. 2000. *Creative collaboration*. Oxford: Oxford University Press.

Landauer, T. K., and S. T. Dumais. 1997. A solution to Plato's problem: The latent semantic analysis theory of acquisition, induction, and representation of knowledge. *Psychological Review* 104, no. 2:211–240.

Lieberman, H., ed. 2001. *Your wish is my command: Programming by example.* San Francisco: Morgan Kaufmann.

Mackay, W. E. 1990. *Users and customizable software: A co-adaptive phenomenon.* Ph.D. diss., Massachusetts Institute of Technology.

MacLean, A., K. Carter, L. Lovstrand, and T. Moran. 1990. User-tailorable systems: Pressing the issues with buttons. In *Proceedings of CHI '90 conference on human factors in computing systems*, edited by J. Carrasco and J. Whiteside. New York: ACM Press.

McKnight, D. H., L. L. Cummings, and N. L. Chervany. 1998. Initial trust formation in new organizational relationships. *Academy of Management Review* 23:383–399.

Meyerson, E. 1994. Human capital, social capital, and compensation: The relative contribution of social contacts to manager's income. *Acta Sociologica* 37:383–399.

Mili, A., R. Mili, and R. T. Mittermeir. 1998. A survey of software reuse libraries. In *Systematic software reuse*, edited by W. Frakes. Bussum, Netherlands: Baltzer Science.

Mockus, A., R. T. Fielding, and J. Herbsleb. 2000. A case study of open source software development: The apache server. In *Proceedings of twenty-second international conference on software engineering.* New York: ACM Press.

Nahapiet, J., and S. Ghoshal. 1998. Social capital, intellectual capital, and the organizational advantage. *Academy of Management Review* 23:242–266.

Nakakoji, K., Y. Yamamoto, Y. Nishinaka, K. Kishida, and Y. Ye. 2002. Evolution patterns of open-source software systems and communities. In *Proceedings of 2002 international workshop on principles of software evolution (IWPSE 2002, Orlando).* New York: ACM Press.

Nardi, B. A. 1993. *A small matter of programming.* Cambridge: MIT Press.

Norman, D. 2001. In defense of cheating. Accessed on 5 March 2003, <http://jnd.org/dn.mss/InDefenseOfCheating.html>.

Oppermann, R., ed. 1994. *Adaptive user support.* Hillsdale, N.J.: Lawrence Erlbaum.

Perens, B. 1999. The open source definition. In *Open sources: Voices from the open source revolution*, edited by C. DiBona, S. Ockman, and Stone. Sebastopol, Calif.: O'Reilly and Associates.

Polanyi, M. 1966. *The tacit dimension.* Garden City, N. Y.: Doubleday.

Preece, J. 2002. Supporting community and building social capital: introduction. *Communications of the ACM* 45, no. 4:37–39.

Putnam, R. D. 1995. Bowling alone: America's declining social capital. *Journal of Democracy* 6, no. 1:65–78.

Raymond, E. S. 1998. Homesteading the noosphere. *FirstMonday* 3, no. 10: http://www.firstmonday.org/issues/issue3_10/raymond/index.html

Raymond, E. S., and B. Young. 2001. *The cathedral and the bazaar: Musings on Linux and open source by an accidental revolutionary.* Sebastopol, Calif.: O'Reilly and Associates.

Reisberg, D. 1997. *Cognition.* New York: Norton and Company.

Resnick, P. 2001. Beyond bowling together: Sociotechnical capital. In *Human-computer interaction in the new millennium*, edited by J. M. Carroll. New York: ACM Press.

Rick, J., M. Guzdial, K. Carroll, L. Holloway-Attaway, and B. Walker. 2002. Collaborative learning at low cost: CoWeb use in English composition. In *Conference on computer-supported collaborative learning.* (CSCL 2002, Boulder). Hillsdale, NJ: Lawrence Erlbaum Associates.

Rittel, H. 1984. Second-generation design methods. In *Developments in design methodology*, edited by N. Cross. New York: John Wiley and Sons.

Rogoff, B., E. Matsuov, and C. White. 1998. Models of teaching and learning: Participation in a community of learners. In *The handbook of education and human development: New models of learning, teaching, and schooling*, edited by D. R. Olsen and N. Torrance. Oxford: Blackwell.

Salomon, G., ed. 1993. *Distributed cognitions: Psychological and educational considerations.* Cambridge: Cambridge University Press.

Scharff, E. 2002a. Applying open source principles to collaborative learning environments. In *2002 Conference on computer-supported collaborative learning.* (CSCL 2002, Boulder). Hillsdale, NJ: Lawrence Erlbaum Associates.

Scharff, E. 2002b. Open source software: A conceptual framework for collaborative artifact and knowledge construction. Ph.D. diss., University of Colorado at Boulder.

Schön, D. A. 1983. *The reflective practitioner: How professionals think in action.* New York: Basic Books.

Schuler, D., and A. Namioka, eds. 1993. *Participatory design: Principles and practices.* Hillsdale, N.J.: Lawrence Erlbaum Associates.

Shipman, F., and R. McCall. 1994. Supporting knowledge-base evolution with incremental formalization. In *Proceedings of 1994 conference on human factors in computing systems.* New York: ACM Press.

Wulf, V. 1999. "Let's see your search-tool!" On the collaborative use of tailored artifacts. In *Proceedings of GROUP '99.* New York: ACM Press.

Ye, Y. 2001. Supporting component-based software development with active component repository systems. Ph.D. diss., University of Colorado at Boulder.

Ye, Y., and K. Kishida. 2003. Toward an understanding of the motivation of open source software developers. In *Proceedings of 25th international conference on software engineering.* (ICSE'03, Portland, OR). Los Alamitos, CA: IEEE Press.

Zuboff, S. 1988. *In the age of the smart machine.* New York: Basic Books.

Contributors

Mark S. Ackerman, School of Information and Department of Electrical Engineering and Computer Science, University of Michigan, Ann Arbor

Eline Aukema, Amsterdam School of Communication Research, University of Amsterdam

Andreas Becks, Fraunhofer Institute for Applied Information Technology, Sankt Augustin Germany

Anita Blanchard, University of North Carolina, Charlotte

Stephen P. Borgatti, Carroll School of Management, Boston College

Mike Bresnen, Warwick Business School, University of Warwick, Coventry, United Kingdom

Robbin Chapman, MIT Media Laboratory, Cambridge, Massachusetts

Rob Cross, McIntire School of Commerce, University of Virginia Charlottesville, VA

Linda Edelman, Bentley College, Waltham, Massachusetts

Gerhard Fischer, Center for LifeLong Learning and Design, Department of Computer Science, University of Colorado, Boulder

Christine Halverson, Social Computing Group, IBM T. J. Watson Research Center, San Jose

Bart van den Hooff, Amsterdam School of Communications Research, University of Amsterdam

Marleen Huysman, Vrije Universiteit Amsterdam, Amsterdam, The Netherlands

Kari Kuutti, Department of Information Processing Science, University of Oulu, Finland

Sue Newell, Royal Holloway, University of London, Egham, Surrey, United Kingdom

Anabel Quan-Haase, University of Toronto

Tim Reichling, Institute for Information Systems, University of Siegen, Germany

Jan de Ridder, Amsterdam School of Communications Research, University of Amsterdam

Markus Rohde, International Institute for Socio-Informatics, Bonn, Germany

Harry Scarbrough, Warwick Business School, University of Warwick, Coventry, United Kingdom

Eric Scharff, Center for LifeLong Learning and Design, Department of Computer Science, University of Colorado, Boulder

Charles Steinfield, Department of Telecommunication, Michigan State University, East Lansing

Anna-Liisa Syrjänen, Department of Information Processing Science, University of Oulu, Finland

Barry Wellman, University of Toronto

Volker Wulf, University of Siegen, Germany, and Fraunhofer Institute for Applied Information Technology, Sankt Augustin Germany

Yunwen Ye, Center for LifeLong Learning and Design, Department of Computer Science, University of Colorado, Boulder, and SRA Key Technology Laboratory, Inc., Tokyo, Japan

Index

Ackerman, Mark S., 401
  information technology and, 5,
    9–11, 269–270
  expertise sharing and, 273–299,
    338
Actors. *See also* Social capital; Trust
  algorithms for, 335, 339–342
  communication and, 335
  filtering systems and, 336–337
  IPoCs and, 337
  matching personal data and,
    339–345
  motivations for, 355, 361–362,
    365–366, 370, 384–387
  networks and, 337 (*see also*
    Networks)
  recognition and, 377, 385
  recommender systems and,
    336–339
  rewards and, 386–387
Ad hoc groups, 277–278
Adler, P. S.
  face-to-face social capital and,
    53–54, 70
  knowledge sharing and, 197
  project-based learning and,
    232–233, 238–240
Ahonen, Heli, 24, 40
Allen, Thomas, 277
Amazon.com, 23, 213
Amsterdam workshop, viii
Answer Garden, 269–270, 282–288
Ariba.com, 213

Arthur, M., 240–241
Artificial intelligence (AI), 336–339
Attitude
  eagerness to share and, 173–180
  efficiency and, 178–179
  motivation and, 355, 361–362,
    365–366, 370, 384–387
Aukema, Eline, 9, 133, 163–186,
  401

Babble, 292–294
Barber, Benjamin, 77
Barlow, John Perry, 117
Basic Support for Cooperative
  Work (BSCW), 89–90, 93–95, 100,
  102
Becks, Andreas, 10–11, 270–271,
  333–354, 401
Bielli, P., 195, 220
Blanchard, Anita, 9, 18, 53–73,
  401
Borgatti, Stephen P., 9–10, 133,
  137–161, 401
Boundary objects, 286–287
Bourdieu, Pierre, 2–3, 6, 22, 42, 87,
  165, 333
Bowling Alone (Putnam), 21
Bresnen, Mike, 6, 9–10, 135,
  231–267, 401
Bressand, A., 194
Brown, John Seeley, 45
Buckingham Shum, S., 338
Burt, Ronald S., 42, 333

Business clusters
  homogeneous/heterogeneous, 224
  IT use in, 218–221
  new vs. established, 224
  proximity issues and, 209–210,
    215–221
  transaction-cost theory and, 218
Business-to-business (B2B) trade,
    224
  categorization of, 211–212
  clusters and, 209–210, 215–221
  digital economy and, 209
  failure rates and, 214
  Internet and, 211–215, 221
  locational aspects and, 221–223
  performance metrics for, 225
  scope of, 211
Business-to-consumer (B2C) trade,
    211, 221–223

Call-tracking, 282–288
Castells, M., 220–221
Chan, Mike, 12
Chapman, Robbin, 11, 270,
    301–331, 401
Chat, 122, 364
Checklists, 194
Chen, C., 177
Chen, X., 177
Choo, C. W., 192–195, 203
Ciborra, C. U., 193
Civic engagement, 3–5
  decline of, 113–114
  FEF and, 88, 92
  free market and, 21–22
  Internet and, 113–131
  Iranian NGOs and, 75–111 (see also
    Iran)
  KBD information system and, 26–47
  Multiple Sport Newsgroup and,
    58–72
  organizational perspective and,
    22–23
Civil society organizations (CSOs),
    76–78
CodeBroker, 370–375, 381

Cognitive ability, 197–198, 217
  project-based learning and,
    247–248, 251–253, 256–258
  sharing expertise and, 273–274
Cohen, Don, 334
Cohen, Ion, 87
Cohen, W., 155
Coleman, James, 4, 86, 164–166, 301
Collective good, 234
Collectivism, 3, 164, 166, 171–172
  eagerness to share and, 173–180
  ICT and, 167–170
CommerceOne, 214
Communitarianism, 3, 6–7
Community. *See also* Networks;
    Virtual communities
  ad hoc groups and, 277–278
  as business model, 221–223
  communities of practice approach
    and, 84–86, 98–99, 104–107
  Computer Clubhouse Network and,
    302–307 (*see also* Computer
    Clubhouse Network)
  computer-supported cooperative
    work (CSCW), 192–195, 274,
    278–279, 336–339
  definition of, 25
  EDC and, 375–378
  electronic services and, 192–195
  information systems (IS) and, 24–47
  infrastructure and, 26
  Internet and, 113–131
  knowledge sharing and, 196–197
    (*see also* Knowledge sharing)
  Marxist tradition and, 2–3
  measurement of, 115
  neighborhoods and, 115
  NGOs and, 75–111 (*see also* Non-
    government organizations (NGOs))
  personal networks and, 137–161
  project-based learning and, 231–267
  psychological safety and, 140–141,
    146–157
  telephone and, 115
  trust and, 23–24, 196–197
Competence traps, 238–239

Computer Clubhouse Network, 270
  benefits of knowledge sharing and,
    324–326
  Clubhouse Council, 303–304
  constructionism and, 302–303
  description of, 302–307
  interaction categorization and,
    305–307
  mentoring and, 304
  Pearls of Wisdom (PoW) and,
    307–309 (*see also* Pearls of
    Wisdom (PoW))
  role of social capital in, 304–307
  technical limitations and, 326–328
Computer-mediated communication
  (CMC), 292
Computers, 10. *See also* Information
  technology (IT)
  Answer Garden and, 282–288, 295
  Babble and, 292–294
  CodeBroker and, 370–375
  impression management and,
    279–280
  Loop and, 292–294
  open source software and, 362–366
  Pearls of Wisdom (PoW) software
    and, 301–331
  repositories and, 282–288, 295
  social-technical gap and, 281–282
  topic-oriented communication and,
    336–339
  virtual communities and, 277 (*see
    also* Virtual communities)
  Zephyr and, 291
Computer-supported cooperative
  work (CSCW), viii, 192–195, 358
  conference on, 21
  expertise sharing and, 273–296,
    336–339
  incentives and, 280–281
  social-technical gap and, 273,
    278–282
Constructionism, 302–303
Constructivism, 302
Contextualization, 287
Contractor, Noshir, 12

Co-optation, 77
Corporatism, 77
Costa, Dora, 114
Courses-as-seeds, 378–381
Covisint, 213
Creativity, 395–399
  CodeBroker and, 370–375, 381
  courses-as-seeds and, 378–381
  design evolution and, 389–391
  Envisionment and Discovery
    Collaboratory (EDC) and,
    375–378, 381
  Experts Exchange and, 366–370,
    381
  externalization and, 356–357
  group concept and, 359–360
  isolation and, 356
  literature on, 356
  managerial awareness and, 142–144
  metadesign and, 358
  motivations for, 355, 361–362,
    365–366, 370, 384–387
  open source software and, 362–366,
    381
  perceived value and, 357–358
  project-based learning and, 231–267
    (*see also* Project-based learning)
  self-application and, 391–393
  SER model and, 360–362, 394
  tacit knowledge and, 356–357
  technical adoption barriers and,
    382–384
  trust and, 387–389
Cronbach's alpha, 171
Cross, Rob, 9–10, 133, 137–161,
  401
Cultural capital, 22

Data matching. *See* Expertise
DeFilippi, R., 240–241
Democracy, 76. *See also* Iran
Design. *See* Creativity
Detlor, B., 192–195, 203
Dieberger, A., 222
Digital economy. *See* E-commerce
Distler, C., 194

Dot-com bubble, 209
Drifting, 193
Duguid, Paul, 45

Ebert, Friedrich, 88
E-commerce
  Answer Garden and, 282–288, 295
  B2B trade and, 209–215
  business clusters and, 215–221
  catalogs and, 213
  cognitive ability and, 217
  digital economy and, 209
  distributors and, 213
  dot.com bubble and, 209
  EDI transactions and, 211–212
  exchanges and, 213
  hunting dog networks and, 26–47
  industry consortiums and, 213–215
  locational aspects and, 221–223
  MRO inputs and, 212–213
  open/closed systems and, 225
  performance metrics for, 225
  procurement services and, 213
  research agenda for, 223–226
  support systems and, 224–225
  trust and, 23
Economic issues, 1–2
  EDI transactions and, 211
  emergence of meaning and, 3–6
  free market and, 21–22
  globalization and, 77–78
  organizational perspective and, 22–23
  religion and, 77–78
  research methods and, 22–23, 223–226
Edelman, Linda, 401
  information technology and, 6, 9–10
  knowledge sharing and, 135
  project-based learning and, 231–267
Education, 78, 95–97. See also Knowledge sharing
Efficiency effects, 167–170
E-lancers, 188

E-learning, 335–336
  e-Qualification platform and, 345–350
Electronic data interchange (EDI) transactions, 211–212
E-mail
  attachments and, 125
  creativity and, 364
  expertise finding and, 348
  face-to-face social capital and, 122–123
  Internet and, 119, 121, 124
  knowledge sharing and, 249–250
E-networks. See Networks
Engagement, 145–146
Engeström, Yrjö, 24, 40
Envisionment and Discovery Collaboratory (EDC), 375–378, 381, 387–388
E-Qualification platform, 345–350
Erickson, T., 222–223, 225
E-steel.com, 213
Ethics, 97–98
Ethnography, 194
Etzioni, Amitai, 2–3
European Conference on Computer Supported Cooperative Work (ECSCW), viii
Evolutionary growth, 360–362, 389–391
Exclusion, 235
Expertise, vii, 273, 296–299, 351–354. See also Knowledge sharing
  Answer Garden and, 282–288
  Babble and, 292–294
  cluster functionality and, 343, 348–350
  CSCW research in, 336–339
  e-Qualification platform and, 345–350
  Experts Exchange and, 366–370, 381, 383
  field studies of, 288–289
  filter systems and, 336–337, 343, 345–350

impression management and,
   279–280
locators and, 276–277
Loops and, 292–294
matching personal data and,
   339–345
Pearls of Wisdom (PoW) and, 304
   (see also Pearls of Wisdom (PoW))
repositories and, 275–276
social-technical gap and, 273,
   278–282
systems for, 289–294, 336–339
virtual communities and, 277
Zephyr and, 291
Experts Exchange, 366–370, 381,
   383
Externalizations, 356–357

Face-to-face (FtF) social capital, 18
causality and, 57
EDC and, 388
e-mail and, 122–123
Internet and, 119
norms and, 65–67
Pearls of Wisdom (PoW) and, 312
trust and, 67–68
virtual communities and, 53–73
Finnish Spitz Club, 26–28, 31, 35, 43
Fischer, Claude, 114
Fischer, Gerhard, 9–11, 271,
   355–399, 401
Food and Agriculture Organization
   (FAO), 76
Ford Foundation, 79
Fraunhofer e-Qualification platform,
   345–350
Free market, 22
Friedrich-Ebert-Foundation (FEF), 88,
   92
Fukuyama, Francis, 21–23, 86

Gender, 139
Germany, 88, 99–100
Ghoshal, Sumantra
creativity and, 359, 384–385
expertise finding and, 334

information technology and, 22
knowledge sharing and, 197,
   234–236
Globalization, 77–78
Grainger.com, 213
Granovetter, Mark, 3–4, 140
Groupthink, 239
Groupware. See Computer-supported
   cooperative work (CSCW)
Grudin, Jonathan, 280–281
Guba, E. G., 60

Halverson, Christine, 401
expertise sharing and, 273–299
information technology and, 9–11,
   269–270
Hampton, K., 57
Hamyaran NGO Resource Center,
   79–80, 89
Hanifan, Lyda Judson, 358–359
Hansen, M., 140
Health issues, 245–248, 251–252,
   254–255
Heath, Christian, 339
Hirschheim, Rudy, 39
Homophilous relationships, 139
Hooff, Bart van den, 9, 133,
   163–186, 401
Huber, G., 138–139
Huberman, A. M., 59
Human capital, 5–6, 164, 196
Human Rights Watch, 78
Hummel, J., 221–222
Huysman, Marleen, vii, 401
business clusters and, 217
information technology and, 1–15
Internet and, 134
knowledge sharing and, 187–207,
   273
Hyperpersonal interactions, 169

IBM Institute for Knowledge
   Management, vii–viii
Ichijo, Kazuo, 44
Iivari, Juhani, 41
Impression management, 279–280

Inclusion, 235
Individualism, 166–167, 170–172.
    See also Creativity
  eagerness to share and, 173–180, 274
  learning trap and, 190–192
  repositories and, 282–288, 295
  tacit knowledge and, 356–357
Info-culture analysis, 194–195
Information
  access and, 139–141
  accuracy and, 150
  channels, 165
  cognitive ability and, 197–198
  eagerness to share and, 173–180, 274
  electronic services and, 192–195
  implicit, 163, 189–190
  managerial awareness and, 142–144
  novel, 140
  personal networks and, 137–157
  problem-solving and, 137
  psychological safety and, 140–141, 146–157
  transactive memory and, 138
Information and communication technology (ICT)
  collectivism and, 167–172
  eagerness to share and, 176–180
  efficiency effects and, 167–170
  future research and, 180–182
  multilevel effects of, 168
  SIDE model and, 169–170
Information systems (IS), 48–51. *See also* Networks
  e-commerce and, 209–229
  emerging role of, 45–47
  hunting dog networks and, 26–47
  infrastructure and, 40–45
  manual, 28
  many-to-many relationships and, 36–39
  as material artifact, 40–45
  objectivity and, 33–34
  one-to-many relationships and, 34–36
  one-to-one relationships and, 31–34
  sense making and, 39–40
  trust and, 46
  virtual communities and, 56–72
Information technology (IT)
  Answer Garden and, 282–288
  Babble and, 292–294
  business clusters and, 218–221
  CodeBroker and, 370–375, 381
  courses-as-seeds and, 378–381
  current research in, 1–15
  e-commerce and, 209–229
  Envisionment and Discovery Collaboratory (EDC) and, 375–378
  Experts Exchange and, 366–370, 381
  human capital and, 5–6
  impression management and, 279–280
  Iran and, 82–84
  knowledge sharing and, 134 (*see also* Knowledge sharing)
  limitations of, 326–328
  Loops and, 292–294
  motivational force behind, 23
  networked society and, 8 (*see also* Networks)
  open source software and, 362–366
  repositories and, 275–276, 282–288, 295
  social-technical gap and, 278–282
  third rationality of, 195
  tool designs for, 187–207
  trap of, 188–191
  trust and, 23–24
  Zephyr and, 291
Institutions. *See* Organizations
Intellectual capital, 22
Intended points of cooperation (IPoCs), 337
International Institute for Socio-Informatics (IISI), 88–89, 94, 101
Internet
  B2B trade and, 211–215
  business clusters and, 220–221
  causality and, 57

chat and, 122, 364
e-commerce and, 117, 209–229
e-mail and, 119, 121–125, 249–250, 348, 364
face-to-face (FtF) social capital and, 119
globalization and, 120
multiplying nature of, 117–120
new relationships and, 119
Pearls of Wisdom (PoW) and, 303
physical environment and, 121
politics and, 124
shared interests and, 117
target groups and, 125
television and, 121–122
transforming effects of, 116–124
Usenet newsgroups and, 118
uses of, 113, 125
Intranet, 192–193
Iran
as axis of evil, 107
BSCW and, 89–90, 93–95, 100, 102
civil society in, 76–81
communication system requirements of, 81–82
demographics of, 78, 81
education and, 95–97
FEF and, 88, 92
German delegation and, 99–100
Mashad conference and, 79
NGO community project, 88–105
(*see also* Non-government organizations (NGOs))
OTD approach and, 75, 82–84
project achievements and, 94–105
technology development and, 82–84
UNDP and, 81
Iranian Civil Society Organizations Resource Center, 92, 103
Iranian Population Council, 79

Java, 309, 338, 366, 368
Johnston, R., 218–219
Jung, Y., 222–223
Junkins, Jerry, 163

Kahn, Matthew, 114
Karelian bear dog IS, 46–51
breeding groups of, 26–28
cultural background of, 29–31
hunting instinct and, 27–31
many-to-many relationships and, 36–39
as material artifact, 40–45
one-to-many relationships and, 34–36
one-to-one relationships and, 31–34
Kellogg, W., 222–223, 225
Khatami, Mohammad, 80
Kiesler, S., 168–169
Klein, Heinz K., 39
Knowledge sharing, vii–viii, 133–135, 158–161
access and, 139–141, 144–145, 153
accuracy and, 150
Answer Garden and, 282–288
awareness and, 143–144
cognitive ability and, 197–202
collectivism and, 166–167, 170–172
community and, 24–47, 196–197
(*see also* Community)
Computer Clubhouse Network and, 302–307
courses-as-seeds and, 378–380
cultural differences and, 154
design analysis and, 198–202
e-commerce and, 210
electronic services and, 192–195
e-mail and, 249–250
emergence of meaning and, 3–6
engagement and, 145–146
expertise and, 273–299 (*see also* Expertise)
Experts Exchange and, 366–370
explicit, 225
gender and, 139
human capital and, 5–6, 139–141
information systems (IS) and, 24–47
infrastructure and, 26
Internet and, 113–131
management and, 195–198 (*see also* Organizations)

Knowledge sharing (cont.)
  matching personal data and, 339–345
  motivations for, 163–186, 197–198
  Pearls of Wisdom (PoW) and, 301–331 (see also Pearls of Wisdom (PoW))
  personal networks and, 137–161
  personal vs. nonpersonal appeals, 144–145
  problem-solving and, 137, 153
  project-based learning and, 241–263
  psychological safety and, 140–141, 146–157
  quality and, 141–148
  quantity and, 148–152
  reciprocity and, 166
  redundancy and, 249–251
  repositories and, 275–276
  sense making and, 39–40
  social-technical gap and, 278–280
  structural dimension of, 196–202
  tacit, 225, 356–357
  technology and, 275–296
  tie strength and, 140
  tool designs for, 187–207
  topic-oriented communication and, 336–339
  trust and, 196, 201
  Usenet newsgroups, 118
  virtual communities and, 277
Kraut, R. P., 5, 219–220
Kumar, K., 195, 220
Kuutti, Kari, 9, 12, 17, 21–51, 401
Kwon, S., 54, 70, 197, 232–233, 238–240

Labor, 2, 78. See also Organizations
Language, 197, 217
Latour, B., 43
Laudon, K., 212–213
Lave, Jean, 84
Lawrence, P., 218–219
Lea, M., 169–170
Leaders, 59
Lechner, U., 221–222

Lee, A., 222–223
Leonard, Dorothy, 35–36
Lesser, Eric, 22, 273–274
Levinthal, D., 155
Lieberman, Henry, 338
Lincoln, Y. S., 60
Linger, Henry, 41
Linux, 366
List servers. See Virtual communities
Logo, 312
Loops, 292–294
Loury, G. C., 4
Luff, Paul, 339
Lurkers, 59
Lyytinen, Kalle, 40

McDermott, Richard, 41
McDonald, David W., 288–289, 338
Mailing lists, 337
Maintenance, operating, and repair (MRO) inputs, 212–213
Management, 334. See also Organizations
  Answer Garden and, 282–288, 295
  awareness and, 142–144, 149, 151–152
  Babble and, 292–294
  expertise sharing and, 273–296
  impressions and, 279–280
  incentives and, 280–281
  individual learning and, 188, 190–192
  knowledge tools and, 187–207
  Loops and, 292–294
  project-based learning and, 241–263
  social capital concept and, 195–198
  trap of, 188–191
  Zephyr and, 291
Markets. See E-commerce
Meindl, J. R., 177
MicroWorlds, 320–321
Miettinen, Reijo, 24
MII Expert Finder, 338
Miles, M. B., 59
MIT Media Lab, 309
Mixed provenance, 286

Moch, M. K., 166
Moore, S., 56
MOOS, 337
MRQAP methods, 150
MSC software company, 288–289
MUDs, 337
Multidisciplinary design teams, 194
Multiple Sport Newsgroup (MSN), 18
  family friendly nature of, 60
  overview of, 60–61
  research on, 58–60, 68–72
  social capital and, 61–68

Nahapiet, Janine
  creativity and, 359, 384–385
  expertise finding and, 334
  information technology and, 22
  knowledge sharing and, 197, 234–236
National Geographic Society, 116–117
NetLab, 116–117
Netville, 120
Networks
  Answer Garden, 282–288
  Babble, 292–294
  cognitive ability and, 197–202, 217
  collectivism and, 166–170
  communities of practice approach and, 84–86, 105–107
  Computer Clubhouse Network and, 302–307 (*see also* Computer Clubhouse Network)
  computer-supported cooperative work (CSCW), viii, 21, 192–195, 273–296, 336–339, 358
  design analysis and, 198–202
  eagerness to share and, 173–180
  e-commerce and, 209–229
  effective sanctions and, 165
  electronic services and, 192–195
  emergence of meaning and, 3–6
  enactment abilities and, 138
  Envisionment and Discovery Collaboratory (EDC) and, 375–378
  homophilous relationships and, 139
  info-culture analysis and, 194–195
  information systems (IS) and, 24–47
  knowledge sharing and, 137 (*see also* Knowledge sharing)
  Loop, 292–294
  Marxist tradition and, 2–3
  Multiple Sport Newsgroup and, 58–72
  NGOs and, 75–111 (*see also* Non-government organizations (NGOs))
  obligations and, 165
  OTD approach and, 75
  Pearls of Wisdom (PoW) and, 301–331 (*see also* Pearls of Wisdom (PoW))
  personal, 137–157
  physical proximity and, 139
  project-based learning and, 231–267 (*see also* Project-based learning)
  psychological safety and, 140–141, 146–157
  relational issues and, 217
  repositories and, 282–288, 295
  situated learning and, 138–139
  solidarity effects and, 238–239
  structural issues and, 197, 217
  sustainability and, 104–105
  tie strength and, 140
  transactive memory and, 138
Newell, Sue, 6, 9–10, 135, 231–267, 401
Newsgroups, 337
Nohria, Nitin, 41
Nonaka, Ikujiro, 44
Non-government organizations (NGOs), 9, 18–19, 108–111
  achievements of, 94–105
  autonomy of, 77
  BSCW and, 89–90, 93–95, 100, 102
  categorization of, 77
  as civil society actors, 76–78
  communities of practice approach and, 84–86, 98–99, 104–107
  community system and, 102–103
  corporatism and, 77

Non-government organizations (NGOs) (cont.)
  definition of, 76–77
  education and, 95–97
  ethics and, 97–98
  FEF and, 92
  globalization and, 77–78
  government and, 80
  Human Rights Watch, 78
  infrastructure and, 75
  international, 80
  Iranian society and, 78–81, 88–90
  networking organization and, 90–92, 99–100
  OTD approach and, 75, 82–84
  social capital and, 86–87
  sustainability and, 100–101, 104–105
  technology and, 79–84, 92–94
  training and, 95
  UNDP and, 81
Nonpersonal appeals, 144–145
Norms, 65–67, 86, 236
  effective sanctions and, 165
  future research for, 180–182
  knowledge sharing and, 164–167, 173–180
  Pearls of Wisdom (PoW) and, 311
  project-based learning and, 256–258
  renegotiation and, 280

Obligations, 165
Open source software, 362–366, 381
Organization and technology development (OTD) approach, 75, 82–85, 98, 100
Organizations. *See also* Networks
  Answer Garden and, 282–288, 295
  B2B trade and, 209–215
  business clusters and, 215–221
  changing boundaries in, 188
  cognitive barriers and, 198–202
  collectivism and, 166–172
  design analysis and, 198–202
  eagerness to share and, 173–180
  e-commerce and, 209–229
  engagement and, 145–146, 148–152, 155–156
  expertise sharing and, 273–296
  groupthink and, 239
  implicit information and, 163
  individualism and, 166–167
  information access and, 139–141, 144–145, 148–149, 151–156
  intranet and, 192–193
  locational aspects and, 221–223
  managerial awareness and, 142–144, 149, 151–152 (*see also* Management)
  project-based learning and, 241–263
  psychological safety and, 140–141, 146–157
  reciprocity and, 166
  resistance to change and, 253–256
  respondent accuracy and, 150
  SIDE model and, 169–170
  sociotechnique domain and, 193–195
  tie strength and, 143
  tool designs for, 187–207
  Zephyr and, 291
Orlikowski, W., 220

Papert, Seymour, 302, 312
Pearls of Wisdom (PoW), 270, 307, 330–331
  architecture of, 309–316
  asynchronous interactions and, 326
  basic units of, 301–302
  benefits of knowledge sharing and, 324–326
  categorization and, 315–316
  constructor tool and, 312–315
  face-to-face (FtF) social capital and, 312
  field studies on, 320–329
  low-floor, high-ceiling approach to, 312–315
  mentoring and, 323
  metalearning and, 308
  norms and, 311

personal area and, 316
search tool and, 316–318
technical limitations and, 326–328
trust and, 311–312, 327–329
Viewer and, 319–320
Peer pressure, 365, 386
Perceived value, 357–358
Personal appeals, 144–145
Pew Internet and Everyday Life Project, 117–120
Pfister, Hans-Rüdig, 337
PhotoShop, 307–308
Piaget, Jean, 302
Pipek, V., 5
Platt, Lew, 163
Polanyi, Michael, 334
Political science, 1. *See also* Iran
axis of evil and, 107
Internet and, 124
Iraq war and, 75
Porter, M., 215–218
Portes, A., 4
Postmes, T., 169–170, 177–178
Privatization, 21
Problem-solving, 137
Project-based learning
cognitive ability and, 247–248, 251–253, 256–258
communities of practice and, 236
environmental context and, 232–233
literature on, 231–233
relational issues and, 247–248, 253–258
resistance to change and, 253–256
solidarity effects and, 258–259
structural dimension and, 247–251, 256–258
UK study on, 241–263
Projecte Internet Catalunya, 117
Proximity, 139
business clusters and, 209–210, 215–221
Prujit, H., 57
Prusak, Laurence, vii–viii, 4, 87, 273–274, 334

Psychological safety. *See* Trust
Putnam, Robert, 86, 301
civic engagement and, 21–22
expertise and, 333–334
Fischer and, 114
information technology and, 2, 4–5, 113–114, 118–119
obligations and, 166
trust and, 23

QAP methods, 150
Quan-Haase, Anabel, 5, 9, 19, 113–131, 401

Reductionism, 1–2
Redundancy, 249–251
Reference systems, 192–193
Referral Web, 338
Regional engineering manager (REM), 244–245, 249, 257, 259–263
Reichling, Tim, 10–11, 270–271, 333–354, 401
Relationship traps, 238–239
Religion, 77–78
Repositories, 192–193
Answer Garden and, 282–288
creativity and, 389–391
networks and, 282–288, 295 (*see also* Networks)
Research
business clusters, 218–221
communities of practice approach, 84–86, 105–107
current debates in, 1–15
cyclic approach, 82–84
design analysis, 198–202
economic, 22–23, 223–226
expertise and, 273–296
info-culture analysis, 194–195
Internet and, 113–131, 221
intranet and, 192–195
iterative data analysis, 59–60
KBD information system, 26–47
knowledge sharing and, 141–152 (*see also* Knowledge sharing)

Research (cont.)
  Multiple Sport Newsgroup, 58–72
  naturalistic inquiry and, 59
  OTD approach, 75, 82–84
  Pearls of Wisdom (PoW), 320–329
  project-based learning, 241–263
  traditional, 22
Reseeding, 360–362
Ridder, Jan de, 9, 133, 163–186, 401
Rohde, Markus, 9, 18–19, 75–111, 401
Role distribution, 364
Ruhleder, Karen, 40

Sanctions, 165
Scarbrough, Harry, 6, 9–10, 135, 231–267, 402
Scharff, Eric, 9–11, 271, 355–399, 402
Schultze, U., 220
Schumpeter, Joseph A., 334
Seeding, Evolutionary Growth, and Reseeding (SER) model, 360–362, 371, 379, 394
Sense making, 39–40
Sensenbrenner, J., 4
Silicon Valley, 216
Social capital, vii–ix
  business clusters and, 209–210, 215–221
  civic engagement and, 17–19 (see also Civic engagement)
  closure and, 251–253
  cognitive ability and, 197–198, 217, 247–248, 251–253, 256–258, 273–274
  collectivism and, 166–172
  communitarianism and, 3, 6–7
  cooperative relationships and, 138–139
  creation of, 54
  creativity and, 355–399 (see also Creativity)
  current research in, 1–15
  darker side of, 237–241
  decline of, 113–114
  definitions of, 1–2, 22–24, 53–54, 86–87, 165, 195–198, 234–237, 301–302, 333, 358
  design analysis and, 198–202
  eagerness to share and, 173–180
  e-commerce and, 209–229
  effective sanctions and, 165
  emergence of meaning and, 3–6
  ethics and, 97–98
  expertise and, 333–354
  face-to-face (FtF), 18, 53–73, 119, 122–123, 312, 388
  FEF and, 88
  free market and, 21–22
  goodwill and, 54
  human capital and, 5–6
  implicit information and, 163, 224–225
  industrialization and, 114
  information systems (IS) and, 24–47
  infrastructure and, 40–45
  Internet and, 113–131
  IT trap and, 188–190
  knowledge sharing and, 133–135 (see also Knowledge sharing)
  Marxist tradition and, 2–3
  Multiple Sport Newsgroup and, 61–68
  NGOs and, 86–87 (see also Non-government organizations (NGOs))
  norms and, 65–67, 164–167, 236
  obligations and, 165
  organizational perspective and, 22–23
  Pearls of Wisdom (PoW) and, 301–331
  personal networks and, 137–161
  project-based learning and, 241–263
  psychological safety and, 140–141, 146–157
  as public good, 86
  redundancy and, 249–251
  relational issues and, 247–248, 253–258, 273–274
  resistance to change and, 253–256
  self-application and, 391–393

situated learning and, 138–139
sociotechnique domain and, 193–195
structural dimension of, 196–202, 247–251, 256–258, 273–274
tie strength and, 143
trust and, 139–141, 196, 201
types of, 333–334
Social capital-sensitive system, 360
Social creativity. *See* Creativity
Social cues, 169–170
Social Identification of Deindividuation Effects (SIDE) model, 169–170
Social Web Cockpit, 339
Socio-technical gap, 273–274, 278–282
Sociotechnique domain, 193–195
Solidarity effects, 238–239, 258–259
Spears, R., 169–170, 177–178
Sproull, L., 168–169
Stahl, Gerry, 326–327, 338
Staples.com, 213
Star, Susan Leigh, 26, 40
Steinfield, Charles, 9, 134–135, 209–229, 402
Straus, Susaan, 35–36
Structural opportunity, 197
Sussman, Gerald, 321
Swan, Jacky, 231–267
Syrjänen, Anna-Liisa, 9, 12, 17, 21–51, 402

Technology
 adoption barriers and, 382–384
 CodeBroker and, 370–375
 education and, 95–97
 e-networks and, 192–195 (*see also* Information technology (IT))
 info-culture analysis and, 194–195
 Internet and, 116 (*see also* Internet)
 limitations of, 326–328
 problem-solving and, 137
 repositories and, 282–288, 295
 sociotechnique domain and, 193–195

transportation, 115
trap of, 188–190
Zephyr, 291
Telephone, 115, 119, 123
Television, 121–122
Teleworkers, 188
Textile industry, 218–219
Tonn, B. E., 56
Transaction-cost theory, 218
Transactive memory, 138
Traver, C., 212–213
Triandis, H. C., 166
Tribalism, 77–78
Trust, 86–87, 335
 activity-based, 24
 checklist for, 60
 community and, 196–197
 creativity and, 387–389
 e-commerce and, 24
 face-to-face (FtF) social capital and, 67–68
 generalized, 86
 information systems (IS) and, 26–47
 knowledge sharing and, 196, 201
 Multiple Sport Newsgroup and, 58–72
 Pearls of Wisdom (PoW) and, 311–312, 327–329
 psychological safety and, 140–141, 146–157
Turnbull, D., 192–195, 203

UN Development Program (UNDP), 81
United Kingdom. *See* Project-based learning
United Nations, 76–78
United States, 3, 211
Usenet newsgroups, 118
Uzzi, B., 4, 6–7, 140

van Dissel, H. G., 195, 220
Video, 192–194
Virtual communities, 10–11
 as business model, 221–223
 causality and, 57

Virtual communities (cont.)
  definition of, 55–56
  dispersed, 55–56
  expertise sharing and, 273–296
  Experts Exchange and, 366–370
  face-to-face (FtF) social capital and, 53–73
  family friendly nature and, 60
  leaders and, 59
  lurkers and, 59
  Multiple Sport Newsgroup, 58–71
  naturalistic inquiry and, 59
  norms and, 65–67
  place based, 55–56
  regular participants and, 59
Vivacque, Adriana, 338
Von Hippel, E., 174
von Krogh, George, 44

Wacquant, L., 165
Wagner, J., 166
Wellman, Barry, 5, 9, 19, 57–58, 113–131, 402
Wenger, Etienne, 84, 85, 202
Wessner, Martin, 337
Who Knows, 338
Williamson, O., 3–4
Witte, J., 57
Wulf, Volker, vii, 402
  expertise finding and, 333–354
  information technology and, 1–15, 270–271

XperNet, 338

Ye, Yunwen, 9–11, 271, 355–399, 402
Yearbook of Global Civil Society 2001, 80–81
Yenta, 338

Zambrano, P., 56
Zephyr, 291, 293
Zero-sum game, 369
Zink, Dan, 12